Neuroethics and Cultural Diversity

SCIENCES

Cognition and Language,
Field Directors – Patrice Bellot

Neuroethics,
Subject Heads – Bernard Reber and Jim Dratwa

Neuroethics and Cultural Diversity

Coordinated by
Michele Farisco

WILEY

First published 2023 in Great Britain and the United States by ISTE Ltd and John Wiley & Sons, Inc.

Apart from any fair dealing for the purposes of research or private study, or criticism or review, as permitted under the Copyright, Designs and Patents Act 1988, this publication may only be reproduced, stored or transmitted, in any form or by any means, with the prior permission in writing of the publishers, or in the case of reprographic reproduction in accordance with the terms and licenses issued by the CLA. Enquiries concerning reproduction outside these terms should be sent to the publishers at the undermentioned address:

ISTE Ltd
27-37 St George's Road
London SW19 4EU
UK

www.iste.co.uk

John Wiley & Sons, Inc.
111 River Street
Hoboken, NJ 07030
USA

www.wiley.com

© ISTE Ltd 2023

The rights of Michele Farisco to be identified as the author of this work have been asserted by him in accordance with the Copyright, Designs and Patents Act 1988.

Library of Congress Control Number: 2023934468

British Library Cataloguing-in-Publication Data
A CIP record for this book is available from the British Library
ISBN 978-1-78945-139-9

ERC code:
LS7 Applied Medical Technologies, Diagnostics, Therapies and Public Health
  LS7_10 Health services, health care research, medical ethics
SH5 Cultures and Cultural Production
  SH5_10 Ethics; social and political philosophy

Contents

Preface xiii
Michele FARISCO

Part 1. Neuroethics as a Field 1

Chapter 1. Examining the Ethics of Neuroscience in Contemporary Neuroethics 3
Cynthia FORLINI

    1.1. Introduction 3
    1.2. A brief history of neuroethics. 4
        1.2.1. Pluralism in the definitions of neuroethics 5
        1.2.2. A fundamental distinction in neuroethics inquiry 6
        1.2.3. Technology and healthcare as drivers of the ethics of neuroscience 7
    1.3. Critiques of the ethics of neuroscience. 9
        1.3.1. Critique 1: reinventing the bioethics wheel. 9
        1.3.2. Critique 2: a dose of "neuroskepticism" 11
    1.4. Responses to critiques of the ethics of neuroscience 12
    1.5. Blind spots in the ethics of neuroscience are opportunities for engagement . . 14
    1.6. Conclusion 16
    1.7. References 16

Chapter 2. Neuroscience of Ethics 21
Georg NORTHOFF

    2.1. Introduction 21
    2.2. Example I: a non-reductionistic and neuro-ecological model of brains 22
        2.2.1. History of neuroscience – passive versus active models of brains 22
        2.2.2. Neuroscience – passive versus active models of the brain 24
        2.2.3. A spectrum model – the hybrid nature of the brain's activity. 24
        2.2.4. The brain's spontaneous activity – constitution of its own
spatio-temporal structure on a functional level 25

2.2.5. Spontaneous activity and mental features – neuro-ecological rather
than neuronal 26
2.2.6. Psychiatric disorders – "spatio-temporal psychopathology" 26
2.3. Example II: from the neural basis of sense of self to relational agency 27
2.3.1. Neuroscience of the self – mapping distinct aspects of the self onto
different brain regions 27
2.3.2. Self and brain – agency is ecological and relational 29
2.4. Example III: enhancement of self – deep brain stimulation 30
2.4.1. Deep brain stimulation – its application in bipolar disorder. 30
2.4.2. Effects of DBS on the self – a quest for neuronal mechanisms 31
2.5. Conclusion 32
2.6. References 33

Chapter 3. Fundamental Neuroethics 37
Kathinka EVERS

3.1. Science and ethics 37
3.2. Neuroethics. 39
3.3. Fundamental versus applied neuroethics. 40
3.4. Fundamental neuroethics as a key component of European research and
innovation in the area of neuroscience 41
3.5. Conceptual analysis in fundamental neuroethics methodology 44
3.6. Fundamental neuroethics connecting neuroscience with "free will" and
social structures 48
3.7. Conclusion 51
3.8. Acknowledgments 51
3.9. References 52

Chapter 4. Diversity in Neuroethics: Which Diversity and
Why it Matters?. 55
Eric RACINE and Abdou Simon SENGHOR

4.1. Background. 55
4.2. Diversity and cultural diversity 57
4.3. Diversity, ethics and neuroethics' uneasy relationship with diversity. 59
4.4. Should neuroethics take cultural diversity into account, and why? 65
4.5. Conclusion 68
4.6. References 70

Chapter 5. Neurofeminism in BCI and BBI Ethics as a Prelude to
Political Neuroethics 77
Mai IBRAHIM and Veljko DUBLJEVIĆ

5.1. Introduction 78
5.2. Brain-to-brain interfaces 79

5.3. Neurosexism	81
5.4. Agential realism	85
5.5. Political perspective in neuroethics	89
5.6. Conclusion	91
5.7. References	91

Chapter 6. Neuroethics as an Anthropological Project 95
Fabrice JOTTERAND

6.1. Introduction	95
6.2. The nature of neuroethics	96
6.2.1. Neuroethics as a second-order discipline	96
6.2.2. Neuroethics, neurology and brain research	97
6.3. Neuroethics as an anthropological project	100
6.4. Protecting the brain and the mind	102
6.4.1. Neuroessentialism and neuroexceptionalism	102
6.4.2. Identity integrity: protecting brain and mind	103
6.5. Concluding remarks	104
6.6. References	104

Part 2. Cultural Influences on Neuroethics 107

Chapter 7. Neuroethics and Culture 109
Arleen SALLES

7.1. Introduction	109
7.2. Neuroethics and the challenge of cultural diversity	111
7.2.1. International neuroethics	111
7.2.2. Culturally aware/engaged neuroethics	112
7.2.3. Towards a global neuroethics	113
7.3. Can neuroethics contribute to the discussion?	114
7.3.1. The brain on culture	114
7.3.2. Culture and the brain: some challenges	116
7.3.3. Is there a role for neuroethics?	118
7.4. Concluding remarks and the way forward	119
7.5. Acknowledgments	120
7.6. References	120

Chapter 8. Globalization of Neuroethics: Rethinking the Brain and Mind "Global Market" 125
Karen HERRERA-FERRÁ

8.1. Introduction	125
8.2. Neuroethics within the global market: the "normality" problem	128

8.2.1. The neuroethical gap: cultural – beliefs and perspectives of the brain and mind 131
8.3. Neuroethics in a consumer country: a narrative from Mexico 132
8.4. Conclusion and future directions 136
8.5. References 137

Chapter 9. The Dilemma of Cross-Cultural Neuroethics 143
Laura SPECKER SULLIVAN and Karen S. ROMMELFANGER

9.1. Framing. 143
9.1.1. Neuroscience and culture? 143
9.1.2. State of the art of cross-cultural neuroethics 145
9.2. Benefits and aims of cross-cultural neuroethics. 146
9.2.1. Intercultural understanding. 147
9.2.2. Self-awareness 147
9.2.3. Mutual interest and cooperation 148
9.2.4. Intracultural creativity 148
9.3. Potential forms of cross-cultural neuroethics 148
9.3.1. Summits and international meetings 149
9.3.2. Historical, sociological and anthropological research 149
9.3.3. Community-based participatory research (CBPR) 150
9.3.4. Cultural competency/critical theory/DEI 150
9.4. Challenges in cross-cultural neuroethics. 151
9.4.1. Intracultural diversity 151
9.4.2. Domestic politics and power 152
9.4.3. International politics and power 152
9.4.4. The dilemma of cross-cultural neuroethics 153
9.5. Conclusion 154
9.6. References 155

Chapter 10. Neuroethics in Religion and Science: Hume's Law and Bodily Value 159
Denis LARRIVEE

10.1. Introduction 159
10.2. Contingency, autonomy and bodily value 165
10.2.1. Identifying in human freedom bodily value: Newtonian regularity and quantum chaos. 165
10.2.2. Biological autonomy and physical contingency – the organism. 167
10.3. Autonomy as a constituent ground of nature: a metaphysical composition 169
10.3.1. Substances and bodily unification 169
10.3.2. Ontological subsistence: the personal subject as a causal origin 171

10.3.3. Newtonian determinism and organismal autonomy: from an extrinsic
to an intrinsic causal order 171
10.4. The personal subject and intrinsic corporal value 173
10.4.1. Wojtyla's metaphysical subject 173
10.4.2. Kant, intrinsic value and the categorical imperative 174
10.5. Conclusion 175
10.6. References 175

**Chapter 11. How Would Neo-Confucians Value Moral
Neuroenhancement?** 179
Jie YIN

11.1. Moral neuroenhancement: the scenario and the conceptual challenge 180
11.2. How would neo-Confucians value moral neuroenhancement? 184
11.3. Concluding remarks: the complementary role of Chinese philosophy
in applied ethics 190
11.4. References 191

Part 3. Illustrative Cases 193

**Chapter 12. How Do Arabic Cultural and Ethical Perspectives Engage
with New Neuro-technologies? A Scoping Review** 195
Amal MATAR

12.1. Background 195
12.2. Methods 196
12.3. Results 197
12.3.1. Culture 198
12.3.2. Ethics literature 202
12.4. Discussion 204
12.5. Conclusion 206
12.6. Acknowledgments 207
12.7. Appendix 207
12.8. References 212

Chapter 13. The Binary Illusion 217
Karin GRASENICK

13.1. A brain is still a brain 217
13.2. Imagine all the people living life in vain 218
13.3. This land is my land, from the asylum to the last island 219
13.4. Little bits of history repeating 221
13.5. What's good for me is good enough for you 222
13.6. They keep saying they have something for you 223
13.7. Sign of the times 226

13.8. Just microscopic cogs for a neuroethics plan (conclusion) 228
13.9. References 229

Chapter 14. What's Next? The Chilean Neuroprotection Initiative, in Light of the Historical Dynamics of Human Rights 235
Manuel GUERRERO

14.1. Introduction. 235
14.2. Battling on the "last frontier": the Chilean neuroprotection legislation 236
14.3. The Chilean neuroprotection initiative, an unfinished project 241
14.4. References 246

Chapter 15. Interrogating the Culture of Human Exceptionalism: Animal Research and the Neuroethics of Animal Minds and Brains 249
L. Syd M JOHNSON

15.1. Introduction. 249
15.2. Brains, minds, consciousness and moral status 250
15.3. Chimeras and humanization: the overexamined problem. 252
  15.3.1. Chimeric non-human primates 257
15.4. Already human-like: the overlooked problem 260
15.5. Implications: justice in neuroscientific research. 260
15.6. Interrogating the anthropocentric culture and human exceptionalism
of neuroethics. 263
15.7. References 266

Chapter 16. Cultural Neuroethics in Practice – Human Rights Law and Brain Death 271
Jennifer A. CHANDLER

16.1. Introduction. 271
16.2. The concept of brain death 273
16.3. Objections to brain death: culture, religion and demographic minorities . . . 275
16.4. What is at stake with the definition of death? 276
16.5. Legal responses to brain death objection. 279
16.6. Accommodation of dissenting views 281
16.7. Conclusion 282
16.8. Acknowledgments 283
16.9. References 283

Chapter 17. Neuroscientific Research, Neurotechnologies and Minors: Ethical Aspects 287
Laura PALAZZANI

17.1. The importance of neuroscientific and neurotechnological research
on minors 287

17.2. Ethical criteria of neuroscientific research on minors in the medical field . . 289
   17.2.1. Justification of scientific relevance of the research. 290
   17.2.2. The benefit/risk balance and the protection of physical and mental integrity 290
   17.2.3. Personal identity 291
   17.2.4. Autonomy: informed consent of minors and informed consent of parents 292
   17.2.5. Incidental findings. 295
   17.2.6. Predictive value of brain images 296
   17.2.7. Single-patient reports 297
   17.2.8. Privacy: the use of neural data 297
   17.2.9. Justice 298
17.3. Research for neuro-enhancement purposes 298
17.4. Use of neurotechnologies in non-medical field without any research 300
17.5. Some conclusive reflections. 301
17.6. References 302

Conclusion 305
Michele FARISCO

List of Authors 311

Index 315

Preface

Michele FARISCO[1,2]

[1] *Centre for Research Ethics and Bioethics, Uppsala University, Sweden*
[2] *Biogem, Biology and Molecular Genetics Research Institute, Ariano Irpino, Italy*

P.1. Why a book on neuroethics and cultural diversity?

There is a growing discussion about the statute of neuroethics as a scientific discipline (Farisco et al. 2018; Johnson and Rommelfanger 2018). There are at least two main reasons why this debate is still open: (1) the discipline is quite young, so there is still the need for clarifying both its methodology and content; (2) since neuroethics is conceived as an interface between academic research and different societal stakeholders, a number of different factors impact the identity of neuroethics (including its methodology and content). Cultural diversity is among the most impactful factors shaping neuroethics, both as a scientific discipline and as a social enterprise.

Addressing the semantic complexity of culture goes far beyond the scope of this book. In fact, a technical, minimalist understanding suffices for its goals. Accordingly, culture can be understood as passing over information from one individual and/or group to another, with an implicit and/or explicit impact on their behavior and possibly their thinking. This definition abstracts away from any explicit reference to specific sets of values, symbols or any details that contribute to defining a collective *Weltanschauung*. Accordingly, this understanding of culture is conceived to be as inclusive as possible. It does not exclude, for instance, other animal species that can also display this kind of cultural behavior. Furthermore, this definition of culture is not limited to sociological and anthropological dimensions: it also includes disciplinary differences, which appear to be quite relevant to the neuroethical debate, which is inter- and multidisciplinary by definition.

Thus, even if limited and minimalistic[1], a cognitive account of culture as shared and socially transmitted information can be assumed as a working definition for the present analysis.

Since the 1980s, there have been several attempts to reconstruct the evolution of culture and cognition, particularly within the field of cognitive anthropology, with the final goal of developing a theoretical framework for cultural evolution (Cavalli-Sforza and Feldman 1981; Lumsden and Wilson 1981; Boyd and Richerson 1985). Emerging from this anthropological research, the concept of a cultural model as elaborated in cognitive anthropology (Bennardo and De Munck 2014; Bennardo 2018) is particularly relevant to the present analysis. Cultural models are "mental representations shared by members of a culture" (Bennardo and De Munck 2014, p. 3). Importantly, cultural models fill in the data of our experience, either at the aware or (mostly) at the unaware level. In this way, cultural models make sense of our experiences, informing our inferences. They eventually facilitate our engagement with the world, allowing us to operate smoothly "on autopilot" and behave in a purposive and communicative way. Importantly, cultural models present both individual and cultural (i.e. collective) variations. This means that a culture can affect the individual through the variation of relevant cultural models.

The following features characterize culture. Some of them are intrinsic to it, while others (namely the latest two), even if not unique to culture, are particularly relevant to the present analysis:

– *Dynamic*: even though it is more or less resistant to change, culture is intrinsically historical. A number of factors, both internal and external to a culture, concur to shape its dynamics.

– *Transformative*: directly dependent on its historical nature, culture is subject to different transformations, that is, significant changes of its identity.

– *Internally differentiated*: cultures are like a framework, that is, they are recognizable but not fully homogeneous.

---

1. For an introduction to the theoretical issues surrounding the definition of culture, including the explanatory value of information, see Sperber's *Explaining Culture: A Naturalistic Approach* (Sperber 1996).

– *Ethically relevant*: cultures are intrinsically connected to ethical values (i.e. evaluative and behavioral standards), and such values inform opinions and behaviors of individuals either explicitly or implicitly.

– *Epistemically relevant*: cultures provide individuals with strategies and tools for interpreting and making sense of the world.

– *Pragmatic*: cultures appear to be intrinsically related to a specific set of actions and behaviors considered key for individual flourishing.

Accordingly, the influence of culture on individuals extends over different dimensions with different degrees of impact. The cultural impact on science and on the public perception of science is particularly relevant to neuroethics, which aims to facilitate the creation of an interface between neuroscience and society at large. In fact, one of the original inspirations of neuroethics is the increasing possibility of exploiting neuroscientific research and related technologies in different contexts and for different purposes, both medical and non-medical, therapeutic and non-therapeutic. The growing societal relevance of neuroscience calls for an ethical reflection that encompasses issues that are both foundational (i.e. what is the impact of neuroscience on fundamental moral notions?) and practical (i.e. what are the criteria for an ethically sound neuroscientific research and use of neurotechnology?).

It is not easy to provide an adequate definition of neuroethics. In short, it can be described as an interdisciplinary field that addresses ethical, legal, social and cultural, as well as philosophical and scientific questions raised by neuroscience and related technologies (Marcus and Charles A. Dana Foundation 2002; Levy 2007; Illes and Sahakian 2011; Farisco et al. 2018; Johnson and Rommelfanger 2018). Its methodology can be conceptual, empirical and normative (or a combination) depending on the perspective we wish to emphasize (Evers et al. 2017b). Since the 2002 Dana Foundation Neuroethics Conference, this field has often been conceived in two main ways: (1) as "ethics of neuroscience", which is a type of applied ethics that aims to provide a repertoire of ethical approaches to address the practical ethical and societal concerns raised by neuroscience research and its applications, for example privacy and the protection of neural data; or (2) as the "neuroscience of ethics", which is an empirical, descriptive approach that focuses on how neuroscientific findings can inform theoretical and practical issues, such as what moral reasoning is, how to understand ethical choices and what the implications of neuroscientific findings for understanding free will are (Marcus and Charles A. Dana Foundation 2002; Roskies 2002a). More recently, a basic research-oriented and conceptual approach, that is, fundamental neuroethics (Evers 2007, 2009), has been gaining traction. Fundamental neuroethics takes as a starting point the view that conceptual analysis plays an important role not only in illuminating key operative notions (e.g. consciousness, the self and human identity), but also in examining basic and foundational issues such as understanding the same notions in different contexts (i.e. ethics and neuroscience) and their mutual relevance, how neuroscientific knowledge is constructed, what its underlying assumptions are and how they are justified, how results can be interpreted, and why or how empirical knowledge of the brain can be relevant to philosophical, social and ethical concerns (Evers et al. 2017b; Farisco et al. 2018; Salles et al. 2019b).

How to address cultural diversity is a challenge for both neuroscience and neuroethics. This is not accidental: neuroethics is engaged in ethical reflection in collaboration with neuroscience, and ethics is *per force* multifaceted and characterized by diversity.

Historically, however, neuroethics originates from culturally specific contexts (i.e. North America and Western Europe), eventually reflecting their theoretical, methodological and practical assumptions (De Vries 2005; see also Chapter 4 of this book). The portability of these assumptions in other contexts is not unproblematic. Beyond methodologies and approaches, it has also been argued that neuroethics shows its Western bias in the topics it chooses to focus on (see Chapter 4 of this book): it has given significant attention to issues related to neurotechnologies that are not a priority in non-Western contexts (Racine 2010; Racine and Sample 2018), while often failing to address those based on the needs of marginalized populations.

The above illustrates the need for a culturally sensitive neuroethics. This book seeks to elaborate a historical and conceptual analysis of neuroethics in relation to cultural diversity, and to provide illustrative models of how to advance in this much-needed neuroethical reflection. This specific analysis of the interaction between neuroethics and cultural diversity is still to be elaborated. This book is a candidate first systematic reflection on the topic: it provides a summary of the state of the art and proposes concrete action plans to advance the debate.

More specifically, three possible strategies for neuroethics to deal with the problems raised by cultural diversity have been identified (see Chapter 7 of this book):

1) Including other cultures in the neuroethical discussion, without much reflection on why, what is really critical, how and for what specific goal in particular. This seems to be the model of so-called "international neuroethics" (Lombera and Illes 2009a).

2) Moving from a simple recognition of cultural differences to actions aimed at engaging with them. This is the "cross-cultural neuroethics" model (see below).

3) Combining the identification of commonalities with building intercultural moral consensus. This is the "global neuroethics" model, which is more inclusive than international neuroethics. In fact, global neuroethics aims to elaborate rules and norms that reflect common values and common general vision, or at least a convergent ethical framework that leads to some type of consensus (Kellmeyer et al. 2019).

The cross-cultural strategy seems particularly relevant to any attempt to build a link among different cultures, including different disciplinary cultures, without eliminating the differences among them. In this sense, cross-cultural neuroethics seems immune to any risk of "ethical colonialism" (i.e. the risk of assuming any cultural model as paradigmatic and dominant) while, for example, global neuroethics may suggest a kind of homologation process. For Karen Rommelfanger and Laura Specker Sullivan, as demonstrated in their joint contribution to the volume, "cross-cultural" properly applies to any project that foregrounds the significance of cultural comparison, whether the goal is to identify similarities against a background of differences, or to highlight differences against a background of similarities. A benefit of this definition is that we do not need to agree on an essential set of criteria for a culture, we only need to be able to identify where there is a difference between two social groups. Broadly, they think that cross-cultural differences often manifest in terms of beliefs, values and practices: both what they are and how they came to be (see Chapter 9 of this book). Thus, cross-culturality does not require an agreed-upon definition of the terms to be compared: rather, it focuses on apparent differences instead. This means shifting the focus *from identity to relation*: since the issues arise at the intersection of two or more terms, the focus should be on their mutual differences, without presupposing a strong, paradigmatic model in order to eliminate or dilute the discrepancies through homologation.

Seeking mutual understanding in a context of tensions between values has been suggested as a possible approach in order to recognize diversity and facilitate mutual understanding (Létourneau 2018). The preliminary condition is a conceptualization of cultural identity that is not as rigid and autonomous, but is instead flexible and relational.

The above holds not only for social and geographical cultures, but also for disciplinary cultures and related diversity. In this case, a cross-cultural goal is also highly desirable. The challenge is how to achieve a true multidisciplinary collaboration. In this respect, embeddedness of neuroethics in scientific research has been advocated and proposed as a possible implementation model. The EU Human Brain Project is a recent and at least in part successful illustration of how neuroethics might do something about disciplinary diversity, not only by reflecting on it, but also by promoting and implementing a concrete methodology in a multidisciplinary environment (Salles et al. 2019a).

Importantly, embeddedness does not mean that neuroethics loses its critical attitudes towards neuroscience and other disciplines, or that it becomes an uncritical advocate of neuroscience, as some have argued (De Vries 2007; Chapter 4 of this book). Uncritical acceptance would be detrimental for the public perception of neuroscience and eventually for neuroscience itself (Racine et al. 2005, 2006, 2017). Embeddedness rather means a shared topic of investigation (e.g. neuronal basis of consciousness), a common goal (e.g. improving the clinical care of consciousness disorders), through the intersection of different methodologies (e.g. clinical, computational, cognitive neurosciences, ethics, philosophy, patients' perspective).

As with any anthology, it is not straightforward to find a shared theoretical stance among the different chapters that follow. If possible, this is even more true of this book, which is conceived as a collection of different, complementary voices coming from diverse cultural (both geographical and disciplinary) backgrounds. Notwithstanding this intrinsic diversity of perspectives and approaches, the different authors share the view of neuroethics as a theoretically solid reflection, at the intersection of empirical and theoretical approaches, upon the conceptual reliability of neuroscience, its relevance to moral reasoning and its ethical impact on society. A panoply of issues arise from this multi- and interdisciplinary model, including: what are its main characteristics? What are the potential advantages for neuroscience and social stakeholders coming from it? How to implement it in practice? How to measure its effectiveness? Etc. While this book provides no final answer to these issues, it illustrates that a convergent approach among different cultural voices in identifying the main issues and getting a consensus on the priorities is possible. This should be taken as the starting point for advancing in the search for shared solutions, and if not final, at least sufficiently reliable to be translated into democratic deliberative processes.

P.2. Plan of the book

This book is structured in three complementary parts: two foundational parts and one that is more practical.

A necessary premise for exploring the relationship between neuroethics and cultural diversity is defining the two terms. Thus, Part 1 is dedicated to the analysis of neuroethics as a field.

In Chapter 1, Cynthia Forlini provides a historical and conceptual analysis of neuroethics understood as the *ethics of neuroscience*. After highlighting the different aims, critiques and potential future directions of neuroethics thus conceived, she argues that the ethics of neuroscience is a good place to anticipate and assess issues in innovation, practice and policy for a balanced adoption of neuroscience and related technology.

In Chapter 2, Georg Northoff provides a critical examination of the other traditional understanding of neuroethics, the so-called *neuroscience of ethics*, arguing that its distinction from the ethics of neuroscience is not always as clear as is commonly assumed. In particular, his analysis focuses on the neuronal activity of the brain in connection with its ecological context, on a neurorelational understanding of the self and on the issue of self-enhancement.

In Chapter 3, Kathinka Evers further problematizes the dichotomy between ethics of neuroscience and neuroscience of ethics, illustrating the *fundamental neuroethics* that she introduced approximately 15 years ago. In fact, the classical dichotomic understanding of neuroethics is not sufficient to cover all the emerging issues at the interface of brain science, philosophy and society at large. For this reason, Evers distinguishes "applied neuroethics" from "fundamental neuroethics" in order to include other areas in the neuroethical discourse in addition to ethics and neuroscience. The main features of fundamental neuroethics as conceptualized by Evers are as follows: it pursues *foundational* analyses within a *multidisciplinary* research domain using an *interdisciplinary* methodology.

In Chapter 4, Eric Racine and Abdou Simon Senghor reflect on the notion of *diversity in relation to neuroethics*. After distinguishing between two main meanings of diversity (human or anthropological vs. socio-cultural diversity), they critically reflect on the cultural identity of neuroethics, which is historically rooted in the Western tradition, concluding by highlighting the important role of human diversity in providing indications about human flourishing.

In Chapter 5, Mai Ibrahim and Veljko Dubljević provide a contribution from the perspective of *political neuroethics*, focusing on the particular topic of sexist and androcentric biases in neuroscience. Taking brain-to-brain interfaces as a case study, the authors conclude by emphasizing the ethical importance of making explicit the socio-political worldviews presupposed by neuroscientific research and the resulting applications.

In Chapter 6, Fabrice Jotterand argues that it is necessary to overcome the traditional dichotomic understanding of neuroethics which, according to him, has impeded a more fundamental reflection on the anthropological implications of neuroscientific research and emerging technologies. For this reason, he claims that neuroethics should be part of a philosophical anthropology project that provides clear reflections on the impact of neurotechnology on how we conceive of our brain, mind and our identity's integrity.

Part 2 is specifically about cultural influences on neuroethics.

In Chapter 7, starting from the distinction between three main understandings of neuroethics (i.e. neurobioethics, empirical neuroethics and conceptual neuroethics) and how they have dealt with culture, Arleen Salles analyzes *how neuroethics can address the challenge of cultural diversity*, arguing that it may play an important role in clarifying some key conceptual issues regarding brain, culture and their mutual interaction.

In Chapter 8, Karen Herrera-Ferrá provides a *reflection on globalization*, starting from its socio-economic features to then outline the potential contribution that neuroethics can make. In particular, she criticizes the "one-size-fits-all" view that ignores cultural specificities, arguing that in order to develop a global neuroethics it is necessary to "(a) [be] mindful of the significant transnational and multidirectional interconnectivity of the diverse human dimensions that results from globalization, (b) [avoid] assuming universality in key concepts of the human mind and essence and (c) [be] inclusive of other perspectives, narratives and neuroethical concerns on the use of neuroscience, neurotechnology and some forms of artificial intelligence".

In Chapter 9, Laura Specker Sullivan and Karen S. Rommelfanger provide an in-depth analysis of *cross-cultural neuroethics*, which undertakes a comparative analysis between different cultural perspectives with several potential advantages, including not essentializing any one culture, not portraying an entire country or geographic region as a monolithic culture and not assuming any one culture as paradigmatic, among others. Yet there is also a potential paradox in the cross-cultural neuroethics project: since it undermines the prominent role of the brain, the cross-cultural model may eventually question the very premise of neuroethics.

In Chapter 10, Denis Larrivee examines aspects of a *Christian understanding of bodily value*, highlighting the connection with relevant metaphysical elements that have recently emerged in scientific reflection. Larrivee acknowledges the divergences between science and religion, especially in relation to their respective normative conclusion, as exemplified by Hume's fallacy principle, but he also argues that clarifying this tension can fruitfully contribute to elaborating a common approach in the many forthcoming challenges raised by neurotechnology and other medical interventions in the brain.

In Chapter 11, Jie Yin provides an illustration of *how Neo-Confucianism evaluates* a neuroethically sensible issue such as *moral neuroenhancement*. She argues that, from a Confucian point of view, moral enhancement through technological innovation is doubtful due to both its technical feasibility and its goal itself. Starting from this particular topic, the author also reflects on the more general question of what the Chinese philosophy can offer to applied ethics, concluding that this may be a privileged place to cultivate mutual understanding and collaboration.

Part 3 presents illustrative cases to provide relevant examples of what has been discussed in the first two parts of the book.

In Chapter 12, the scoping review about *Arabic cultural and ethical perspectives on new neuro-technology*, presented by Amal Matar, is an original view on a topic that still lacks sufficient attention. Two research questions guide the analysis: How are novel neurotechnologies harnessed to investigate cultural elements in the Arab region? And what are the discussions, if there are any, pertaining to neuroethics in the Arab region? The results show that the cultural elements of language, religious practice and behavior deserve the main attention, and that there is a wide gap between the current international debates on neuroethics and the relevant discussion in the Arab region.

Chapter 13 by Karin Grasenick provides a reflection on the *societal embeddings of neuroscience*, with particular focus on *its implications for how gender differences are conceived, articulated and converted in social practice*. Combining a fictional and an actual historical example, the author explores the issue of whether and how neuroethics can contribute to contextualizing both the methodology and the results of neuroscientific investigation in order to maximize societal benefits and avoid artificial discrimination. The chapter ends with seven open questions for neuroethics in order to encourage a much-needed interdisciplinary dialogue.

Chapter 14 by Manuel Guerrero is an attempt to *read the neurorights movement as part of the evolution of human rights*. Starting from the case of the Chilean Constitution, where the notion of neuroprotection has been recently introduced, the author provides a historical and conceptual analysis of human rights, arguing that neurorights should be interpreted as part of this broader framework. In addition, to make the claim for new neurorights to be more effective and productive, it is necessary to elaborate and promote them in direct collaboration with a number of societal stakeholders.

Chapter 15 by L. Syd M Johnson is an in-depth reflection on the *anthropocentric speciesism and human exceptionalism* that, according to the author, have prevailed in the neuroethical discussion so far. Her reflection spans from reviewing the brain-based capacities considered ethically relevant in humans and also found in different animal species, to analyzing the problem of humanizing animals, to then paying attention to justice in research, outlining the need for consistency in the scientific and ethical justification of research. This reflection eventually leads the author to acknowledge the neuroethics commitment to both rigorously explore the implications of neuroscience and to critically interrogate research that threatens important and fundamental ethical values.

Chapter 16 by Jennifer A. Chandler focuses on the specific issue of *brain death in relation to culture* and specifically to human rights law. The topic is quite challenging, and several dimensions of cultural diversity concur with its controversy, including social, ethnic and racial diversity. Another important dimension analyzed in the text is mistrust towards medical practice, which often informs the resistance to the notion of brain death and related criteria, especially by racial minorities in some Western countries. The challenge for democratic societies is whether and how diverging views on this question can be accommodated. Cultural neuroethics can make an important contribution to this discussion in order to clarify the issues and to elaborate effective action models.

Chapter 17 by Laura Palazzani reflects on an often-neglected topic: the *neuroscientific and neurotechnological research on minors*. These raise plenty of ethical and legal issues, including informed consent, self-determination and the minimization of risks, among others. The author starts from a justification of neuroscientific research on minors in the medical field and the definition of relevant ethical criteria for it, including the scientific relevance, the benefit/risk balance, the protection of personal identity and autonomy. Then different emerging issues are outlined, such as the need for an appropriate handling of incidental findings and the predictive value of brain images, privacy, justice and non-medical application of research. This chapter is one of the few examples so far that specifically reflects on the ethics of neuroscientific research on minors.

P.3. Acknowledgments

This research was funded by the European Union's Horizon 2020 Framework Programme for Research and Innovation under the Specific Grant Agreement No. 945539 (Human Brain Project SGA3).

Special thanks go to my mentor, Kathinka Evers, who supported the project since the beginning; to my colleague Arleen Salles for inspiring discussions about the topic of neuroethics and culture; to the colleagues from the Centre for Research Ethics and Bioethics, Uppsala University, and from Workpackage 9 of the Human Brain Project for a very fruitful exchange of ideas.

January 2023

P.4. References

Bennardo, G. (2018). Cultural models theory. *Anthropology News*, 59(4), e139–e142.

Bennardo, G. and De Munck, V.C. (2014). *Cultural Models: Genesis, Methods, and Experiences*. Oxford University Press, New York.

Boyd, R. and Richerson, P.J. (1985). *Culture and the Evolutionary Process*. University of Chicago Press, IL.

Cavalli-Sforza, L.L. and Feldman, M.W. (1981). *Cultural Transmission and Evolution: A Quantitative Approach*. Princeton University Press, NJ.

De Vries, R. (2005). Framing neuroethics: A sociological assessment of the neuroethical imagination. *American Journal of Bioethics*, 5, 25–27.

De Vries, R. (2007). Who will guard the guardians of neuroscience? Firing the neuroethical imagination. *EMBO Reports*, 8, 65–69.

Evers, K. (2007). Towards a philosophy for neuroethics. An informed materialist view of the brain might help to develop theoretical frameworks for applied neuroethics. *EMBO Reports*, 8 Spec No. S48-51.

Evers, K. (2009). *Neuroéthique. Quand la matière s'éveille*. Odile Jacob, Paris.

Evers, K., Salles, A., Farisco, M. (2017). Theoretical framing of neuroethics: The need for a conceptual approach. In *Debates About Neuroethics: Perspectives on its Development, Focus and Future*, Racine, E. and Aspler, J. (eds). Springer International Publishing, Dordrecht.

Farisco, M., Salles, A., Evers, K. (2018). Neuroethics: A conceptual approach. *Cambridge Quarterly of Healthcare Ethics*, 27, 717–727.

Illes, J. and Sahakian, B.J. (2011). *The Oxford Handbook of Neuroethics*. Oxford University Press, Oxford, UK, and New York.

Johnson, L.S.M. and Rommelfanger, K.S. (2018). *The Routledge Handbook of Neuroethics*. Routledge, Taylor & Francis Group, New York.

Kellmeyer, P., Chandler, J.A., Cabrera, L., Carter, A., Kreitmar, K., Weiss, A., Illes, J. (2019). Neuroethics at 15: The current and future environment for neuroethics. *American Journal of Bioethics Neuroscience*, 10(3),104–110.

Létourneau, A. (2018). Differing versions of dialogic aptitude. Bakhtin, Dewey, and Habermas. In *Dialogic Ethics*, Arnett, R.C. and Cooren, F. (eds). John Benjamins Publishing, Amsterdam.

Levy, N. (2007). *Neuroethics*. Cambridge University Press, Cambridge, UK, and New York.

Lombera, S. and Illes, J. (2009). The international dimensions of neuroethics. *Developing World Bioethics*, 9, 57–64.

Lumsden, C.J. and Wilson, E.O. (1981). *Genes, Mind, and Culture: The Coevolutionary Process*. Harvard University Press, Cambridge, MA.

Marcus, S. and Charles A. Dana Foundation (2002). *Neuroethics: Mapping the Field: Conference Proceedings, San Francisco, California, 13–14 May 2002*. Dana Press, New York.

Racine, E. (2010). *Pragmatic Neuroethics: Improving Treatment and Understanding of the Mind-Brain*. MIT Press, Cambridge, MA.

Racine, E. and Sample, M. (2018). Two problematic foundations of neuroethics and pragmatist reconstructions. *Cambridge Quarterly of Healthcare Ethics*, 27, 566–577.

Racine, E., Bar-Ilan, O., Illes, J. (2005). fMRI in the public eye. *Nature Reviews Neuroscience*, 6, 159–164.

Racine, E., Bar-Ilan, O., Illes, J. (2006). Brain imaging: A decade of coverage in the print media. *Science Communication*, 28, 122–142.

Racine, E., Dubljevic, V., Jox, R.J., Baertschi, B., Christensen, J.F., Farisco, M., Jotterand, F., Kahane, G., Muller, S. (2017). Can neuroscience contribute to practical ethics? A critical review and discussion of the methodological and translational challenges of the neuroscience of ethics. *Bioethics*, 31, 328–337.

Roskies, A. (2002). Neuroethics for the new millenium. *Neuron*, 35, 21–23.

Salles, A., Bjaalie, J.G., Evers, K., Farisco, M., Fothergill, B.T., Guerrero, M., Maslen, H., Muller, J., Prescott, T., Stahl, B.C. et al. (2019a). The human brain project: Responsible brain research for the benefit of society. *Neuron*, 101, 380–384.

Salles, A., Evers, K., Farisco, M. (2019b). The need for a conceptual expansion of neuroethics. *AJOB Neuroscience*, 10, 126–128.

Sperber, D. (1996). *Explaining Culture: A Naturalistic Approach*. Blackwell, Oxford, UK, and Cambridge, MA.

PART 1

Neuroethics as a Field

1

Examining the Ethics of Neuroscience in Contemporary Neuroethics

Cynthia FORLINI
School of Medicine, Faculty of Health, Deakin University, Geelong, Australia

1.1. Introduction

In 2002, a group of more than 150 academics in bioethics, neuroscience, philosophy, law and public policy, as well as professionals in medicine, law and journalism, gathered in San Francisco (USA) for a meeting entitled "Neuroethics: Mapping the Field" (Marcus 2002). Their common interest was to define the remit of a scholarly and practical field that studies the ethical implications of advances in brain science. Hence, the birth of contemporary neuroethics. The impetus for such a field was famously articulated by the late William Safire as the need for "a distinct portion of bioethics, which is the considerations of good and bad consequences in medical practice and biological research. But the specific ethics of brain science hits home as research on no other organ does" (Safire 2002). Part of what Safire suggested "hits home" is the millennia long project of understanding the biological and philosophical underpinnings of how the brain determines who and how we are as humans and individuals (Bennett and Hacker 2022). That project has inspired significant ethical issues concerning how we intervene in the brain and mind and how we communicate that knowledge. Studying and resolving these issues required the concerted effort of a field of study.

Neuroethics and Cultural Diversity, coordinated by Michele FARISCO. © ISTE Ltd 2023.

In the 20 years since the *Mapping the Field* meeting, neuroethics has become more mainstream in neuroscience and bioethics. It has all the trappings of an established field: an expansive literature (Leefmann et al. 2016), journals dedicated to neuroethics scholarship (i.e. Neuroethics, AJOB Neuroscience, Journal of Cognition and Neuroethics), an International Neuroethics Society, higher education training and research programs (Buniak et al. 2014) as well as dedicated funding schemes. These features came about through the tireless efforts of academic and professional pioneers who demonstrated the relevance of neuroethics to scientific, clinical and public discourses and its proven ability to bring about change in policy and practice. The aim of this chapter is to present the evolution of a branch of neuroethics (i.e. the ethics of neuroscience) by highlighting the diversity of aims, critiques and future directions of the field. The presence of an ethics of neuroscience creates a place where issues in innovation, practice and policy are foreseen and discussed to ensure that technology and neuroscience are not uncritically adopted or rejected.

1.2. A brief history of neuroethics

The endeavor that neuroethics now represents existed before its moniker. A few early references to neuroethics arose in clinical contexts grappling with innovation in neuroscience and neurology (Racine 2008). In the 1970s, Pontius, a physician, questioned the long-term implications of intervening in neurodevelopment to accelerate walking in newborns (Pontius 1973). Pontius proposed that "by raising such questions, attention is focused on a new and neglected area of ethical concern – neuroethics. In the present context, this concept stresses the importance of being aware of the neurological facts and implications while experimenting with the newborn's mobility" (Pontius 1973). This proposal foreshadowed our contemporary debates about human performance enhancement across the life span, and hasty (and sometimes uncritical) implementation of innovation in neuroscience (Racine 2010). Most importantly, her inquiry signaled a blind spot in the ethics of healthcare at the time from which neuroethics could emerge.

Later, in the 1980s, Cranford, a neurologist, proposed that a "neuroethicist" is "a neurologist who has taken a specific interest in bioethical issues and becomes an active member of [their] institutional Ethics Committee or becomes an individual consultant" (Cranford 1989). Cranford explained that a neurologist is best placed to occupy this role because "[h]e or she understands the neurological facts and has extensive clinical experience in dealing with these neuroethical dilemmas at the bedside, serves in a significant educational and consultative capacity by clarifying the neurologic facts and integrating them with the ethical and legal issues" (Cranford 1989). Where Pontius opened the door to paying attention to novel issues, Cranford proposed the need for a novel role (scholarly and practical) and the required competencies for doing so.

Contemporary neuroethics has evolved into a field that draws from many disciplines (i.e. multidisciplinary) and integrates knowledge to produce new conceptual and empirical scholarships (i.e. interdisciplinary). A neurology or psychiatry practice is not a prerequisite for engaging in this endeavor. However, the role of a *neuroethicist* is still one that creates and reframes knowledge that may come from, or needs to be applied to, a different disciplinary or practical context to inform what ought to be done in a given situation (Forlini 2017).

1.2.1. *Pluralism in the definitions of neuroethics*

The absence in this chapter of a formal definition of neuroethics is quite deliberate. Neuroethics enjoys plural perspectives on what it is and its raison d'être. One feature of neuroethics definitions is how they relate to bioethics. Neuroethics has been described as a "field of contemporary bioethics" (Racine 2010) as well as one that "intersects with biomedical ethics" (Illes and Bird 2006). The former description implies that neuroethics sits under the goals, values, theories and methods of bioethics as a home discipline. The latter description implies that, while there are some common features, neuroethics extends beyond the confines of bioethics. The mere existence of neuroethics in the orbit of bioethics is the topic of multiple critiques that will be addressed in a later section.

Neuroethics is also often defined as a function of areas of practice. Four areas emerged from the *Mapping the Field* meeting: "(i) the implications of neuroscience for notions of the self, agency and responsibility; (ii) social policy applications that make new resources such as healthcare and education available to society; (iii) therapeutic intervention through advances in clinical practice; and (iv) public discourse and training" (Illes and Bird 2006). These areas are consistent with the early focus of the neuroethics literature on neuroimaging and its impact on various domains of health, law and society (Leefmann et al. 2016). Racine (2010) revised these original areas of practice in terms of four general domains of neuroethics which address (1) the responsible conduct of *research* in neuroscience and related fields, (2) *clinical* issues in neurological and psychiatric care, (3) *public and cultural* aspects of health policy and understanding of neuroscience by diverse audiences and (4) theoretical and reflective work that examines the foundation of neuroethics, including moral reasoning and decision-making. These four domains are inclusive of the diverse topics and perspectives that have become a part of the neuroethics landscape over the past two decades. The breadth of neuroethics scholarship is extensively scoped in Buniak et al.'s (2014) bibliography of neuroethics as well as

Leefman et al.'s (2016) bibliometric analysis of the neuroethics literature. Both works highlight a fundamental distinction in the knowledge sought through neuroethics inquiry.

1.2.2. *A fundamental distinction in neuroethics inquiry*

General definitions of neuroethics do not tend to capture a distinction made early in the emergence of neuroethics as a field of study, which gives rise to different perspectives in identifying the driving forces of neuroethics inquiry. In the landmark paper "Neuroethics for the new millennium", Roskies (2002) separates *the neuroscience of ethics* from *the ethics of neuroscience*. This distinction is generally supported, but not without critique about its potential to promote essentialist and reductionist views of neurological and psychological traits (Racine and Sample 2017). Racine (2010) refers to this distinction as the *knowledge-driven perspective* of neuroethics. This perspective positions neuroethics as a truly interdisciplinary endeavor because these two parts draw from different disciplines and pose questions that require public deliberation (Roskies 2002a; Levy 2008; Racine 2010).

The neuroscience of ethics is the feature that aligns best with a definition of neuroethics that intersects with bioethics, rather than being considered a sub-field. Buniak et al. (2014) refer to the neuroscience of ethics as "first tradition neuroethics". It reflects studying "traditional philosophical notions such as free will, self-control, personal identity and intention […] from the perspective of brain function" (Roskies 2002a). This endeavor recruits findings from neuroscience broadly writ to explain *how* humans make decisions to enact morality, rather than discuss *what* a moral agent ought to do. Understanding the underlying neurocognitive mechanisms can impact the way in which we attribute intention, responsibility, and fault in legal proceedings (i.e. neurolaw), make value judgments about self-control relevant to topics like addiction and obesity, and appreciate factors that influence rational thought such as affect (Levy 2008). Roskies acknowledges that technology is a rate-limiting step in answering the myriad questions about how our brains determine who we are and whether those fundamental traits can be modified. Perhaps that explains why despite being one of the defining features of the field of neuroethics that makes neuroethics a truly unique field of study, the neuroscience of ethics is not a dominant topic in terms of the number of neuroethics publications that address it (Leefmann et al. 2016; Vidal 2018).

1.2.3. *Technology and healthcare as drivers of the ethics of neuroscience*

Roskies' designation of the ethics of neuroscience is closer to what Safire's early definition of neuroethics refers to and the description of neuroethics as a sub-field of bioethics. Roskies (2002) proposes two mandates for the ethics of neuroscience:

> (1) the ethical issues and considerations that should be raised in the course of designing and executing neuroscientific studies and (2) evaluation of the ethical and social impact that the results of those studies might have, or ought to have, on existing social, ethical and legal structures (Roskies 2002a).

The first mandate designates an ethics of practice that aligns with the traditional bioethical analysis of what ought to be done in research, clinical and public spheres. The second focuses on the ethical implications of neuroscience to reconcile new knowledge about the brain with social systems such as expectations of "normal" behavior and performance, personal and legal responsibility, as well as policies and practices that support neurocognitive health in the population. Buniak et al.'s (2014) taxonomy of topics in this "second tradition neuroethics" is extensive and includes: issues in neuroscience research, dual-use research, neuroimaging, neurogenetics, neurobiomarkers, predictive technologies, treatment-enhancement debate, neuro-psychopharmacology, invasive and non-invasive brain stimulation, neural stem cells, neural tissue and gene transplants, brain–machine interfaces, neurofeedback, synthetic impacts and posthuman issues. The top three areas of activity from a publishing perspective are brain stimulation, pharmacology and addiction, and neuroimaging (Leefmann et al. 2016). These topics are consistent with the other two neuroethics perspectives which are *technology-driven* and *healthcare-driven* (Racine 2010).

In a retrospective of neuroethics' accomplishments since its inception, Farah notes that, "You [neuroethics] are at your best when you are scanning the horizon for new scientific and technical developments that intersect in new ways with ethics and law" (Farah 2021). Farah provides examples of scientific and technical developments such as open- and closed-loop deep brain stimulation and the neuroimaging of pain resulting in neuroethics work that addresses the implications of the application of a neurotechnology. This work is consistent with the technology-driven perspective of neuroethics, which states that the field is "[...] defined by the technologies it examines rather than any particular philosophical approach" (Wolpe 2004). Those technologies might be new or emerging such as neuroimaging, brain stimulation, brain organoids – all examples of innovation from the past few decades. Neuroethics issues might also arise due to repurposing or dual use of existing technology as is the case with the non-medical use of stimulants for cognitive enhancement. That innovation is the impetus for examining what is right and just, and whether any of the existing mechanisms (e.g. policies, laws, regulation) upholding those concepts are fit for purpose. For example, early neuroethics scholarship focused heavily on neuroimaging, discussing topics such as incidental findings and the potential for mind reading (Glannon 2017). Those issues have evolved into a debate about whether there are specific rights (i.e. neurorights) related to how data and information about the brain and mind are (or ought to be) protected and preserved (Ienca 2021; Rommelfanger et al. 2022). However, proponents of the technology-driven perspective tend to exclude the neuroscience of ethics on the basis of the latter being a different topic (Farah and Wolpe 2004). The reason for this exclusion is not entirely clear, given that many of the topics in the neuroscience of ethics derive from capabilities of technology such as neuroimaging, which purports to provide visual evidence of moral reasoning or deep brain stimulation that might modify fundamental traits.

Commitment to issues about neurotechnology instead of a philosophical approach distinguishes the technology-driven perspective of neuroethics from bioethics where there is ongoing contention between normative and empirical ethicists about which approach yields the best course of action (Hurst 2010). Yet, the open-ended methodology of this perspective also serves as a point of comparison for Evers et al. (2017), who call the ethics of neuroscience "neurobioethics". The authors highlight a focus on solving moral concrete problems in the ethics of neuroscience by sometimes drawing upon philosophy, as well as other fields (Evers et al. 2017b). The open-endedness of neuroethics scholarship extends to empirical methods. In demonstrating the diversity of empirical methods that neuroethics draws upon (e.g. participant observation to qualitative interviews, from quantitative surveys to participatory action research), Paravini and Singh (2018) argue that "methodological flexibility within bioethics turns into an opportunity for generating novel methods that cross disciplinary boundaries". In this fashion, neuroethics has committed to maintaining interdisciplinarity while fostering innovation.

The *healthcare-driven perspective* of neuroethics mirrors medical and clinical ethics closely. Some authors have suggested that: "The single most important factor supporting [neuroethics] is the opportunity for an increased focus and integration of the ethics of related research to improve patient care" (Racine and Illes 2008). Given the focus on the brain, the patient care referred to here is in neurology and psychiatry. Functional neurosurgery, deep brain stimulation and disorders of consciousness were prominent topics in the evolution of neuroethics (Glannon 2017). However, Fins (2017) argues that the clinical roots of neuroethics laid by Pontius and Cranford have been forgotten because few of the pioneers in neuroethics as we know it were clinicians. Instead, neuroethics evolved to focus on the application of technology and its impact on understanding human behavior. Some neuroethics scholars are working to reorient neuroethics scholarship towards a health-driven perspective by committing to a pragmatic methodology that prioritizes the lived experiences of patients and their context (Fins 2017; Pavarini and Singh 2018). Indeed, this commitment to the reality of patients is evident in recent work that examines the long-term effects of deep brain stimulation to illustrate "how extending the life span without improving quality of life may introduce a burden of harms for patients and families" (Gilbert and Lancelot 2021). There is likely more of this type of critical analysis work to do as the long-term effects of some health-related neurointerventions become apparent.

1.3. Critiques of the ethics of neuroscience

The emergence and existence of neuroethics invite significant critiques about the originality of its scholarship, legitimacy of the brain as the sole focus and required stakeholders. It is difficult to pull apart the critiques of neuroethics that pertain specifically to the ethics of neuroscience, as is the topic of this chapter, from those pertaining to the neuroscience of ethics as some position the ethics of neuroscience as contingent on what we believe the neuroscience of ethics can show us (Vidal 2018). However, many of these critiques have been answered and served to fortify the justification of a new field of study, confirming that there is indeed need for the enterprise of neuroethics. The critiques also provide a roadmap for the development of neuroethics.

1.3.1. *Critique 1: reinventing the bioethics wheel*

Positioning neuroethics as a function of bioethics is a common feature in many definitions of neuroethics, albeit one that is often contested. One of the main critiques is that neuroethics content and methods are already accommodated by the established field of bioethics. Parens and Johnston (2007) draw parallels with other sub-specialties such as "nanoethics" and "genethics" to show that issues focused on healthcare and technology always converge upon the fundamental bioethical principles of autonomy, justice, beneficence and nonmaleficence. The same applies to the conceptual and empirical methods that neuroethics scholarship draws upon. They are the same as those used in bioethics. As a result, neuroethics could be perceived to be recreating bioethics inquiry. Indeed, "reinventing the bioethics wheel" may include overpromising on what neuroethics can deliver and invertedly contributing to hype on a topic in academic and public spheres (Parens and Johnston 2007).

One reason for this overpromising is the anticipatory nature of neuroethics scholarship, especially in the technology-driven perspective. Vidal (2018) posits that "[t]he proleptic structure of neuroscience helps define research topics, shape communication and public understanding, inspire expectations, and identify expertise" based on what neuroethicists *believe* will be the future impact of neuroscience. Parens (Parens and Johnston 2007) provides an autobiographical example of overpromising by describing an early bioethics project in which the decisive advice on how to regulate human enhancement promised to the funder was impossible to deliver due to the unforeseen complexity of distinguishing enhancement from other salient practices (i.e. treatment). Indeed, there is a similar story to tell on the topic of cognitive enhancement, which serves as a cautionary tale about anticipating ethical issues independently of their real-world context. Cognitive enhancement was a pillar of early neuroethics, spurring numerous empirical studies and much hand-wringing about potential ethical issues in competitive environments (Erler and Forlini 2020). However, despite abundant data and multiple possible ethical perspectives to choose from, neuroethics still has not produced any actionable insight into cognitive enhancement (Forlini 2020). Some have argued that the speculative nature of the cognitive enhancement debate is the cause of this lack of insight because putative cognitive enhancement technologies have not demonstrated the characteristic safety and efficacy that would cause imminent ethical dilemmas (Fins 2008). They are neither so efficacious as to encourage widespread use, nor so harmful as to warrant new forms of regulation (Erler and Forlini 2020). Indeed, there is emerging evidence showing that stakeholders in academic environments where cognitive enhancement is discussed as being rife and problematic do not consider the non-medical use of prescription medication an ethical problem, but rather a health-related one (Dunn et al. 2021). Fins (2008) argues that this type of speculation "comes at the cost of therapeutic engagement of patients who have heretofore been marginalized and sequestered from the fruits of neuroscience and the possibility of therapeutics". While there may be facets of cognitive enhancement that provoke serious ethical dilemmas in the future, current neuroethical resources might be more appropriately allocated to other areas of need.

The attempt to distinguish neuroethics from bioethics has other unintended consequences. Vidal (2018) posits that the work of distinguishing neuroethics from bioethics is performative because it helps to bring about the reality it describes, that is, a distinct field. Furthermore, Parens and Johnston warn that focusing on neurotechnology can exaggerate the progress of research and contribute to hype about emerging or existing technology and their ethical, social and legal implications (Parens and Johnston 2007). Contention about hype in neuroethics has arisen in the literature. Mind reading and lie detection have long been topics of interest in neuroethics (Farah and Wolpe 2004). This interest has been revived and discussed through the expectation that brain–computer interfaces, non-invasive brain stimulation and portable electroencephalograms can gather brain-derived data that pose threats to privacy (Ienca et al. 2018). Wexler (2019) contested the imminence of those threats based on Ienca et al.'s misrepresentation of the state of the technology in question by conflating theoretical feasibility with real-world issues. The critique suggested that this type of anticipatory work contributes to "neurohype" and precipitous regulation that may not be viable or needed (Wexler 2019). Recent bioethics scholarship on the ethics of artificial intelligence has also addressed the challenges of engaging in anticipatory ethics without the key details about specific applications (Rogers et al. 2021).

1.3.2. Critique 2: a dose of "neuroskepticism"

Safire's statement about the brain "hitting home" like no other organ is the foundation of another major critique of neuroethics. The critique is described by the term "neuroskepticism", coined by Whitehouse, which reflects critical appraisal of the foundation and perpetuation of neuroethics (2012). Here, the critique is not about novelty, but about whether or not neuroscience is poised to provide the expected insights. Neuroethics is predicated on the assumption that technologies like neuroimaging are "a window into the mind and into the constitutive features of personhood" (Vidal 2018). Here, Vidal refers to the "neuroexceptionalism" that bolsters neuroethics' autonomy as a field of inquiry. However, Vidal adds that, "valuable as they are, neuroethics discussions of myths and misunderstandings about neuroimaging are constructed so as to strengthen that postulate, never to call it into question" (Vidal 2018). Indeed, it has been proposed that neuroethics is too much aligned with the promise of neuroscience to provide anything other than an optimistic perspective on the development and implementation of neuroscience (Brosnan 2011).

The lack of critical distance manifests in two ways. First, Whitehouse notes that the "neuro" prefix is used to "enhance the credibility of the various terms it modifies" in professional and public fora (Whitehouse 2012). Indeed, there is empirical evidence to suggest that neuroscientific explanations of psychological phenomena increase the credibility of statements even when neuroscience is irrelevant (Weisberg et al. 2008). We only have to look towards the numerous burgeoning neuro-fields such as neurolaw, neuropolitics, neuroeducation, neuromarketing, neuroaesthetics, neurotheology and neuroleadership to see the extent of the promise neuroscience has made "to reconfigure these fields of human activity to a significant degree" (Whitehouse 2012). However, Whitehouse argues that the neuro prefix is actually promoting a neuroessentialist view of all these fields, a view that proposes "overly simplistic scientific solutions to the complex human and natural system failures"

(Whitehouse 2012). It is also a reductionist view that conceptualizes the brain as an independent organ, ignoring that it is part of the rest of the body (i.e. embodied) and an external physical environment (i.e. embedded) (Glannon 2009). Whitehouse adds that the focus on neuroscience ignores the psychosocial aspects of cognition and behavior that neuroscience appropriates in addition to de-prioritizing issues of more pressing importance such as climate change.

Second, some have suggested that the lack of critique is caused by a skew in neuroethics stakeholders towards professions that have an optimistic view of "neuro-progress" (Whitehouse 2012). Leefman et al.'s (2016) bibliometric analysis of neuroethics literature showed that "a large part of neuroethics is the result of the engagement of neuroscientists, psychologists and medical doctors with normative and conceptual issues, for which they have not sufficiently trained in their academic education". The work of these major stakeholders in neuroethics depends on the promise and progress of neuroscience. Their participation and influence create a "simple" mechanism according to Vidal (2018): "neuroethics first predicts tensions that neuroscientific progress will generate, and then offers to manage them". The result is a self-referentiality that ensures the longevity of the field but creates major challenges for neuroethics. It calls into question what type of training is required to walk the fine line between neuroscientific optimism and productive critical appraisal, especially given the neuroethical commitment to interdisciplinarity (Forlini 2017; Forlini et al. 2017). This concentration of neuro-optimistic stakeholders also makes neuroethics an insular environment. Neuroscientists are not sufficiently aware of the salient ethics issues related to neuroscience or to neuroethics scholarship to develop solutions (Brosnan and Cribb 2014). Brosnan and Cribb (2014) suggest that this gap is due to divergence in the respective substantive ethical agendas of ethics and the neurological sciences. The former often focuses on the permissibility of neuroscience (i.e. the ethics of neuroscience broadly writ) and the latter on its day-to-day implementation (i.e. project-specific research ethics).

1.4. Responses to critiques of the ethics of neuroscience

The myriad critiques of the endeavor, conceptual underpinnings and empirical methods of neuroethics have not stymied work in the area. This section presents a cursory overview of some ways in which those working in neuroethics have responded to critiques about the novelty of the field and its relationship to neuroscience. These responses have helped to not only establish neuroethics as a field of study but also expand its engagement with other disciplines and the public.

There are several ways in which neuroethics has distinguished itself from bioethics. In a direct response to Parens and Johnston's question "Does it make sense to speak of neuroethics?" (Parens and Johnston 2007), Racine outlined eight distinct domains of neuroethics scholarship, noting that critiques are based on the technology-driven perspective of neuroethics, which does not represent the whole of neuroethics scholarship (Racine 2008). Racine's pragmatist approach to neuroethics puts the welfare of patient populations at the forefront of neuroethics work and the eight domains. However, there are two important points that arise from those domains that represent key contributions of neuroethics. One is bioethics' lack of attention to basic and clinical neuroscience. This lack of attention is often cited as the origin of the need for neuroethics but has also given rise to conceptual innovation. One example is Giordano's (2017) proposal for an *Operational Neuroethical Risk Analyses and Management Paradigm* (ON-RAMP), which is designed specifically for the study of neuroethicoloegal and social issues. ON-RAMP goes beyond the traditional anticipatory ethics approach by querying the type and objectives of the science and technology, framing the intended applications in their context of use, defining the domains (mostly socio-cultural) that are likely to be impacted by the neuroscience and technology, and modeling and plotting trajectories of effect (Giordano 2017). Another example is the development of a conceptual approach to neuroethics, which emerged from a need to challenge "neuroscientific objectivity in the production and interpretation of 'facts' about the brain" (Evers et al. 2017). At a philosophical level, "fundamental neuroethics" designates the analysis of "the meaning of neuroscientific terms, theories, and interpretations, as well as their relationship to how the same or similar terms are used in other disciplines and in ordinary, nonscientific discourse" (Farisco et al. 2018). These neuroethical innovations demonstrate that neuroethics has "stretched" classical ethics in different ways by questioning concepts and evidence that are often taken for granted (Buller 2018). They also disprove the neuroskeptic perspective of neuroethics, which is uncritically optimistic about neuroscience by demonstrating a rare self-reflection on the nature, integrity and application of neuroethics scholarship.

The other area that is highlighted by Racine's domains of neuroethics is its intentional interdisciplinarity. Neuroethics was conceived of as an interdisciplinary field that requires diverse input and reflexive practice to foster useful real-world solutions (Racine and Sample 2017). It draws from as well as informs other epistemological domains. While academic structures might require it, disciplinary exclusivity is incompatible with this endeavor. Engaging in multidirectional communication to create common ground and shared understanding of issues, key concepts, empirical data and expertise is a key mandate of neuroethics (Racine et al. 2005; Illes et al. 2010). This "neurotalk" can help to engage stakeholders and guard against the biases and hype that neuroskeptics caution against (Illes et al. 2010). However, the hype is not exclusively an enemy to good neuroscience and neuroethics. It might be regarded as an ally, a canary in the coal mine of sorts. "Neurohype" can be an indicator of which areas of neuroscience are likely to be prioritized by research, healthcare or policy (van de Werff et al. 2016). Paying attention to neurohype allows researchers to anticipate and reflect on the ethical implications of research and how it supports or refutes current norms and values.

There is still much work to be done to bridge some longstanding gaps between neuroethics and other stakeholders. Namely, from a neuroscience perspective, neuroethicists are seen as the "ethics police", the result of a conflation with traditional research ethics and engaging in "science fiction", a euphemism for the unrealistic discourses of anticipatory or speculative ethics (Wexler 2020). One attempt to clarify the neuroethics endeavor for neuroscience stakeholders and create common ground is the "Neuroethics Questions to Guide Ethical Research in the International Brain Initiatives" (NeQN) (Global Neuroethics Summit Delegates et al. 2018). The NeQN cover aspects of the ethics of neuroscience and the neuroscience of ethics that lead to ethically informed neuroscience such as considering the impact of a neuroscientific model of a disease or the moral significance of neural models that are developed in laboratories. These series of questions are inspired by the numerous national brain projects founded across the globe. They aim to unpack not only the scientific and social values that are driving neuroscience research but also the cultural influences. Ultimately, the overarching aim of the interdisciplinary work of neuroethics is to ensure that the voices and values relevant to an area of research, scholarship, practice or policy are heard and examined. In this fashion, stakeholders representing these areas are mutually accountable to one another, sharing the responsibility for ensuring that neuroscience and its impact are portrayed and conducted to the benefit of all involved (Forlini et al. 2015).

1.5. Blind spots in the ethics of neuroscience are opportunities for engagement

During two decades of formal scholarship, neuroethics was founded, debated, defended and developed. Indeed, this work has led to the consideration and integration of neuroethics scholarship and experts in large-scale brain projects (Global Neuroethics Summit Delegates et al. 2018). That progress is remarkable, but there are still some blind spots in both content and methods that neuroethics needs to address. The Emerging Issues Task Force of the International Neuroethics Society proposed three topics that are poised to shape the future of neuroethics (Emerging Issues Task Force International Neuroethics Society 2019). First, "rapid and continuous increases in knowledge and technical capability" in neuroscience as well as related fields such as data science, communications and information technology.

These innovations will require the ability to reach across disciplines to foster shared understanding and action on their application and implications. Second, "expanding [the] global landscape of large-scale neuroscience research that generates increasingly diverse perspectives and greater access to knowledge" denotes the need for a neuroethics that is inclusive of different socio-cultural value systems and transparent with respect to assumptions and biases. Lastly, "[i]ncreases in commercial, military, and government applications of neurotechnologies" especially direct-to-consumer technologies, are inspiring a recalculation of associated benefit and risk, especially privacy. These topics are revised extensions of contemporary neuroethics issues that are accompanied by important mandates for changing how we "do" neuroethics.

Investigating emerging content areas will require action on two methodological fronts. The aim is to ensure that neuroethics is equipped to translate scholarship into practice, a concept called "translational neuroethics" by Wexler and Specker Sullivan (2021). Their suggestion is to reshape the current speculative and technocentric neuroethics by aligning it more closely to real-world contexts by being "integrated, inclusive and impactful" (Wexler and Specker Sullivan 2021). Stakeholder engagement, defined broadly to include other disciplines, experts, industry, communities and the public, is a key feature of neuroethics (Specker Sullivan and Illes 2017). Specker Sullivan and Illes (2017) argue that engagement in neuroethics will need to be fortified to tackle the types of issues discussed above. They advise that engagement will need to be early on in projects and innovation, ideally integrated through the life span of the project, fit for purpose and actionable by the stakeholder involved (e.g. scientists or communities) and sustainable through the creation of pathways and funding that support interdisciplinary collaboration. The Human Brain Project, a European multidisciplinary brain project, has been enacting these recommendations as neuroethics is integrated into the project guided by the "reflection" category of the Responsible Research and Innovation framework (Salles and Farisco 2020). This approach to neuroethics engagement and integration has also led to conceptual innovation in neuroethics, which reinforce its foundations as a field of study (Farisco et al. 2018).

One of the ongoing issues that stakeholders of the Human Brain Project are still grappling with is "how to acknowledge, understand, and manage cultural considerations in brain research in itself and its implications" (Salles and Farisco 2020), which constitutes the second methodological blind spot. There is a glaring lack of diversity in neuroscience and neuroethics (Matshabane 2021). Inclusion of culturally and linguistically diverse communities in these domains is essential for socially just and impactful research and public discourses (Matshabane 2021). Ray illustrates how people of color are underrepresented in many major neuroethics topics and provides an example of the difference more representative samples would make to the cognitive enhancement debate (Ray 2020). The same can be said for the need to engage with the populations of low- and middle-income countries to address global mental health so as to avoid applying concepts, frameworks and technologies developed in other contexts that will not resonate (Singh 2020). Operating on the scale of national brain projects and global mental health means that neuroethics must do more to ensure that neuroethics inquiry is appropriately representative and fit for the purposes of the people and communities it is meant to benefit.

1.6. Conclusion

The pluralistic definitions of the ethics of neuroscience (as well as neuroethics more broadly) and its driving forces might be confusing to those expecting a scholarly and practical field of study to have exclusive disciplinary commitments. Admittedly, the lack thereof was a challenge for early trainees in neuroethics (Forlini 2017). Neuroethics could be perceived as a boundless catch-all field with insufficient guiding principles to know what questions and scholarship fall within its remit and who can call themselves a "neuroethicist". Racine and Sample (2017) highlight some outstanding questions about the goals of neuroethics for training, research, disciplinary and organizational status, public engagement, ethics and mentoring to help more clearly situate neuroethics scholarship in the academic endeavor and, in turn, neuroethics knowledge in social and public discourse. However, they caution that reconciling divergent perspectives in neuroethics has an opportunity cost because "[a]ny time and effort spent on new syntheses are resources that could be spent applying extant theory" (Racine and Sample 2017). Instead, pluralism can be viewed as an evolutionary advantage for neuroethics which accommodates organic growth and innovation in the field according to contemporary issues, and an inclusive approach to expertise that addresses the appetite for interdisciplinary work in research (Forlini 2017). As a result, neuroethics is an agile field, which ensures its relevance and longevity in academic and public discourses.

1.7. References

Bennett, M.R. and Hacker, P.M.S. (2022). *The Philosophical Foundations of Neuroscience*, 2nd edition. Blackwell Publishing, Hoboken, NJ.

Brosnan, C. (2011). The sociology of neuroethics: Expectational discourses and the rise of a new discipline. *Sociol Compass*, 5(4), 287–297.

Brosnan, C. and Cribb, A. (2014). Between the bench, the bedside and the office: The need to build bridges between working neuroscientists and ethicists. *Clin Ethics*, 9(4), 113–119.

Buller, T. (2018). The new ethics of neuroethics. *Camb Q Healthc Ethics*, 27(4), 558–565.

Buniak, L., Darragh, M., Giordano, J. (2014). A four-part working bibliography of neuroethics. Part 1: Overview and reviews – Defining and describing the field and its practices. *Philos Ethics Humanit Med*, 9(9). doi: 10.1186/1747-5341-9-9.

Cranford, R.E. (1989). The neurologist as ethics consultant and as a member of the institutional ethics committee. The neuroethicist. *Neurol Clin*, 7(4), 697–713.

Dunn, M., Dawson, P., Bearman, M., Tai, J. (2021). "I'd struggle to see it as cheating": The policy and regulatory environments of study drug use at universities. *High Educ Res Dev*, 40(2), 234–246.

Emerging Issues Task Force International Neuroethics Society (2019). Neuroethics at 15: The current and future environment for neuroethics. *AJOB Neurosci*, 10(3), 104–110.

Erler, A. and Forlini, C. (2020). Neuroenhancement [Online]. Available at: https://www.rep.routledge.com/articles/thematic/neuroenhancement/v-1.

Evers, K., Salles, A., Farisco, M. (2017). Theoretical framing of neuroethics: The need for a conceptual approach. In *Debates About Neuroethics: Perspectives on Its Development, Focus, and Future*, Racine, E. and Aspler, J. (eds). Springer International Publishing, Cham.

Farah, M.J. (2021). Checking in with neuroethics. *Hastings Cent Rep*, 51(1), 3.

Farah, M.J. and Wolpe, P.R. (2004). Monitoring and manipulating brain function: New neuroscience technologies and their ethical implications. *Hastings Cent Rep*, 34(3), 35–45.

Farisco, M., Salles, A., Evers, K. (2018). Neuroethics: A conceptual approach. *Camb Q Healthc Ethics*, 27(4), 717–727.

Fins, J.J. (2008). A leg to stand on: Sir William Osler and Wilder Penfield's "neuroethics". *AJOB*, 8(1), 37–46.

Fins, J.J. (2017). Toward a pragmatic neuroethics in theory and practice. In *Debates About Neuroethics: Perspectives on Its Development, Focus, and Future*, Racine, E. and Aspler, J. (eds). Springer International Publishing, Cham.

Forlini, C. (2017). Growing up with neuroethics: Challenges, opportunities and lessons from being a graduate student at a disciplinary crossroads. In *Debates About Neuroethics: Perspectives on the Field's Development, Focus, and Future*, Racine, E. and Aspler, J. (eds). Springer International Publishing, Cham.

Forlini, C. (2020). Empirical data is failing to break the ethics stalemate in the cognitive enhancement debate. *AJOB Neurosci*, 11(4), 240–242.

Forlini, C., Partridge, B., Lucke, J., Racine, E. (2015). Popular media and bioethics scholarship: Sharing responsibility for portrayals of cognitive enhancement with prescription medications. In *Handbook of Neuroethics*, Clausen, J. and Levy, N. (eds). Springer International Publishing, Cham.

Forlini, C., Bell, E., Carter, A. (2017). Throwing the ethics (hand) book at professional organizations in the neurological sciences. *AJOB Neurosci*, 8(4), W1–W2.

Gilbert, F. and Lancelot, M. (2021). Incoming ethical issues for deep brain stimulation: When long-term treatment leads to a new form of the disease. *J Med Ethics*, 47(1), 20.

Giordano, J. (2017). Toward an operational neuroethical risk analysis and mitigation paradigm for emerging neuroscience and technology (neuroS/T). *Exp Neurol*, 287(Pt 4), 492–495.

Glannon, W. (2009). Our brains are not us. *Bioethics*, 23(6), 321–329.

Glannon, W. (2017). The evolution of neuroethics. In *Debates About Neuroethics: Perspectives on Its Development, Focus, and Future*, Racine, E. and Aspler, J. (eds). Springer International Publishing, Cham.

Global Neuroethics Summit Delegates, Rommelfanger, K.S., Jeong, S.J., Ema, A., Fukushi, T., Kasai, K., Ramos, K.M., Salles, A., Singh, I. (2018). Neuroethics questions to guide ethical research in the international brain initiatives. *Neuron*, 100(1), 19–36.

Hurst, S. (2010). What empirical turn in bioethics? *Bioethics*, 24(8), 439–444.

Ienca, M. (2021). On neurorights. *Front Hum Neurosci*, 15(485). doi: 10.3389/fnhum. 2021.701258.

Ienca, M., Haselager, P., Emanuel, E.J. (2018). Brain leaks and consumer neurotechnology. *Nat Biotechnol*, 36(9), 805–810.

Illes, J. and Bird, S.J. (2006). Neuroethics: A modern context for ethics in neuroscience. *Trends Neurosci*, 29(9), 511–517.

Illes, J., Moser, M.A., McCormick, J.B., Racine, E., Blakeslee, S., Caplan, A., Hayden, E.C., Ingram, J., Lohwater, T., McKnight, P. et al. (2010). Neurotalk: Improving the communication of neuroscience research. *Nat Rev Neurosci*, 11(1), 61–69.

Leefmann, J., Levallois, C., Hildt, E. (2016). Neuroethics 1995–2012. A bibliometric analysis of the guiding themes of an emerging research field. *Front Hum Neurosci*, 10(336). doi: 10.3389/fnhum.2016.00336.

Levy, N. (2008). Introducing neuroethics. *Neuroethics*, 1(1), 1–8.

Marcus, S.J. (eds) (2002). *In Neuroethics: Mapping the Field*. Dana Press, Washington D.C.

Matshabane, O.P. (2021). Promoting diversity and inclusion in neuroscience and neuroethics. *EBioMedicine*, 67(103359).

Parens, E. and Johnston, J. (2007). Does it make sense to speak of neuroethics? Three problems with keying ethics to hot new science and technology. *EMBO Rep*, 8 (Spec Issue), S61–S64.

Pavarini, G. and Singh, I. (2018). Pragmatic neuroethics: Lived experiences as a source of moral knowledge. *Camb Q Healthc Ethics*, 27(4), 578–589.

Pontius, A.A. (1973). Neuro-ethics of "walking" in the newborn. *Percept Mot Skills*, 37(1), 235–245.

Racine, E. (2008). Comment on "Does it make sense to speak of neuroethics?" *EMBO Rep*, 9(1), 2–3.

Racine, E. (2010). *Pragmatic Neuroethics: Improving Treatment and Understanding of the Mind-Brain*. MIT Press, Cambridge, MA.

Racine, E. and Illes, J. (2008). Neuroethics. In *The Cambridge Textbook of Bioethics*, Singer, P.A. and Viens, A.M. (eds). Cambridge University Press, Cambridge.

Racine, E. and Sample, M. (2017). The competing identities of neuroethics. In *The Routledge Handbook of Neuroethics*, L. Syd M Johnson and Karen S. Rommelfanger (eds). Routledge, Taylor & Francis Group, New York.

Racine, E., Bar-Ilan, O., Illes, J. (2005). FMRI in the public eye. *Nat Rev Neurosci*, 6(2), 159–164.

Ray, K.S. (2020). When people of color are left out of research, science and the public loses. *AJOB Neurosci*, 11(4), 238–240.

Rogers, W.A., Draper, H., Carter, S.M. (2021). Evaluation of artificial intelligence clinical applications: Detailed case analyses show value of healthcare ethics approach in identifying patient care issues. *Bioethics*, 35(7), 623–633.

Rommelfanger, K.S., Pustilnik, A., Salles, A. (2022). Mind the gap: Lessons learned from neurorights. *Science & Diplomacy*, February 2022.

Roskies, A. (2002). Neuroethics for the new millenium. *Neuron*, 35(1), 21–23.

Safire, W. (2002). Visions for a new field of neuroethics. In *Neuroethics: Mapping the Field*, Marcus, S.J. (ed.). Dana Press, Washington, DC.

Salles, A. and Farisco, M. (2020). Of ethical frameworks and neuroethics in big neuroscience projects: A view from the HBP. *AJOB Neurosci*, 11(3), 167–175.

Singh, I. (2020). Neuroscience for global mental health. *Cerebrum: The Dana Forum on Brain Science*, September–October 2022.

Specker Sullivan, L. and Illes, J. (2017). Models of engagement in neuroethics programs: Past, present, and future. In *Debates About Neuroethics: Perspectives on Its Development, Focus, and Future*, Racine, E. and Aspler, J. (eds). Springer International Publishing, Cham.

Van De Werff, T., Slatman, J., Swierstra, T. (2016). Can we "remedy" neurohype, and should we? Using neurohype for ethical deliberation. *AJOB Neurosci*, 7(2), 97–99.

Vidal, F. (2018). What makes neuroethics possible? *Hist Hum Sci*, 32(2), 32–58.

Weisberg, D.S., Keil, F.C., Goodstein, J., Rawson, E., Gray, J.R. (2008). The seductive allure of neuroscience explanations. *J Cogn Neurosci*, 20(3), 470–477.

Wexler, A. (2019). Separating neuroethics from neurohype. *Nat Biotechnol*, 37(9), 988–990.

Wexler, A. (2020). The urgent need to better integrate neuroscience and neuroethics. *AJOB Neurosci*, 11(3), 219–220.

Wexler, A. and Specker Sullivan, L. (2021). Translational neuroethics: A vision for a more integrated, inclusive, and impactful field. *AJOB Neurosci*, 1–12. doi:10.1080/21507740.2021.2001078.

Whitehouse, P.J. (2012). A clinical neuroscientist looks neuroskeptically at neuroethics in the neuroworld. In *The Neuroscientific Turn*, Littlefield, M.M. and Johnson, J.M. (eds). University of Michigan Press, Ann Arbor, MI.

Wolpe, P.R. (2004). Neuroethics. In *The Encyclopedia of Bioethics*, Post, S.G. (ed.). MacMillan Reference, New York.

2
Neuroscience of Ethics

Georg NORTHOFF[1,2,3]

[1] Faculty of Medicine, Centre for Neural Dynamics, The Royal's Institute of Mental Health Research, Brain and Mind Research Institute, University of Ottawa, Ontario, Canada
[2] Mental Health Centre, Zhejiang University School of Medicine, Hangzhou, China
[3] Centre for Cognition and Brain Disorders, Hangzhou Normal University, China

2.1. Introduction

Neuroethics is concerned with the relationship between neuroscientific findings and ethical concepts including free will, moral judgment, self, among others. On the one hand, it focuses on the investigation of the psychological and neural conditions of these ethical concepts and, on the other, on ethical problems arising from neuroscientific advances.

Roskies (2002) distinguishes correspondingly between the ethics of neuroscience and the neuroscience of ethics. The former deals with ethical problems arising in neuroscience, such as the validity of informed consent in psychiatric patients, enhancement of cognitive functions by neuroscientific interventions and coincidental findings in neuroimaging. The latter investigates neural mechanisms that may underlie ethical concepts such as informed consent, moral judgment and will (Figure 2.1).

The purpose of this chapter is to demonstrate the relevance of empirical findings for issues in the neuroscience of ethics and the ethics of neuroscience. While the conceptual distinction between the two holds firm, empirical reality often provides a more blurred picture. Three examples at the interface between neuroscience and ethics demonstrate that the line between the neuroscience of ethics and the ethics of neuroscience can become unclear, with particular relevance for psychiatry.

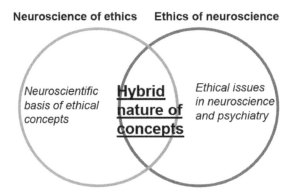

Figure 2.1. *Distinction between ethics of neuroscience and neuroscience of ethics*

Firstly, we explore how data clearly shows that the brain's neuronal activity aligns to its ecological context, implying a relational and spatio-temporal model of brains. Secondly, we examine the concept of self in a neurorelational way, on the premise that the self as the basis of agency cannot be reduced to the brain, but instead to the relationship between the external world and the brain. Thirdly, we discuss the issue of self-enhancement in the context of deep brain stimulation.

2.2. Example I: a non-reductionistic and neuro-ecological model of brains

2.2.1. *History of neuroscience – passive versus active models of brains*

The way neuroscience researchers approach the study of the brain can have a significant impact on their empirical investigations, as well as on the interpretation of their philosophical implications. One model, favored by the British neurologist Sherrington (1906), proposed that the brain and the spinal cord were primarily reflexive; that is, the brain reacts in predefined and automatic ways to sensory stimuli. Those stimuli from outside the brain, originating in either the body or the environment, are assumed to determine subsequent neural activity. The resulting activity, and more generally any neural activity in the brain, is then traced back to the stimuli to which the brain passively reacts. We may therefore speak of the passive model.

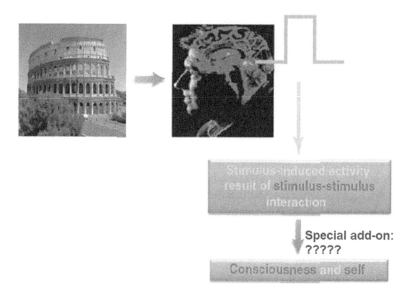

Figure 2.2a. *Passive model of the brain: Neural activity resulting sufficiently from extrinsic stimuli*

Thomas Brown, one of Sherrington's students, advanced an alternative view, namely that neural activity in the spinal cord and brain stem is not driven and sustained by external stimuli, but by spontaneous activity originating in the brain itself. Hans Berger, who introduced the electroencephalogram (EEG), also observed spontaneous activity within the brain that remained independent of any external stimuli (Berger 1929). Other neuroscientists agreed with Brown, proposing that the brain actively generates operational–behavioral activity known as spontaneous activity (Northoff 2014a, 2014b; Raichle 2015). The idea of central activity has gained traction in neuroscience with the observation of spontaneous oscillations, as well as connectivity between different regions of the brain and what is referred to as the default-mode network DMN (Raichle et al. 2001; Greicius et al. 2003; Raichle 2015); the DMN is a network that mainly includes regions in the middle of the brain that have been shown to be related to our experience or sense of self (see below for details). These observations highlight the central role of the brain's spontaneous activity, including both resting state and stimulus-induced activity; the implication is an active model. This is well illustrated in a passage by Kurt Goldstein:

> The nervous system has often been considered as an organ at rest, in which excitation arises only as a response to stimuli. [...] It was not recognized that events that follow a definite stimulus are only an

expression of a change of excitation in the nervous system, that they represent only a special pattern of the excitation process (Goldstein 2000, pp. 95–96).

2.2.2. Neuroscience – passive versus active models of the brain

The question of which model of the brain is valid has gained increased attention with the discovery of the DMN, a neural network that covers various regions in the so-called cortical midline structures (CMS) (Northoff and Bermpohl 2004; Northoff et al. 2006; Qin and Northoff 2011; Andrews-Hanna et al. 2014). The DMN shows particularly high levels of metabolism (when compared to the rest of the brain) and neural activity in the absence of any external stimuli.

The DMN's high levels of resting state activity are associated with diverse mental features, including sense or experience of self, consciousness, inner thoughts which are also described as mind wandering, episodic memory retrieval, time perception of both past and future, and random thoughts. Given this wide range of functions, the DMN's role remains unclear.

It is clear that the nature of the DMN supports the concept of an active model of the brain. Put more philosophically, the active model has been regarded as similar to Immanuel Kant's argument against a passive model of the mind (Kant 1998; Northoff 2018). David Hume proposed the opposite view, namely that impressions of external stimuli completely determined mental activity. This dispute has resurfaced in the context of theoretical neuroscience.

2.2.3. A spectrum model – the hybrid nature of the brain's activity

An empirically plausible model of brain activity that takes into account the relationship between spontaneous and stimulus-induced activities would be highly desirable (see Northoff (2018) for details). The brain neither generates its neural activity in a completely passive way, driven by external stimuli, nor in an exclusively active way, driven by spontaneous activity. Given the evidence, we need to accept a model of the brain that undermines the passive/active dichotomy and integrates both in a spectrum that allows for categorizing different forms of neural activity according to the degree of the brain's participation in generating that activity.

A spectrum model assumes that different sorts of neural activity involve various levels of resting state, some more active, others more passive. This is relevant when placing the brain in the context of body and environment. Neural activity is thus both intrinsic to the brain, the body and the environment – with the three usefully referred to as a "trinity" (Edelman et al. 2011).

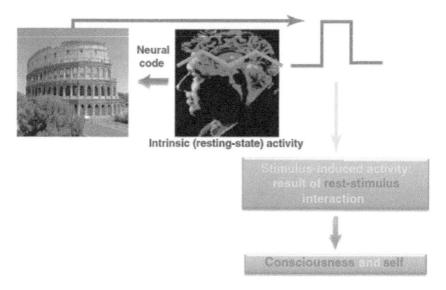

Figure 2.2b. *Spectrum model of the brain: Neural activity resulting from the interaction between intrinsic resting state activity and extrinsic stimuli*

2.2.4. The brain's spontaneous activity – constitution of its own spatio-temporal structure on a functional level

The brain's spontaneous activity is relevant for mental features such as consciousness and self-awareness. They also appear to relate, albeit in an ill-defined way, to the way the brain's spontaneous activity constructs its own "space–time" at a functional (rather than anatomical–structural) level (Northoff et al. 2020). On the spatial side, this concerns the constitution of a particular topography with different networks being related to each other in a hierarchical way (Golesorkhi et al. 2021). On the temporal side, the spontaneous activity of the brain can also be characterized by a complex temporal structure whose neural activity fluctuates in different bands of frequency. These are coupled; for example, slower frequency bands with higher ones. The result is that complex temporal structures in the brain's intrinsic activity relate, albeit unclearly, to spatial structures, as well as to a range of neural networks.

2.2.5. *Spontaneous activity and mental features – neuro-ecological rather than neuronal*

Spontaneous activity in the spatio-temporal structure extends beyond the brain and is aligned to the body (e.g. the heart and stomach) and to the external world (Bab-Rabello et al. 2016; Richter et al. 2017; Tallon-Baudry et al. 2018), suggesting a spatio-temporal alignment of the brain to the body (Northoff and Huang 2017; Northoff 2018). Alignment to the world is especially obvious when listening to music and dancing; that is, we align our brain's temporal features and rhythms of its neural activity (its frequencies and synchronization) to the temporal feature and rhythm of the music or, more generally, to the world (Schroder and Lakatos 2008; Schroeder et al. 2008).

The above forms of alignment are central for the state of consciousness; the better the alignment the more we can become conscious of the body and the world. We therefore posit a spatio-temporal model of both consciousness and mental features in general (Northoff 2014a, 2014b, 2018; Northoff and Huang 2017). The brain and its spatio-temporal features must be related to those of the world to make consciousness possible; if they are not related, consciousness is lost, as in disorders of unresponsive wakefulness, sleep and anesthesia. Most importantly, the extension of the spatio-temporal structure beyond the brain and body to the world signifies spontaneous activity as intrinsically neuro-ecological and relational, also entailing a non-reductionistic view of the brain. This is of utmost importance to psychiatric disorders, where the mental changes can be traced to the impact of social and developmental factors such that both neuronal and social changes cannot be separated.

2.2.6. *Psychiatric disorders – "spatio-temporal psychopathology"*

We are now ready to confront the relevance of the spectrum model in the case of psychiatric disorders. In schizophrenia and bipolar disorder (BD), for example, major changes occur in the brain's spontaneous activity (Martino et al. 2016; Northoff and Duncan 2016). This is manifested in the networks' resting states, as well as in the variability of their neural activity. At the same time, patients with these conditions react abnormally or differently in stimulus-induced or task-evoked activity. The abnormal neuronal speed of spontaneous activity in these patients, as reflected in neuronal variability, is related to the experience of abnormal consciousness of time speed which, in turn, manifests as symptoms such as abnormal speed of thought or movement, like psychomotor agitation and thought racing in manic episodes of BD (Martino et al. 2016; Northoff and Duncan 2016).

The spectrum model can thus be invoked when linking spontaneous and task-evoked activities to mental features (see Figure 2.2(c)).

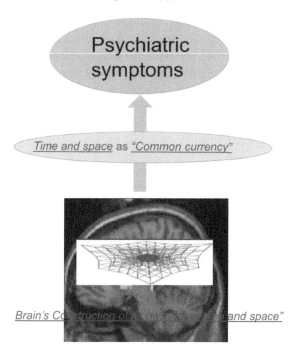

Figure 2.2c. *Spatio-temporal psychopathology*

2.3. Example II: from the neural basis of sense of self to relational agency

2.3.1. *Neuroscience of the self – mapping distinct aspects of the self onto different brain regions*

The question for the constitution of self-awareness has been one of the most salient problems in philosophy, psychology and neuroscience. William James distinguished between three selves: physical, mental and spiritual; similar concepts of self have been discussed by neuroscientists (Northoff and Panksepp 2008). Panksepp (1998) and Damasio (1999), among others, suggest that what has been referred to as the "proto-self" in sensory and motor domains, resembles James's account of the physical self. Similarly, what has been termed as the "minimal self" (Gallagher 2005) or "core or mental self" (Damasio 1998) may correspond to James's concept of mental self. Finally, Damasio's (1998) "autobiographical self" and Gallagher's (2005) "narrative self" strongly rely on linking past, present and future, and is akin to James' spiritual self.

These selves are related to activity in distinct brain regions. For instance, the "proto-self", outlining our body in emotional and sensory-motor terms, is associated with subcortical regions like the peri-aqueductal gray matter (PAG), colliculi and tectum (Northoff and Panksepp 2008), and the "core or mental self" building on the "proto-self", with the thalamus and cortical regions, such as the ventromedial prefrontal cortex (see for instance Northoff and Panksepp 2008). Finally, the "autobiographical or extended self" is associated with cortical regions such as the hippocampus and cingulate gyrus.

Neuro-imaging studies reveal that various cortical regions, especially midline structures, integrated with subcortical regions like the thalamus, are involved in self-related processing (SRP) to yield an integrated subcortical–cortical midline system (SCMS). The assumed existence of this system is consistent with research findings that show that core self-related functioning involves both cortical and subcortical regions.

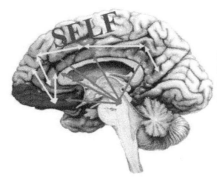

Rest-self overlap and containment: High spontaneous activity and its spatiotemporal structure overlap with and predict self-specific activity

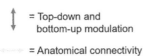

Regions
Red = Orbitomedial prefrontal cortex
Blue = Anterior cingulate cortex
Yellow = Dorsomedial prefrontal cortex
Green = Posterior cingulate and retrosplenium

Figure 2.3. *Cortical and subcortical midline structures*

Cortical regions have been studied in particular since cognitive components, including evaluative judgments, are involved (see Northoff and Bermpohl 2004;

Northoff et al. 2006); this is reflected in the experimental paradigms of most imaging studies of the self, which compare the evaluation of self-related and non-self-related stimuli. This raises the question of whether neural activity in the SCMS is associated with the cognitive functions implicated in the evaluation of stimuli as self-related, or in the self-relatedness of the stimuli themselves.

2.3.2. Self and brain – agency is ecological and relational

The data have major implications for ethical concepts like agency (e.g. the ability to voluntarily initiate an action) and moral responsibility, since both are based on the sense of self. If that same self is based on the brain's spontaneous activity and its spatio-temporal structure, the two concepts need to be defined in a spatio-temporal rather than cognitive content-based way. Consequently, agency and responsibility are not about specific content generated by cognition, but about spatio-temporal features and scales, offering a novel line of study for ethicists and philosophers. The closer a person's self is to the spatio-temporal, the more robust their agency and responsibility. Furthermore, since the self is strongly based on spontaneous activity, it is more appropriately viewed as neuro-ecological, rather than being located in the brain and its neuronal activity. The same applies to agency and moral responsibility.

Assuming a sense of moral responsibility is ecological in nature, an individual is only the symptom bearer, and not the cause, of moral failure. Analogously, the same can be said for the brain and its neuronal activity, which is associated with agency and moral responsibility. That, in turn, is relevant in the case of psychiatric disorders, which are relational and neuro-ecological, and thus spatio-temporal disorders of the world–brain relation (Northoff 2018). Given a patient's altered relationship with the world, and sense of self, they may manifest a changed sense of agency and moral responsibility. Psychiatrists therefore should not impose their values onto their patients. Instead, we may want to try to understand the world as perceived by the patient. This is important since what we need to treat are not the objective symptoms, as diagnosed by the psychiatrist, but the subjective experiences and perception of the patients themselves: it is because of the latter that they suffer and come to the psychiatrist, not because of the former. For instance, it is possible that subjects with delusions and hallucination function perfectly well and do not subjectively experience any suffering. Accordingly, experience and symptoms may dissociate from each other for which reason it is important to consider both.

2.4. Example III: enhancement of self – deep brain stimulation

Deep brain stimulation (DBS) is a potential form of treatment for severe forms of conditions such as anorexia nervosa, major depression and obsessive-compulsive disorder (Lozano and Lipsman 2013). Whether and how DBS impacts the self in terms of effectiveness and unwarranted side effects generates ethical concerns. Gilbert et al. (2017) discuss different notions of self in the context of DBS, like a predominant definition of the self in terms of cognitive contents rather than in terms of relation. Gilbert and Goering observe a relationship between the pre-operative effects of the disease on self and post-operative effects on self-estrangement.

We may then set up a specific hypothesis that the pre-operative spontaneous activity's CMS and their fluctuations in different frequencies including the association between slow and fast frequencies predict effects of DBS on self. Clinically, there is the possible option of measuring the power spectrum in the brain's spontaneous activity (via functional MRI and EEG) in relation to the self, in order to predict the risk of the DBS-induced experience of being estranged from the own self.

Gilbert et al. (2017) argue correctly that DBS is no different from other forms of treatment, such as psychotropic medications and psychological therapies, in that all of them change the spatio-temporal structure and, in turn, the self. The interaction possibly restores the self rather than replacing it. My colleagues and I have shown that inserting fetal cells when transplanting the brain tissue in Parkinson's disease does not adversely affect the patient's sense of self. On the contrary, the fetal tissue improves abnormal movement, and also restores their original sense of self and personal identity. Specifically, we asked subjects for the subjective experience of their identity before and after the fetal cell implantation (Northoff 1999). None of the five patients experienced any changes in their personal identity, that is, the experienced temporal continuity of their own self after the implantation relative to before. This may be explained by the fact that the fetal cells only impacted those regions related to movements, that is, dopaminergic cells in basal ganglia and motor cortex, but not those networks like the default-mode network and its cortical midline structure that are key for constituting the experience of self-continuity (Northoff 2017).

2.4.1. Deep brain stimulation – its application in bipolar disorder

Goering et al. (2017) point out how DBS potentially affects the self. They suggest that it registers neural activity and delivers stimulation when the neural activity related to the target symptom occurs. Although Goering et al. welcome the introduction of DBS, they are careful, if not skeptical, about its potential effects on the self, and specifically agency. They believe these effects may be mitigated by the support of family and friends, leading to their concept of "relational agency".

Can neuroscience, in tandem with neuro-philosophy, contribute to better understand the mechanisms and effects of DBS? We need to be clear about the neuronal mechanisms underlying specific psychopathological symptoms. For example, fMRI studies in BD (26) demonstrate elevated neuronal activity and variability in the resting state in the somatomotor cortex in manic patients. These findings may be associated with a hallmark of manic BD, namely spontaneous initiation of movement and subsequent psychomotor agitation. The opposite pattern is observed in depressed BD patients manifesting psychomotor retardation (see Northoff et al. (2018) for the latter).

2.4.2. Effects of DBS on the self – a quest for neuronal mechanisms

What about the effects of DBS on the self? There should be none since only motor cortical activity is altered. However, as mentioned earlier, DMN neural variability changes occur that are the opposite of those in the motor network; namely depressed patients show increased neural variability, whereas in manic patients neural variability declines.

What do these findings imply for a potential influence of DBS on the self and agency? First, "normalization" of motor neuronal variability should also re-balance its relationship to DMN/CMS; this, in turn, should "normalize" the self. We know that if we do not move, ruminations can develop and we can get "stuck" in our self. In contrast, if we are overactive, our sense of self can evaporate. In short, psychomotor behavior and self are closely linked. That is, DBS may restore the "original self" by replacing the "disease self". However, the findings may also be reversed. There may be self-estrangement from the own self, as indicated by Gilbert. This may be related to the fact that the restoration of the movement, including their subcortical–cortical motor circuit, may affect their relation to the default-mode network/cortical midline structure that mediates self-awareness in a negative way: for instance, functional connectivity between subcortical–cortical motor circuit and DMN/CMS may decrease after DBS, which, in turn, may be manifested in the experience of self-estrangement, that is, the movements are no longer properly related to the self. This is further supported by the fact that DBS may also affect agency.

DBS may also affect agency (Goering et al. 2017). Thus, movements either initiated or suppressed by external stimulation may depend on the degree to which the effects of DBS are integrated with the brain's spontaneous activity. The more effective the response to the stimulation, the more likely it is to be attributed to the self rather than to an external agent. However, we have to be aware that it is a continuum between the internal self and external environment – this may be shifted towards either end with both increased self-awareness and self-estrangement. Albeit hypothetically, this continuum of internal self-awareness and self-estrangement may be mediated by the functional connectivity between different neural networks in the brain like between the DMN/CMS (self) and sensory regions (non-self). That remains speculative though.

The introduction of DBS for individually tailored therapy is a welcome development; as a neuroscientist, I remain wary since the neuronal mechanisms underlying psychopathological symptoms remain unknown. As a neurophilosopher, I am equally wary of DBS; I suggest a need to re-define the notion of self in spatio-temporal terms that cross boundaries between neuronal and social dimensions, and thus between the brain and the world, thereby facilitating a novel view of self.

As a clinical psychiatrist, I will judge the therapeutic effects of DBS by its ecological and social manifestations concerning the perception and behavior of the respective subjects. For that, DBS must manipulate those mechanisms that account for spatio-temporal alignment while, at the same time, provide ecological and social stimuli that activate the brain's spatio-temporal alignment. Accordingly, DBS alone is not sufficient; complimentary and individually tailored ecological and social therapies (e.g. contact with others, music therapy) are necessary to fully exploit potential benefits of DBS.

2.5. Conclusion

The convergence of neuroscience and ethics has major implications for psychiatry. Neuroethics cannot be reduced to an exclusively neuronal view of concepts, such as agency and moral responsibility. A truly neuro-ecological approach is indicated where the brain's neural activity is intrinsically related and aligned to its respective environmental (and bodily) context (Northoff 2014, 2016, 2018). Such neuro-ecological and neuro-social determination is, for instance, manifested in the shaping of both neural dynamic (Duncan et al. 2016) and psychological time perception (Wang et al. 2020) by early traumatic life events. This, in turn, may impact emotion and ultimately, depression severity (Wang et al. 2020), as well as decision-making (Nakao et al. 2013). Abnormal shaping of neural dynamics and psychological time perception by strong traumatic events may, in turn, lead to psychopathological symptoms like depression (Wang et al. 2020). Accordingly, psychiatric disorders may require a neuro-ecological rather than merely neuronal model of the brain. Finally, the history of psychiatry informs us that empirical findings and conceptual issues are inseparable and highly relevant to the future study of psychiatric disorder.

2.6. References

Andrews-Hanna, J., Smallwood, J., Spreng, R. (2014). The default network and self-generated thought: Component processes, dynamic control, and clinical relevance. *Annals of the New York Academy of Sciences*, 1316, 29–52.

Babo-Rebelo, M., Wolpert, N., Adam, C., Hasboun, D., Tallon-Baudry, C. (2016). Is the cardiac monitoring function related to the self in both the default network and right anterior insula? *Philosophical Transactions of the Royal Society B: Biological Sciences*, 371, 1708.

Berger, H. (1929) About the electroencephalogram of huamns/Über das Elektrenkephalogramm des Menschen. *Archiv für Psychiatrie und Nervenkrankheiten*, 87, 527–570.

Damasio, A. (1999). *The Feeling of What Happens. Body and Emotion in the Making of Consciousness*. Houghton Mifflin Harcourt, Boston, MA.

Gallagher, S. (2005). *How the Body Shapes the Mind*. Clarendon Press, Oxford.

Gilbert, F., Goddard, E., Viaña, J., Carter, A., Horne, M. (2017). I miss being me: Phenomenological effects of deep brain stimulation. *American Journal of Bioethics Neuroscience*, 8, 96–109.

Goering, S., Klein, E., Dougherty, D. (2017). Staying in the loop: Relational agency and identity in next generation DBS for psychiatry. *American Journal of Bioethics Neuroscience*, 8(2), 59–70.

Goldstein, K. (2000). *The Organism: A Holistic Approach to Biology Derived from Pathological Data in Man*. Zone Books/MIT Press, New York.

Greicius, M., Krasnow, B., Reiss, A., Menon, V. (2003). Functional connectivity in the resting brain: A network analysis of the default mode hypothesis. *Proceedings of the National Academy of Sciences*, 100, 253–258.

Huang, Z., Zhang, J., Wu, J., Northoff, G. (2015). Decoupled temporal variability and signal synchronization of spontaneous brain activity in loss of consciousness: An fMRI study in anesthesia. *Neuroimage*, 124, 693–703.

Kant, I. (1998). *Critique of Pure Reason*. Cambridge University Press, Cambridge [Original published in 1781].

Lozano, A.M. and Lipsman, N. (2013). Probing and regulating dysfunctional circuits using deep brain stimulation. *Neuron*, 77, 406–424.

Martino, M., Magioncalda, P., Huang, Z., Northoff G. (2016). Contrasting variability patterns in the default mode and sensorimotor networks balance in bipolar depression and mania. *Proceedings of the National Academy of Sciences of the United States of America*, 113, 4824–4829.

Menon, V. (2011) Large-scale brain networks and psychopathology: A unifying triple network model. *Trends in Cognitive Sciences*, 15, 483–506.

Northoff, G. (2014a). *Unlocking the Brain. Volume I: Coding.* Oxford University Press, New York.

Northoff, G. (2014b). *Unlocking the Brain. Volume II: Conciousness.* Oxford University Press, New York.

Northoff, G. (2014c). *Minding the Brain. A Guide to Neuroscience and Philosophy.* Palgrave MacMillan, London.

Northoff, G. (2016). *Neuro-Philosophy and the Healthy Mind. Learning from the Unwell Brain.* Norton, New York.

Northoff, G. (2018). *The Spontaneous Brain. From the Mind-Body Problem to the World-Brain Problem.* MIT Press, Cambridge, MA.

Northoff, G. and Bermpohl, F. (2004). Cortical midline structures and the self. *Trends in Cognitive Sciences*, 8, 102–107.

Northoff, G. and Duncan, N. (2016). How do abnormalities in the brain's spontaneous activity translate into symptoms in schizophrenia? From an overview of resting state activity findings to a proposed spatiotemporal psychopathology. *Progress in Neurobiology*, 145(146), 26–45.

Northoff, G. and Huang, Z. (2017). How do the brain's time and space mediate consciousness and its different dimensions? Temporo-spatial theory of consciousness (TTC). *Neuroscience and Biobehavioral Reviews*, 80, 630–645.

Northoff, G. and Panksepp, J. (2008). The trans-species concept of self and the subcortical-cortical midline system. *Trends in Cognitive Sciences*, 12, 259–264.

Northoff, G., Heinzel, A., de Greck, M., Bogerts, B. (2006). Self-referential processing in our brain – A meta-analysis of imaging studies on the self. *Neuroimage*, 31, 440–445.

Northoff, G., Magioncalda, P., Martino, M., Lane, T. (2018). Too fast or too slow? Time and neuronal variability in bipolar disorder – A combined theoretical and empirical investigation. *Schizophrenia Bulletin*, 44, 54–64.

Panksepp, J. (1998). *Affective Neuroscience*. Oxford University Press, Oxford.

Qin, P. and Northoff, G. (2011). How can the brain's resting state activity generate hallucinations? A "resting state hypothesis" of auditory verbal hallucinations. *Schizophrenia Research*, 127, 202–214

Raichle, M.E. (2015). The brain's default mode network. *Annual Review of Neuroscience*, 38, 433–447.

Raichle, M.E., Snyder, A.Z., Gusnard, D., Simpson, A. (2001). A default mode of brain function. *Proceedings of the National Academy of Sciences of the United States of America*, 98, 676–682.

Richter, C., Babo-Rebelo, M., Schwartz, D., Tallon-Baudry, C. (2017). Phase-amplitude coupling at the organism level: The amplitude of spontaneous alpha rhythm fluctuations varies with the phase of the infra-slow gastric basal rhythm. *Neuroimage*, 146, 951–958.

Roskies, A. (2002). Neuroethics for the new millenium. *Neuron*, 35, 21–23.

Schroeder, C.E. and Lakatos, P. (2008). Low-frequency neuronal oscillations as instruments of sensory selection. *Trends in Neurosciences*, 32, 9–18.

Schroeder, C.E., Lakatos, P., Kajikawa, Y. (2008). Neuronal oscillations and visual amplification of speech. *Trends in Cognitive Sciences*, 12, 106–113.

Sherrington, C.S. (1906). *The Integrative Action of the Nervous System*. Yale University Press, New Haven, CT.

Tallon-Baudry, C., Campana, F., Park, H., Richter, C., Babo-Rebelo, M. (2018). The neural monitoring of visceral inputs, rather than attention, accounts for first-person perspective in conscious vision. *Cortex*, 102, 139–149.

3

Fundamental Neuroethics

Kathinka EVERS
Centre for Research Ethics and Bioethics, Uppsala University, Sweden

3.1. Science and ethics

Science and ethics[1] have traditionally been uneasy companions; they have a history of troubled relations. The ethical tradition in modern science can in Europe be traced back to the English statesman and philosopher of science, Francis Bacon, prophet and protector of the dawning scientific revolution. To Bacon, science was more than an academic quest for knowledge; it was a systematic study aiming for mastery over nature with the purpose of enabling human beings to improve their living conditions (Bacon 1624). In Bacon's vision, science should have a social and ethical agenda. However, it was not Bacon's view that came to dominate science in the centuries to come but rather the opposite, official view at the time, famously expressed by Robert Hooke in his proposal for the Statutes of the Royal Society 1663:

> The business and design of the Royal Society [is] to improve the knowledge of natural things, and all useful Arts, Manufactures, Mechanick practices, Engynes, and Inventions by Experiments – (not meddling with Divinity, Metaphysics, Moralls, Politicks, Grammar, Rhetorick, or Logick).

By virtue of dealing essentially with subjective human interests, ethical debate was long banished from traditional science by its norms of "disinterestedness" and

---

1. I here understand the term "ethics" in the traditional philosophical sense to mean the analysis of moral reasoning, that is, the branch of knowledge that deals with moral principles.

Neuroethics and Cultural Diversity,
coordinated by Michele FARISCO. © ISTE Ltd 2023.

"objectivity"[2], ruling that all research results should be conducted, presented and discussed quite impersonally, as if produced by androids or angels (Ziman 1998). Many scientists were insensitive regarding ethical problems arising from their research, and refused to acknowledge that science had any responsibility for dealing with them. This was the classic image of the "ivory tower", in which scientists should work like extra-terrestrials in supreme isolation from human affairs and rest content in their conviction that truth was their sole legitimate aim. This "no ethics" principle was not merely an obsolete model that could be uninstalled by a keystroke, but an integral part of a complex cultural form (Ziman 1998) that required much time and effort to change. Nevertheless, during the 20th century, the culture of science did at last become more in line with Bacon's utopia: the traditionally individualistic and socially secluded science was increasingly substituted for project-oriented and often interdisciplinary teamwork science, requested to justify itself in terms of potential human consequences.

In the 20th century, after a long and hard struggle and having secured a place in the scientific community, notably in science education[3], ethicists and moral philosophers today find themselves facing an interesting *reversed* situation when natural sciences, in particular neuroscience, enter "their" domains purporting to shed scientific light on the phenomenon of moral thought. This bold move is received with varying levels of enthusiasm ranging from optimistic faith in finding final solutions to age-old philosophical problems, to downright rejection[4]. Moral attitudes are often thought to be the last domains that stubbornly resist scientific understanding, and attitudes differ over whether that is a cause for celebration or scandal. We may compare this to the reactions that Jacques Monod (1970) received when he suggested that life could be explained in biochemical terms: many readers were outraged; others rejoiced. Vitalism (the doctrine that living bodies have some feature that prevents them from being entirely explained in physical and chemical terms) was still propagated at the time but has since been abandoned. Today, Monod's view is not considered so provocative.

Neuroscience is a young science that has developed considerably in the last decades. Neuroscientist Gerald Edelman maintained in 1992 that "what is now going on in neuroscience may be looked at as a prelude to the largest possible scientific revolution, one with inevitable and important social consequences" (Edelman 1999, p. xiii). One ultimate goal of that scientific revolution is to gain, at last, a deeper

---

2. These are two of the scientific virtues in Robert Merton's classic list (Merton 1973).
3. For example, courses in research ethics form an unalienable part of science education in many countries, for example, Sweden.
4. See, for example, Didier Sicard (2006) for a skeptical discussion.

understanding of the nature and functioning of the human mind, including its development of morality. What does it mean for an animal (whether human or not) to act as a "moral agent"? Why did the evolution of higher cognitive functions produce moral beings rather than amoral ones? As neuroscientist Jean-Pierre Changeux interrogates: From where does "the natural predisposition (mainly neural) of humans to pass moral judgments" come from? (Changeux 1997, p. 114).

As we understand more about the details of the regulatory systems in the brain and how emotions, ideas and decisions emerge in neural networks, it is increasingly evident that feelings, thoughts and preferences originate in our neurobiology, and that our neurobiology is profoundly shaped by our evolutionary history. Since the genesis of functional brain imaging, there has been a significant increase in the number of neuroscientific studies of consciousness; complex behaviors and emotion, that began to reveal the neural basis for cognitive/affective functions (e.g. Illes 2002). These included studies on volition and self-control or self-monitoring (e.g. Libet et al. 1999), moral judgment (e.g. Greene et al. 2001), decision-making, (e.g. Hanna Damasio et al. 1994), racial attitudes (e.g. Hart et al. 2000), fear (e.g. Ledoux 1999) and lying and deception (e.g. Langleben et al. 2002). The increasing ability of the human species to understand and even design its own brain would, as was suggested (Farah et al. 2004), shape history as powerfully as the development of metallurgy in the Iron Age, mechanization in the Industrial Revolution, or genetics in the second half of the 20th century.

3.2. Neuroethics

These neuroscientific advances and the challenges they encounter inspired new academic disciplines, one of which is *neuroethics*. Neuroethics is a young area of intellectual and social discourse that deals with our consciousness, identity, sense of self and the values that we develop; it is an interface between the empirical brain sciences, philosophy of the mind, moral philosophy, ethics and the social sciences. It is the study of the questions – for example, theoretical, conceptual or normative – that arise when scientific findings about the brain are carried into philosophical analyses, medical practice, legal interpretations, health and social policy, and can, by virtue of its interdisciplinary character, be seen as a sub-discipline of, notably, neuroscience, philosophy or bioethics, depending on which perspective we wish to emphasize.

Many of the questions that modern neuroethics raises are not new; they were raised already during the French Enlightenment, notably by Diderot who stated in his *Eléments de Physiologie*: *"C'est qu'il est bien difficilie de faire de la bonne métaphysique et de la bonne morale sans être anatomiste, naturaliste, physiologiste et médecin*"[5]. Moreover, ethics committees throughout the world have long dealt with ethical problems arising from advances in neuroscience, though not necessarily under the "neuroethics" label[6].

However, as an academic discipline *labeled* "neuroethics", it is a very young discipline. The first "mapping conference" on neuroethics was held in 2002 (Marcus 2002) and references to neuroethics in the literature were made little more than a decade earlier. These early articles described, for example, the role of the neurologist as a neuroethicist faced with patient care and end-of-life decisions (Cranford 1989), and philosophical perspectives on the brain and the self (Churchland 1986). Today, the pioneers of modern neuroethics have developed an entire body of literature and scholarship in the field of neuroethics that is rapidly expanding.

3.3. Fundamental versus applied neuroethics

So far, researchers in neuroethics have focused on applications, that is, on practical ethical questions that arise through neuroscientific advances such as those involved in neuroimaging techniques, cognitive enhancement or neuropharmacology[7]. This applied neuroethics is what Adina Roskies (2002) has labeled "ethics of neuroscience", distinct from "neuroscience of ethics", which is concerned with whether and how neuroscientific knowledge can offer explanations of the nature and development of moral thought and judgment. This oft-quoted distinction is indeed useful, but it is not exhaustive in the sense that the field of neuroethics does not only comprise neuroscience and ethics but also philosophy, notably philosophy of science, and a wide range of issues that the distinction fails to cover.

If, in contrast, "applied neuroethics" is distinguished from "fundamental neuroethics", the distinction may include further areas than ethics and neuroscience and become more far-reaching if not exhaustive.

Three core features, related to topic and to methodology, distinguish fundamental neuroethics (a concept introduced by Evers 2007, 2009): it pursues

---

5. This work is a collection of writings by Diderot starting in 1774 but the book was not published until 1875. The above quote is from p. 136 in the 1875 edition. The quote is, which translates to: "It is very difficult to make good metaphysics and good morals without being an anatomist, naturalist, physiologist and physician", well in line with "fundamental neuroethics" presented in this article.

6. See, for example, opinions by the *Comité Consultatif National d'Éthique*, France, in the 1980s.

7. A review of numerous such issues is provided in Illes (2006).

foundational analyses within a *multidisciplinary* research domain using an *interdisciplinary* methodology. Topically, fundamental neuroethics pursues basic research and analyses foundational concepts and methods used in the neuroscientific investigations of notions like, for example, simulation, identity or consciousness. These analyses necessarily involve both empirical scrutiny of the science in question and philosophical analyses of the concepts involved. Fundamental neuroethics is accordingly multidisciplinary because – in line with the above quote from Diderot – it involves elements from different disciplines, including natural and social sciences as well as philosophy of science, philosophy of language, philosophy of mind and moral philosophy. In other words, fundamental neuroethics cannot be subsumed under any particular classical discipline. Methodologically, fundamental neuroethics is ipso facto interdisciplinary because it combines a variety of methods, for example, empirical and conceptual methods related to the different disciplines.

These features distinguish fundamental neuroethics from other approaches to neuroethics (see below). Yet, since all forms of neuroethics require some foundational analyses in order to be viable, it could also be said that all forms of neuroethics must somehow involve, or be developed on a basis of fundamental neuroethics (whether or not they use this label).

3.4. Fundamental neuroethics as a key component of European research and innovation in the area of neuroscience

Because of its indisputable importance, since 2013, the European Commission has significantly supported science in the European Research Area to deepen our understanding and improve our knowledge of the brain. Fundamental neuroethics has played an essential role in that process, embedding it as a critical component from the very start of different projects, such as the Horizon 2020 Future and Emergent Technologies (FET) Flagship "The Human Brain Project (HBP)" and the FET Proactive "Digital twins for model-driven non-invasive electrical brain stimulation (Neurotwin)" project. Fundamental neuroethics has also been a key component in strengthening responsible research and innovation practices.

The Human Brain Project (HBP), which was launched on October 1, 2013, is a large 10-year scientific research project based on exascale supercomputers that aims to build a collaborative ICT-based scientific research infrastructure to allow researchers globally to advance knowledge in the fields of neuroscience, computing and brain-related medicine[8]. Fundamental to the HBP approach is to investigate the brain on different spatial and temporal scales (i.e. from the molecular to the large

---

8. For more information, see: https://www.humanbrainproject.eu.

networks underlying higher cognitive processes, and from milliseconds to years). To achieve this goal, the HBP relies on the collaboration of scientists from diverse disciplines, including neuroscience, social science, philosophy and computer science, to take advantage of the loop of experimental data, modeling theories and simulations. The idea is that empirical results are used to develop theories, which then foster modeling and simulations, which result in predictions that are in turn verified by empirical results and further clarified by conceptual/theoretical analyses.

From the earliest planning stages and the beginning of the project, the HBP's core research has comprised social science and humanities, including philosophy and neuroethics, joined in what was called "Ethics & Society Subproject" (one of 11 research subprojects in the HBP)[9]. Other major Brain Initiatives (in, for example, the US, China, Japan, Canada and Australia) have ethics monitoring, ethics regulations and ethical advisory boards but they comprise no dedicated research in those areas, which makes the HBP unique in this regard (Evers 2017).

Research in philosophy and neuroethics[10] has been largely devoted to fundamental neuroethics. It has inter alia focused on the following topics[11]:

a) simulation (e.g. Dudai and Evers 2014; Farisco et al. 2018);

b) neuronal epigenesis and cultural imprinting on brain architecture (e.g. Evers 2016a, 2020; Evers and Changeux 2016; Salles 2016);

c) human identity (e.g. Salles 2021);

d) consciousness, brain disorders and neurotechnological mind-reading (e.g. Evers and Sigman 2013; Farisco et al. 2015; Evers and Giordano 2017; Pennartz et al. 2019; Northoff et al. 2020);

---

9. In financial terms: social science and humanities research has received ≈5% of the HBP's total budget. A quarter of that was dedicated to philosophy and neuroethics, which accordingly received ≈1.25% of the HBP's total budget.
10. The neuroethics and neurophilosophy group at the Centre for Research Ethics & Bioethics (CRB), Uppsala University, is in charge of this research in the HBP and related teaching. More details about the neuroethics and neurophilosophy group at Uppsala University can be found at: https://www.crb.uu.se/research/neuroethics/.
11. Other topics that have been addressed are: Responsible Research and Innovation (RRI) (e.g. Salles et al. 2018, 2019), Dual Use (e.g. Giordano and Evers 2018) and Data Protection and Privacy (e.g. Farisco and Evers 2016; Salles 2016). The research has been complemented by teaching and seminars, for example, a series of higher seminars by M. Guerrero on "Social praxeology and neuronal epigenesis" (2018), "Sound, listening practices and music" (2018), "Towards and ethics of Neurorobotics" (2019) and "The socialized body" (2020).

e) digital twins and virtual brains (e.g. Evers and Salles 2021; Evers et al. 2021);

f) artificial intelligence (e.g. Salles et al. 2020a, 2020b).

Importantly, this focus on fundamental neuroethics research does not imply that applied neuroethics research is ignored: on the contrary, it is considered to be an important task for fundamental neuroethics to provide the theoretical foundations that can be clearly and adequately communicated and thereby enable informed analyses of applied issues and help ensure beneficial applications (see Evers 2007). For example, the philosophical analyses of consciousness are directly connected to clinical applications to benefit, notably, patients with disorders of consciousness (Evers 2016c; Farisco and Evers 2017), as well as people who suffer from drug addiction (Farisco et al. 2018a). The analyses of cultural imprinting on brain architecture have, for example, been used in connection to understanding the effects of poverty on brain development (Lipina and Evers 2017).

Thus, in fundamental neuroethics, theory and practice complement each other. An advanced and refined theoretical framework combining clearly described scientific data and hypotheses with well-defined terminology helps to ensure the best possible use of the results. Vice versa, a successful application may provide the theoretical framework with crucial substance and justification.

The mutual relevance of theory and practice, diverging from the fact that fundamental research and technological applications feed into each other, has been expressed not least in the *Opinion* documents that the HBP's Ethics and Society Subproject published on a variety of topics: Opinion and Action Plan on "Data Protection and Privacy" (2016), Opinion on "Responsible Dual Use" Political, Security, Intelligence and Military Research of Concern in Neuroscience and Neurotechnology (2018), and Opinion on Trust and Transparency in Artificial Intelligence (2021). These opinions all combine conceptual analyses of key concepts with social science and ethics studies, forming a solid basis for recommendations on how to approach and manage the topics in question. They are also combined with concrete action plans to guide their implementation[12].

A second international brain research initiative in which fundamental neuroethics is embedded from its onset is the new European "Neurotwin project". Initiated in January 2021, Neurotwin aims to develop personalized therapy for neurodegenerative disorders through the construction of hybrid brain models able to

---

[12]. The opinions can be read and downloaded from the HBP's website: https://www.humanbrainproject.eu/en/social-ethical-reflective/publications/.

represent the effects of non-invasive electrical brain stimulation appropriate in the context of large-scale connectivity alterations and oscillatory deficits that are characteristic of some brain disorders, such as Alzheimer's disease[13]. This type of work raises important ethical and philosophical questions that need to be addressed and handled responsibly, and so fundamental neuroethics has been included in the core of the project[14].

3.5. Conceptual analysis in fundamental neuroethics methodology

The methodology of fundamental neuroethics has been further developed and enriched since its introduction. Generally speaking, there are three different methodological approaches in neuroethics, identified by Evers et al. (2017), Farisco et al. (2018b) and Salles et al. (2019): (i) "neurobioethics" (which mirrors bioethics) applies ethical theory and reasoning to address the practical issues raised by brain research, its clinical applications and communication; (ii) "empirical neuroethics" uses empirical data to inform theoretical issues (e.g. what is moral reasoning?) and practical issues (e.g. who is really a moral agent?) and (iii) "conceptual neuroethics" uses conceptual analysis to address topics such as how neuroscientific knowledge is constructed and why or how empirical knowledge of the brain can be relevant to philosophical, social and ethical concerns.

The neuroethics and neurophilosophy group at Uppsala University have developed the third methodological conceptual approach.

The conceptual analyses can heuristically be described as comprising (notably but not exclusively) three closely interconnected foci: scientific descriptions, philosophical analyses, and ethical and social considerations. Although distinct, each

---

13. Personalized hybrid brain models uniting the physics of electromagnetism with physiology – Neurotwins or NeTs – are poised to play a fundamental role in understanding and optimizing the effects of stimulation at the individual level. Benefiting from newly emerging physical and physiological modeling techniques, Neurotwin's ambition is to develop advanced individualized whole-brain models that predict the physiological effects of transcranial electromagnetic stimulation at the individual level and use them to characterize pathology design and test optimal brain stimulation protocols in Alzheimer's disease. This new sophisticated approach can lead to a significant breakthrough in personalized therapy for neurodegenerative disorders, delivering disruptive solutions through model-driven, individualized therapy, where physics and computational neuroscience come together. More details about the project can be found on Neurotwin's website: https://www.neurotwin.eu/.
14. The philosophy and ethics research and ethics monitoring is led from the neuroethics and neurophilosophy group at CRB, Uppsala University, by Evers and Guerrero.

focus feeds into the others and should to that extent not be understood as strictly chronologically distinguished[15]:

1) Description of the scientific research domain in terms of:

a) goals and purposes;

b) methods;

c) theoretical underpinnings;

d) results.

2) Conceptual analyses of the science[16] in question:

a) meaning of key concepts and terminology;

b) selection of methods;

c) comparative assessment of theoretical underpinnings including analyses of their conceptual and logical consistency;

d) possible scientific interpretations of results.

3) Identifying ethical and social relevance and implications, questioning, for example:

a) Are the goals realistic, justified and adequately communicated?

b) Is the research well carried out, for example, is it in line with good scientific practice and codes of conduct?

c) Is the presupposed theoretical framework socially and ethically reasonable and acceptable, for example, compliant with the relevant ethics regulations?

d) Are the results beneficial, and do they involve risks?

Adequate scientific descriptions are crucial, since they help ensure that the subsequent philosophical and ethical or social analyses do not lead astray, for example, focus on misconceptions or fake problems, that is, straw men[17]. This is

---

15. In other words, while it may often be natural to begin the analyses by describing the research in question in scientific terms, that description may include elements of philosophical reasoning or ethical considerations. For example, if the scientific research is motivated by the latter.

16. These analyses are largely but not exclusively focused on the philosophy of science, but they may also include philosophy of, for example, language, the mind or moral philosophy.

17. This, we may note, was a charge made against philosophy of science, in the philosophical discussions of materialism in the 20th century. Gaston Bachelard complained that philosophers of science "identify materialism with a crude conception of matter [...] lacking

perforce an interdisciplinary quest, for although philosophy and neuroethics may contribute to the formulations and assessments of these descriptions, they require collaborations with scientists from the relevant domains.

Philosophical analyses are essential in at least three ways linking to both science and ethics. Advanced conceptual clarifications may (i) enrich the scientific descriptions, (ii) create an overview of the scientific and philosophical challenges versus promises that arise and present hypothetical scenarios of how to deal with them, and (iii) provide a basis that enables realistic and clear discussions of ethical and social implications.

The work on simulation in fundamental neuroethics (e.g. Dudai and Evers 2014) may serve to illustrate these three points.

Simulation is a powerful method in science and engineering; however, simulation is an umbrella term, and its meaning and goals differ among disciplines. Rapid advancements in neuroscience and in computing draw increasing attention to large-scale brain simulations, but what is the meaning of simulation, and what should the method expect to achieve? Aiming to enrich the scientific descriptions and create an overview of the scientific and philosophical challenges versus promises that arise, Dudai and Evers discuss the concept of simulation from an integrated scientific and philosophical vantage point in terms of meaning, goals and medium, and pinpoint selected issues that are specific to brain simulation identifying realistic hopes and caveats. On the basis of these scientific descriptions and overviews, hypothetical scenarios are constructed. Science and society should aim to benefit from contemplating the future and prepare for it, even if this future is not necessarily around the corner. Suppose, for the sake of argument, that the brain and computer sciences combined will indeed be able one of these days to come up with a simulated human brain. What questions will we face? Some of these questions are primarily theoretical. For example: How similar will the simulation be to the original? Will consciousness emerge? How would we recognize whether a future brain simulation is conscious or not? Is realistic human brain simulation possible in the absence of consciousness? Other questions are more practically oriented towards what we stand to gain (or lose) in social, ethical and clinical terms. The latter type of questions are more fully addressed in a subsequent study by Farisco et al. (2018), mapping technical and conceptual issues in large-scale brain simulation and disorders of consciousness.

---

any experimental basis" and set up a straw man: a "matterless materialism" (Bachelard 1953, Introduction; Changeux 2004, p. 7).

Concerning the identification of ethical and social relevance and implications of a specific research area (the third focus listed above), it is important to note that ethics is not merely an after-thought, something to be added post factum, but should be an integrated part of the scientific process itself. Science can inherently raise ethical and social issues as, for example, research on racial or gender-related distinctions may illustrate. Studies of race and gender can be tremendously important to, for example, deepen the understanding of structures of inequality between groups, yet they can also be explicitly or implicitly conducted in racist or sexist terms, not least if the society in which they are pursued is permeated with racist or sexist ideologies. These inherent values are not always overt but can be both hidden and quite resistant to transparency. This importantly connects ethics to conceptual analyses: inherent values may be more hidden if the concepts are unclear which in turn can make their social consequences more insidious. Advancing conceptual clarity may bring inherent norms to the surface, which better allows discussions on how to deal with them.

History offers numerous illustrations from normative as well as scientific discourses of how language and the meaning assigned to terms influence the contexts in which they are used. The meanings assigned to concepts may change (and sometimes they are changed deliberately) to mirror or drive social changes, such as, for example, in the quests to abolish or reduce racism and misogyny. To illustrate, in Sweden, non-white and female humans are not considered in the same way today in comparison to how they were viewed in the 19th century. At that time, the racist and misogynistic views of the scientific community reflected the attitudes of the dominating society, and so white male superiority was considered to be a biological fact.

That said, conceptual analyses alone are far from sufficient to bring out and explain inherent normativity in scientific research: advanced studies of, notably, history and social science are unalienable in order to not only identify the norms but also to understand their background, context, reasons and possible scenarios for dealing with, and perhaps influence them. Thus, we should also note the relevance of these academic areas to fundamental neuroethics.

Identifying ethical and social relevance and implications is important in terms of risk/benefit analyses aiming to ensure that possible benefits are identified and reaped, while risks are clearly anticipated and managed. In this context, it is crucial to adequately communicate the research to distinct audiences, engage with them and apply the principles of responsible research and innovation (RRI). RRI implies that societal actors (researchers, citizens, policy-makers, business, third sector organizations, etc.) work together during the whole research and innovation process in order to better align both the process and its outcomes with the values, needs and expectations of society.

In practice, RRI is implemented as a package that includes multi-actor and public engagement in research and innovation, enabling easier access to scientific results, the take-up of gender and ethics in the research and innovation content and process, and formal and informal science education[18].

The above description of the conceptual analyses that fundamental neuroethics involves hopefully clearly illustrates how the distinctions into three domains are merely heuristic: they are closely interconnected, enrich each other and develop together.

3.6. Fundamental neuroethics connecting neuroscience with "free will" and social structures

One area of research in fundamental neuroethics that illustrates well the interdisciplinary methodology described above as well as the bridge it can provide between neuroscience and social structures is the classical problem of "free will". Below is a very brief summary of this research and some suggested social implications. The description is brief because the aim is merely to illustrate the fundamental ethics approach, not to give a full account of this rich research area[19].

The problem of "free will" partly arises from the fact that any system of norms for adequate behavior presupposes, in order to make practical sense, that human beings have some capacity voluntarily to control, or in some measure influence, their behavior. In other words, "ought" implies "can", in some sense and measure. In total absence of such a capacity, recommendations or prescriptions become meaningless in practice if not in theory. This freedom to voluntarily influence our nature and destiny lies at the heart of human identity: to be human means for many people to have "free will", to be able to choose what we do, think and say – and to improve and develop as individuals. We experience that, given the practical opportunity, we do in some sense control our lives. Even though we are always under some constraints and our identities are results of our upbringing and past experiences, we believe, or want to believe, that in a situation where we stand faced with a range of alternative actions, we are in some measure at liberty to choose among them.

---

18. See the home page of Horizon (2020).
19. The research is more extensively presented in Evers (2009, 2016b, 2021).

This experience of free will also functions as a social axiom. Having the freedom of choice is, as a rule, regarded as a necessary condition for personal responsibility around which all societies are built. In its social applications, personal responsibility for a given action presupposes that the action was not constrained, or forced, but voluntary, so that the agent could in some sense have chosen to act differently. We are only considered to be morally or legally responsible for actions that we could in principle have omitted.

The neuroethical problem of free will consists of explaining how this socially crucial conception of human beings as free and responsible individuals can be combined with the neuroscientific views on us and on our behavior. Is it reasonable to believe in free will when what we experience as a free choice is the result of electrochemical interactions in the brain, and a sort of biological decision-making program shaped by evolution? Or is free will simply an illusion? But if that is so, what happens to the notion of personal responsibility? This problem has distinct facets: ontological (Does such a thing as "free will" exist?), epistemological (What might it mean to have a "free will"? Is the belief that we possess free will rational? What do or can we know about this?), as well as ethical and social (How should we behave and interact in the light of our beliefs in this regard?).

The epistemological debate over whether the belief in possessing "free will" can rationally be maintained in light of modern science has often been presented as a choice between two extremes. On the one hand, we have the conception of strict determinism, the philosophical view that all events are completely determined by previously existing causes, that appears to exclude free will; on the other hand, we have a conception of freedom of choice unfettered by preceding causes that appears to exclude determinism. However, it has been argued (Evers 2009) that neither position is scientifically realistic since the introduction in neuroscience of the "theorem of variability" (Changeux et al. 1973), which suggests that we can be simultaneously determined and "free".

The neuroscientific theorem of variability can simply be described as follows. The human brain is a motivated neural system, genetically equipped with a predisposition to explore the world and to classify what it finds there. This motivated system is intensely and spontaneously active independently of and in interaction with its environment, as an autonomous system that constantly produces representations that it projects onto the world as it tests its physical, social and cultural environments. This testing involves continuous processes of evaluation and narration[20] that make emotionality and emotional choice essential neuronal features:

---

20. The brain's intense and spontaneous activity constantly producing representations that it projects onto the world as it tests its physical, social and cultural environment independently

when different values oppose one another, one of them will sooner or later be selected (unless both are discarded). Cerebral connections are organized in a gigantic network wherein distribution in space depends on a species-specific, organizational arrangement of connections, the distribution of which depends on a reserve of random variations that is sufficient to assure the network's plasticity and its accessibility to physical, social and cultural inputs. New combinatory models are continually produced and tested as a neural embodiment of creative activity, where the same afferent message may stabilize different arrangements of connections; causal antecedents can produce variable outcomes. This variability suggests a non-necessary, that is, contingent relationship between cause and effect without denying that there is a causal relation.

The philosophical reasoning (as developed in Evers 2009) goes, summarily, as follows. Plasticity – the general capacity of the neuron and its synapses to change properties as a function of their state of activity – ensures our cerebral capacity to exert a non-zero influence over events. So long as there is variability in the causal relations (i.e. determinism including variability), there may be a fundamental dimension of voluntary influence in our choices. This measure of variability and control is sufficient to safeguard the possibility of "free will" under these particular aspects, leaving open the extent of the freedom that we actually possess (an extent that is probably very small). According to this neuroethical position, human beings might act as free and responsible agents while being contingently causally determined. Rather than posing a threat to our unalienable notions of free will and personal responsibility, neuroscience can be interpreted to offer empirical support to maintain them, reconceptualized in accordance with empirical data and related models.

The possible ethical and societal implications of this scientific-cum-philosophical reasoning are numerous. De facto, our cerebral variability notwithstanding, our capacity for choice is strongly limited by causal mechanisms and antecedent events over which we in fact have no or very limited influence: our natural, social and cultural environments shape us. The important questions before us concern how to understand this balance and how to structure our societies accordingly. Arguably, social institutions need to be adapted so as to take science into account; they need to be based on biological realism and understanding of the larger contexts. The models of the brain revealed by neuroscience and philosophy need to be more broadly integrated into our social structures and general world-

---

and in interaction with these environments suggests that the brain is a narrative organ motivated by a continuous process of narration; spinning its own neuronal tale. At some point in the mother's womb, the infant brain begins the tale that will in due course develop into the neuronal history of its entire life (Evers 2009, Chapter 2).

views, shaping our legal and educational systems, child-care and moral upbringing, furthering their biological realism in a socially beneficial manner (Evers and Changeux 2016; Salles 2017).

3.7. Conclusion

Three core features, related to topic and to methodology, distinguish fundamental neuroethics: the pursuit of foundational analyses within a multidisciplinary research domain using an interdisciplinary methodology. Topically, fundamental neuroethics pursues basic research and analyzes foundational concepts and methods used in the neuroscientific investigations. Since these analyses necessarily involve elements from different disciplines, including natural and social sciences as well as the humanities, fundamental neuroethics is multidisciplinary. Methodologically, fundamental neuroethics is ipso facto interdisciplinary because it combines a variety of methods related to the different disciplines. On the one hand, these core features can be said to distinguish fundamental neuroethics from other approaches to neuroethics (e.g. neurobioethics, or empirical neuroethics). On the other hand, since all forms of neuroethics require some foundational analyses in order to be viable, it could also be argued that all forms of neuroethics should somehow, explicitly or implicitly, involve, or be developed on a basis of fundamental neuroethics (as the label suggests).

The theoretical sophistication of successful fundamental neuroethics research enhances its practical utility. As the levels of conceptual, scientific and normative clarity and transparency within a given research area increase as the result of fundamental neuroethics analyses, the uses and applications of this research may also become all the more realistic and well guided. This is a key idea driving fundamental neuroethics: refining onto ever higher levels how the abstract and the concrete, thought and action, theory and practice, may enrich and enhance each other to the benefit of our understanding and interaction within the natural, social and cultural contexts we inhabit.

3.8. Acknowledgments

I thank Michele Farisco, Manuel Guerrero and Arleen Salles for their valuable comments regarding earlier versions of this manuscript.

This research was funded by the European Union's Horizon 2020 Framework Programme for Research and Innovation under the Specific Grant Agreement No. 945539 (Human Brain Project SGA3).

3.9. References

Bachelard, G. (1953). *Le Matérialisme rationnel*. Presses Universitaires de France, Paris.

Bacon, F. (2010). *The New Atlantis*. Watchmaker Publishing, Astoria, OR.

Changeux, J.-P. (1997). *Neuronal Man. The Biology of Mind*. Princeton University Press, NJ.

Changeux, J.-P. (2004). *The Physiology of Truth: Neuroscience and Human Knowledge*. Harvard University Press, Cambridge, MA.

Changeux, J.-P., Courrège, P., Danchin, A. (1973). A theory of the epigenesis of neural networks by selective stabilization of synapses. *Proceedings of the National Academy of Sciences of the United States of America*, 70, 2974–2978.

Churchland, P.S. (1986). *Neurophilosophy: Toward a Unified Science of the Mind-Brain*. MIT Press, Cambridge, MA.

Cranford, R.E. (1989). The neurologist as ethics consultant and as a member of the institutional ethics committee: The neuroethicist. *Neurologic Clinics*, 7, 697–713.

Damasio, H., Grabowski, T., Frank, R., Galaburda, A.M., Damasio, A.R. (1994). The return of Phineas Gage: Clues about the brain from the skull of a famous patient. *Science*, 264, 1102–1105.

Diderot, D. (1875). *Œuvres complètes de Diderot*, Volume IX, edited by J. Assézat and M. Tourneux. Garnier, Paris.

Dudai, Y. and Evers, K. (2014). To simulate or not to simulate: What are the questions? *Neuron*, 84(2), 254–261.

Edelman, G.M. (1992). *Bright Air, Brilliant Fire. On the Matter of the Mind*. BasicBooks, New York.

Evers, K. (2009). *Neuroéthique. Quand la matière s'éveille*. Éditions Odile Jacob, Paris.

Evers, K. (2016a). Can we be epigenetically proactive? In *Open Mind: Philosophy and the Mind Sciences in the 21st Century*, Metzinger, T. and Windt, J.M. (eds). MIT Press, Cambridge, MA.

Evers, K. (2016b). The responsible brain: Free will and personal responsibility in the wake of neuroscience. *Revista Latinoamericana de Filosofía*, 42(1), 33–44.

Evers, K, (2016c). Neurotechnological assessment of consciousness disorders: Five ethical imperatives. *Dialogues in Clinical Neuroscience*, 18(2), 155–162.

Evers, K. (2017). The contribution of neuroethics to international brain research initiatives. *Nature Reviews Neuroscience*, 18, 1–2.

Evers, K. (2021). Variable determinism in social applications: Translating science to society. *Intellectica*, 2021/2, 75, 73–89.

Evers, K. and Changeux, J.-P. (2016). Proactive epigenesis and ethical innovation. A neuronal hypothesis for the genesis of ethical rules. *EMBO Reports*, 17(10), 1361–1364.

Evers, K. and Giordano, J. (2018). Then utility – and use – of neurotechnology to recover consciousness: Technical and neuroethical considerations in approaching the "hard question" of neuroscience. *Frontiers in Human Neuroscience*, 11(564), 2–5. doi: 10.3389/fnhum.2017.00564.

Evers, K. and Sigman, M. (2013). Possibilities and limits of mind-reading: A neurophilosophical perspective. *Consciousness and Cognition*, 22, 887–897.

Evers, K., Salles, A., Farisco, M. (2017). Theoretical framing of neuroethics: The need for a conceptual approach. In *Debates About Neuroethics. Perspectives on Its Development, Focus, and Future*, Racine, E. and Aspler, J (eds). Springer International Publishing, Dordrecht.

Farah, M.J., Illes, J., Cook-Deegan, R., Gardner, H., Kandel, E., King, P., Parens, E., Sahakian, B., Wolpe, P.R. (2004). Neurocognitive enhancement: What can we do and what should we do? *Nature Reviews Neuroscience*, 5(5), 421–425.

Farisco, M. and Evers, K. (eds) (2016). *Neurotechnology and Direct Brain Communication. New Insights and Responsibilities Concerning Speechless but Communicative Subjects*. Routledge, Taylor & Francis Group, London and New York.

Farisco, M. and Evers, K. (2017). The ethical relevance of the unconscious. *Philosophy, Ethics, and Humanities in Medicine*, 12(1), 11. doi: 10.1186/s13010-017-0053-9.

Farisco, M., Laureys, S., Evers, K. (2015). Externalization of consciousness. Scientific possibilities and clinical implications. In *Current Topics in Behavioural Neurosciences*, Geyer, M., Ellenbroek, B., Marsden, C. (eds). Springer, Berlin, Heidelberg. doi: 10.1007/7854_2014_338.

Farisco, M., Evers, K., Changeux, J.P. (2018a). Drug addiction: From neuroscience to ethics. *Frontiers in Psychiatry*, 9, 595. doi: 10.3389/fpsyt.2018.00595.

Farisco, M., Salles, A., Evers, K. (2018b). Neuroethics: A conceptual approach. *Cambridge Quarterly of Healthcare Ethics*, 27(4), 717–727.

Farisco, M., Hellgren Kotaleski, J., Evers, K. (2018c). Large-scale brain simulation and disorders of consciousness: Mapping technical and conceptual issues. *Frontiers in Psychology*, 9, 585. doi: 10.3389/fpsyg.2018.00585.

Giordano, J. and Evers, K. (2018). Dual use in neuroscientific and neurotechnological research: A need for ethical address and guidance. In *Ethics and Integrity in Health and Life Sciences Research*, Koporc, Z. (ed.). Emerald Publishing Limited, Bingley.

Greene, J.D., Sommerville, R.B., Nyström, L.E., Darley, J.M., Cohen, J.D. (2001). An fMRI investigation of emotional engagement in moral judgment. *Science*, 293, 2105–2108.

Hart, A.J., Whalen, P.J., Shin, L.M., McInerney, S.C., Fischer, H., Rauch, S.L. (2000). Differential response in the human amygdala to racial outgroup vs ingroup face stimuli. *NeuroReport*, 11(11), 2351–2355.

Langleben, D.D., Schroeder, L., Maldjian, J.A., Gur, R.C., McDonald, S., Ragland, J.D., O'Brien, C.P. (2002). Brain activity during simulated deception: An event-related functional magnetic resonance study. *Neuroimage*, 15, 727–732.

Ledoux, J. (1999). *The Emotional Brain*. Phoenix, New York.

Libet, B., Freeman, A., Sutherland, K. (eds) (1999). The volitional brain. Towards a neuroscience of free will. *Journal of Consciousness Studies*, 6(8–9).

Lipina, S. and Evers, K. (2017). Neuroscience of childhood poverty: Evidence of impacts and mechanisms as vehicles of dialog with ethics. *Frontiers in Psychology*, 8, 61.

Marcus, S.J. (ed.) (2002). *Neuroethics: Mapping the Field. Conference Proceedings. San Francisco, California, 13–14, May 14, 2002*. The Dana Press, New York.

Merton, R. (1973). The normative structure of science, In *The Sociology of Science: Theoretical and Empirical Investigations*, Merton, R.K. (ed.). University of Chicago Press, IL.

Monod, J. (1970). *Le Hasard et la nécessité : essai sur la philosophie naturelle de la biologie moderne*. Le Seuil, Paris.

Racine, E. and Illes, J. (eds) (2002). *Ethical Challenges in Advanced Neuroimaging. Brain and Cognition*. Academic Press, New York.

Roskies, A. (2002). Neuroethics for the new millenium. *Neuron*, 35, 21–23.

Salles, A. (2017). Proactive epigenesis and ethics. *EMBO Reports*, 18(8), 1271.

Salles, A. (2021). What is the human? A call for conceptual reflection before normative discussion. In *Regulating Neuroscience: Translational Legal Challenges*, Hevia, M. (ed.). Elsevier, Amsterdam.

Salles, A., Evers, K., Farisco, M. (2018). Neuroethics and philosophy in responsible research and innovation: The case of the human brain project. *Neuroethics*, doi: 10.1007/s12152-018-9372-9.

Salles, A., Bjaalie, J., Evers, K., Farisco, M., Fothergill, T., Guerrero, M., Maslen, H., Muller, J., Prescott, T., Stahl, B.C. et al. (2019a). The human brain project: Responsible brain research for the benefit of society. *Neuron*, 101(3) [Online]. Available at: https://doi.org/10.1016/j.neuron.2019.01.005.

Salles, A., Evers, K., Farisco, M. (2019b). The need for a conceptual expansion of neuroethics. *AJOB Neuroscience*, 10(3). DOI: 10.1080/21507740.1632972.

Sallin, K., Lagercrantz, H., Evers, K., Engström, I., Hjern, A., Petrovic, P. (2016). Resignation syndrome: Catatonia? Culture-Bound? *Frontiers in Behavioral Neuroscience*, 10, 1–18.

Sicard, D. (2006). *L'Alibi éthique*. Plon, Paris.

Ziman, J. (1998). Why must scientists become more ethically sensitive than they used to be? *Science*, 282, 1813–1814.

4

Diversity in Neuroethics: Which Diversity and Why it Matters?

Eric RACINE[1,2,3] and Abdou Simon SENGHOR[1,3]
[1] *Pragmatic Health Ethics Research Unit, Institut de recherches cliniques de Montréal, Quebec, Canada*
[2] *Université de Montréal, Quebec, Canada*
[3] *McGill University, Montreal, Quebec, Canada*

4.1. Background

Neurological and psychiatric conditions do not spare any cultural or political communities (World Health Organization 2001, 2006). While their prevalence varies within populations, depending on a range of economic, political, social and cultural factors, all groups of humans are concerned at different levels, in different ways, with these conditions.

Unsurprisingly, because of the complexity of these conditions and their impact on human functions, they create a host of situations which defy engrained social habits (e.g. how do we react to, and integrate, changes in personality induced by conditions such as Alzheimer's disease?) and available medical treatments (e.g. what can Western-style or traditional medicine do for these people?). It is safe to say that no human culture has all the best clinical or ethical solutions to respond to these conditions and the difficult situations they generate.

In reaction to challenging clinical, research and policy situations involving neurological patients and neuroscience, the field of neuroethics has taken form as a specific kind of response to the moral issues generated by various existing and novel treatments of neurological and neuropsychiatric conditions. For many neuroethicists, the fundamental, and sometimes the only legitimate considerations for this field are advances in neurotechnology (see, for example, Wolpe (2004)). To a lesser extent, neuroethics also tackles issues raised by clinical care and public policy issues (as reviewed in Racine (2010) and Leefmann et al. (2016)). Much like bioethics (Messikomer et al. 2001; Moreno 2004; Fox and Swazey 2005; Chattopadhyay and De Vries 2008), neuroethics originates primarily from North America and Western Europe and is embedded in, and reflects, a number of theoretical, methodological, and practical assumptions rooted in these contexts (De Vries 2005, 2007; Doucet 2005; Racine and Sample 2018). In particular, the question of whether and how neuroethics has integrated diversity, in all its forms, remains open. Is neuroethics embedded in neuroscience to the point of reflecting its agenda and uncritically promoting the values of this area of biomedical science (Turner 2003; De Vries 2007)? Is it open to questioning some of the deeper assumptions about the worth of neurotechnologies based on non-Western worldviews? Is it inherently committed to forms of biomedical reductionism that make the brain the source of the self? (Ortega 2009; Vidal 2009). Does neuroethics promote reflectivity about the diagnostic and therapeutic solutions offered by Western-style biomedical science, or rather does it defend them? Now, after almost 20 years of scholarly and practical activity, it is worth asking whether neuroethics has reflected a diversity of human situations, to what extent, and why. We must also reflect on why and how neuroethics should reflect this diversity. That is, what cultural diversity is, what it means in the context of ethics which is directly concerned with human values and why neuroethics should care about it. Tackling this issue involves scrutinizing our views and practices in matters of diversity and ethics, a difficult and engaging topic.

In this chapter, we ask questions about neuroethics' relationship to a diversity of human experiences and moral practices. We discuss how the exploration of this issue involves asking questions about neuroethics' relationship to various human moralities and its understanding of the task of ethics in relation to these. For example, how does ethics relate to the diversity of moral values and human practices surrounding neurological and neuropsychiatric conditions? Does it aim to provide a universal ethical framework? If not, is it condemned to ethical relativism, which respects all human moralities and their practices?

We first unpack the notion of diversity and highlight some of its assumptions and orientations. This leads us to reflect critically on what diversity is. We distinguish between human or *anthropological diversity*, and cultural diversity – that is, *socio-cultural diversity*. These aspects of diversity, one more personal, and the other more collective, cannot be separated but, for the purpose of this chapter, they help us stage the ongoing tension between an individual's values and existence and their relationship to the group(s) which they are a part of, including the moral values and model of existence promoted by these groups.

Secondly, we reflect critically on where neuroethics stands with respect to socio-cultural diversity and anthropological diversity. We explain that, given the historical roots of neuroethics as a Western (mainly North American and Western European) movement, it has inescapably only partly reflected the diversity of moral concerns that neurological and psychiatric conditions give way to.

Finally, we make an argument for the important role of human diversity as a set of insights about human flourishing. Indeed, from a pragmatic ethics standpoint, human flourishing and the diversity of ways of pursuing a flourishing life are the end goals of ethics, although this is less a fixed goal than an actual process and pursuit to achieve such a goal (Pekarsky 1990).

4.2. Diversity and cultural diversity

Diversity, including cultural diversity, has been discussed for several decades and has been the object of extensive social science and public policy work, especially as a cornerstone of anthropology and a prompt for discussions of ethics in face of divergence. In ethics, diversity often evokes issues of health disparities, notably in the influential North American context because therein cultural diversity is the root cause of fundamental differences in the treatment of individuals in essentially all aspects of life (e.g. education, politics, work, economics and of course health, including brain health) (Sullivan 2017). Moreover, because of its cultural dominance, American history and ways of thinking about and acting in the face of diversity, including racial segregation and systemic racism and corresponding correcting responses, tend to be viewed expediently as universal history, applying to all societies and cultures across time and space (Gerstle 1993; Schaub 2017). Accordingly, given this context, diversity is often discussed in terms of the differential, discriminatory treatment of people based on American socio-cultural categories, notably those rooted in multicultural approaches (Kymlicka 1995), as well as racial categories common to the USA. This is obviously a very important consideration and represents a contribution of American, and other liberal societies of Anglo-Saxon influence, to ideas of socio-cultural diversity. However, to go further, we must recognize this context for what it is and step back and ask what diversity is more fundamentally, beyond this common socio-political understanding, experience and response to diversity.

Diversity, in a sense, simply alludes to the plural, the non-monolithic, the non-singular. However, human diversity often signals that there are rather deep differences in ways of acting, behaving, thinking and feeling – differences rooted in quasi-instinctive entrenched patterns of behaviors which are hard to discuss openly. Diversity, notably in ethics, is frequently encountered and thought of as a problem because of an underlying assumption driving humans and human collectivities to think in terms of oneness, the singular universal and the same. It may also be true that when humans with profoundly different worldviews try to cooperate and communicate (e.g. to determine which treatment is preferable, to determine which policy is best), they strive to agree on the best and most appropriate course of action. However, in such exchanges, dominant views tend to be idealized as the objective way of doing things, such that other ways of thinking are cast aside without due consideration. In reality, no one is the same and differences permeate societies and human groups (within and between them), even those that consider themselves to be rather homogeneous.

Specifying that diversity, or certain differences, originate in culture (cultural diversity) raises the question of what we mean by culture and how universal any account of culture can be. Does it represent acquired (non-innate) habits and differences? Does culture represent differences which are the result of customs such as religion, languages and shared narratives about the past and future? These are profound questions which cultural anthropologists – and to some extent biological anthropologists, by default – have been dealing with for over a century. Here, it may be safer to embrace a broad understanding of diversity and a correspondingly broad understanding of human culture that is not dependent on the overly specific accounts of what contribution biology or culture make to our individualities and differences. This is why we propose distinguishing between diversity, understood as a socio-cultural phenomenon (*socio-cultural diversity*) such as language, ethnicity and gender, and *anthropological diversity*. The latter describes the deeper, more personal, experiential and existential nature of human diversity as a fundamental fact of human existence, despite the appearance of socio-cultural uniformity in certain contexts. There is also an important number of human biological differences such as sex, skin color and ability–disability that create *biological diversity*.

Biological diversity has often been essentialized to be considered as quasi-sociocultural categories (e.g. group identities based on race alone, gender roles based on sex). However, the meanings of different forms of biological diversity are socially constructed (Cohan and Howlett 2017), and – as the phenomena of sexism, racism and ableism illustrate – the equation of aspects of human biology with various social and individual attributes is often undertaken in discriminatory and unfounded ways. Freeing human identities and existences from reductionist biological categories necessarily involves taking issue with sweeping equations between biological diversity and socio-cultural diversity. In reality, we need to recognize that socio-cultural identities are constantly socially biologized (e.g. sexualized, racialized or ablelized), such that we become accustomed to seeing ourselves through the lenses of these fallacious categories. By focusing the analysis on biological functions, biological reductionism easily gives way to the essentialization of inequalities of race, gender, etc. It opens the door to a hegemonic attitude of certain cultures or countries over others based on a claimed biological superiority. Indeed, there has been a constant desire to essentialize and biologize cultural differences towards such goals (Lanre-Abass 2010). This biological essentialization process is a cornerstone of European colonialism and its racial ideology (Meltzer 1993).

4.3. Diversity, ethics and neuroethics' uneasy relationship with diversity

Diversity – anthropological and socio-cultural – is everywhere. This fundamental fact is of importance because ethics is concerned with the role and place of moral habits (plural, and therefore implying differences) in thinking about how to best deal with a whole range of human situations, including those concerning people with neurological and neuropsychiatric conditions. In the face of the diversity of moral habits, humans and societies cannot hold onto the idea that all things are equal or that all human practices, ways of being, acting, etc., are of equal worth, for very long. Simply politely respecting the diversity of ways of being can only go so far in trying to develop forms of human cooperation and relationships. Humans have interests and value certain outcomes and experiences, such that they need to find, for themselves as individuals and as human groups, what goals and practices are worth pursuing or not. Traditions and acquired habits of all sorts can seemingly guide us, as long as they are not challenged by the novelty of situations or by conflicts with other traditions, where the need to readjust and rethink the valuations and interests at stake becomes apparent. So, in spite of being pervasive, diversity represents a fundamental condition of humans who need to act and decide what to do.

Perhaps the simplest and most elegant response to diversity in Western scholarship has been the appeal of human freedom and the self-construction of identity. This orientation appears to settle the issue of diversity, since everyone is granted freedom to live a life of their own, according to their preferences as an individual. This is the line of thinking developed by the classic liberal thinkers such as Locke, Voltaire and Mill. This intellectual and political tradition has emphasized the importance of human liberties and the value of diverse manifestations of individual personalities and preferences, as long as those are expressed within the confines of public and social order, such that no prejudice or harm is caused to others (Mill 1956). Implicitly and explicitly, liberalism (and now neoliberalism) still dominates the regulation of major social systems in Western societies, such as the economy, politics, education, health and research in neuroscience and beyond (Racine et al. 2017a).

The intersection between liberal thinking – at least a narrowed-down utilitarian version of it – and modern science warrants much more attention given their juncture in capitalism. Science provides new powers (*scientia potentia est*), while liberal economics is inherently built on the development of new needs and markets (Harari 2016). Profits generated by capitalist modes of production are reinvested in science and technology to develop new products and profits, often using the natural and intellectual resources of less powerful countries (Kokwaro and Kariuki 2001). To its proponents, this "virtuous" circle appears to be a pure reflection of the true universal nature of humankind, such that it justified much of European colonialism, starting in the Renaissance. Here, the search for wealth and power had the support of European science and technology and a vanguard ideology of bringing "civilization" to a wide array of human communities. In this process, a tremendous amount of anthropological and socio-cultural diversity was destroyed across the world, sometimes explicitly genocided to impose European languages, religions, economics and politics. Although this may appear as a distant past, colonialism still structures much of the world order and is pervasive in attitudes towards cultural and racial diversity (Acemoglu et al. 2002), and, in many respects, it has been fronted as simple and "natural" liberalism whose colonial aspects are simply silenced or ignored[1]. The colonizing attitude was manifested, for example, by a control of Black subjectivity through education and the centrality of the Euro-American knowledge promoted by Western settlers, ignoring African epistemologies and bringing their perceived beneficial modernity over African modes of knowing (Ndlovu-Gatsheni 2013).

---

1. Racism, slavery, colonialism and the use of genocidal tactics are not European inventions. It is a current mainstay to believe that violence against humans and cultures chiefly originates from the Western world, and is even something that is intrinsically Western. It is true that to understand our contemporary situation, we must look to events explaining our current situation. However, human history as far as we can read – and even before as far as we can reconstruct archeologically – is very much the history of oppression between people, of imposition of languages, cultures and religions. Europe of the Renaissance and Modern area expanded its power through technology and the pursuit of a liberal-style economic development based on the exploitation of off-shore colonies and its people. The history of European racism is interwoven with this massive development, which still structures international relationships and our everyday contemporary lives. In this chapter, we are primarily concerned with the relationship between science, economics and racism, as developed in the Western world.

In matters of science and ethics, European and North American medicine and biomedical science, including basic and applied neurosciences, were part of this broader process of capitalist colonization and investment in science as enlightened salvation (Abiodun 2019; Rollins 2021). In the United States of the 19th century, phrenologists argued that their psychology and neuroscience would bring about a philosophical, educational and social revolution. They invested their understanding of the brain with the greatest potentials and stated that:

> To know ourselves is a matter of greatest importance and there is no other means by which we can acquire this knowledge so well, as by the aid of Phrenology. It teaches us for what occupation in life we are by nature best qualified, and in what pursuit we may be most successful (cited in Racine 2010, p. 112).

In that period and the following decades, biological- and neuroscience-based essentialization and racism were also prevalent (Abiodun 2019). Hagner (2001) brought to light how, in Europe, neuroscience was sometimes envisioned – as illustrated in German neuroscience – to promulgate neurocentric ideals of human development and flourishing. For example, the influential neuroscientist Theodor Meynert envisioned that neuroscience could "improve culture and man as a whole" (Hagner 2001, p. 543). Followers of Meynert, like Oscar Vogt, soon migrated to social hygiene when discussing the neuroscience of morality. In 1912, the leading neuroscientist Vogt wrote that:

> Man will increasingly become a brain animal. In our further development, the brain will play an increasingly important role. But this development will bring ever increasing health dangers with it. Thus, a fortuitous future of our species depends significantly on the expansion of brain hygiene (Vogt 1912, cited in Hagner 2001, pp. 553–554).

Although the specifically racist and discriminatory use of neuroscience and the neuroscience of morality has been overwhelmingly condemned (Shevell 1999), neuro-essentialism and neuroscientism have endured (Racine and Zimmerman 2012). More recently, the idea of a neurophilosophy, a form of philosophical thinking using neuroscience knowledge as a form of foundational knowledge – as developed notably by Patricia Churchland – has made neuroscience the primary discipline that would provide a sound scientific basis for the construction of human ethics (Churchland 1986, 1995, 2012). For example, the philosopher and neuroscientist Greene claims that "Social neuroscience is, above all else, the construction of a metaphysical mirror that will allow us to see ourselves for what we are and, perhaps, change our ways for the better" (Greene 2011). In the context of ethics, the cognitive neuroscientist Gazzaniga (2005) has called for a "brained-based ethics", while the neurobiologist Changeux (1996) has appealed to neuroscience to provide a universal foundation for human ethics. These claims reflect the very real promise that neuroscience can shed light on a wide array of human social and moral habits. But they sometimes verse in excess when describing neuroscience as the ultimate form of self-knowledge, rather than as one of its many contributions alongside other disciplines such as psychology, anthropology and other philosophical and ethical approaches. In contrast to reductionist neurophilosophy, neuropragmatism links experience and culture in the sense that reasoning permitting the cognition of the environment is based on contexts (Solymosi and Shook 2013). Also, global neuroethics is interested in addressing the cross-cultural problems of neuroscience so that it can have beneficial effects for a global society (Rommelfanger et al. 2019). In summary, neuroscientism – the authority afforded to neuroscience in vetting moral ideals and describing the ultimate insights into human nature – has granted neuroscience epistemic supremacy over all other discourses and, by cascade, over other cultures. Neuroscience functions as a cultural force as it embodies its sources and perpetuates certain values associated with its origins (Choudhury and Slaby 2011; Vidal and Ortega 2017). This supremacy of neuroscience needs to be questioned by neuroethics (Jotterand and Ienca 2017). If neuroethics accepts the dominant influence of liberalism and mainstream neuroscience uncritically, it offers limited critical resources with respect to understanding and responding to diversity. Neuroethics – like genetics before it in the context of genetics – may become the advocate of neuroscience, failing to engage in genuine dialogue, as critics of neuroethics have cautioned (De Vries 2005, 2007; Pitts-Taylor 2010). This is a worry which is substantiated with some uncritical acceptance of the promises of neuroscience and neurotechnology by some neuroethicists (Racine and Forlini 2010; Gilbert et al. 2018), thereby echoing the media's enthusiasm for neurotechnology and neuroscience-based solutions to almost everything (Racine et al. 2005, 2006, 2007, 2010, 2017c; Dubljević et al. 2014). Thus, an important question, albeit one which cannot be answered easily, is whether neuroethics has reflected anthropological diversity and socio-cultural diversity, and why. Again, this depends on what we mean by diversity.

In a more general and superficial sense, neuroethics has in principle been open to diversity as grants and journal publications are peer-reviewed, although peer review has limitations. Some compilations of edited volumes have explicitly opened their pages to international perspectives on neuroethical topics (Jotterand and Dubljević 2016). The Neuroethics Society, created in 2007, became the International Neuroethics Society a few years later to explicitly reflect and encourage diversity and international representation. However, it is clear that some geographic regions have not participated equally in the development of this field (Lombera and Illes 2009). Furthermore, the most salient topics discussed in neuroethics tend to center around neurotechnologies, with more limited attention on topics that concern the deeper sources of inequalities of neurological and neuropsychiatric conditions which could be of much broader relevance globally speaking (Racine 2010; Racine and Sample 2018). If there is a determination to integrate other perspectives based on socio-cultural realities, there must be a commitment to the development of methodologies able to reflect how moral values are shaped by human cultures. Kaya and Seleti (2013) write that African scholars often fail to devise methodological and theoretical approaches to value their realities. They also point to the fact that African educational systems are designed on the basis of other cultures that are different from African cultures. In the context of neuroethics, this situation and others should be kept in mind to avoid the imposition of a single account of neuroethics and the construction of genuine international dialogue. Moreover, Black scholarships are still racialized and related to a narrow and biological understanding of race (Dei and Atweneboah 2014), thereby demonstrating pervasive misunderstandings which impede genuine dialogue. This is one of the reasons for the worthwhile development of an international neuroethics summit to foster dialogue and promote diversity (Rommelfanger et al. 2019).

It is quite possible that these limitations of neuroethics in addressing socio-cultural diversity – which is often discussed under the topic of health and economic disparities – lead to additional difficulties in representing the broad spectrum of ethical questions (Zizzo et al. 2016), topics and situations that occur in everyday experiences of healthcare and research (Racine and Sample 2018). There is data showing how, for example, the ethics literature about a common neurological disease like Parkinson's disease is heavily centered on the sensational issues raised by advanced neurotechnologies (e.g. deep brain stimulation, stem cells, tissue transplantations), rather than on ordinary healthcare and research experiences and challenges created by health inequalities (Zizzo et al. 2017). This echoes bioethics' own troubled relationship with socio-cultural diversity (Hedgecoe 2010) and its ability to reflect the root social and political causes of a number of important lived ethical problems (Turner 2003; Hedgecoe 2004; Turner 2004). As a cultural creation, neuroethics has adopted dominant views about socio-cultural diversity which limit its connection to problems experienced internationally, as well as within the Western world, where a host of marginalized people and communities are excluded from common preoccupations. To move beyond these barriers, neuroethics needs to metamorphose and grow beyond the common understanding that the field is chiefly concerned with the ethics of neurotechnology and ask difficult questions about who cares and why we care about neurotechnology.

In a deeper, anthropological sense, diversity (anthropological diversity) with respect to moral habits may have received greater attention in what is often considered to be the other component of neuroscience; the neuroscience of morality. This field is erroneously called the neuroscience of ethics, implying an unhelpful conflation of ethics and morality. Indeed, the term "ethics" designates a discipline, and the "neuroscience of ethics" is sometimes wrongly applied by taking this discipline as its object. It would be more accurate to understand the neuroscience of ethics as a "neuroscience of morality", since the neuroscience of ethics does not investigate ethics per se (i.e. the academic discipline that takes morality as its object), but rather certain aspects of morality such as moral judgments, moral emotions, moral intuitions and moral behaviors (Racine et al. 2017b). By examining the neurological mechanisms underlying phenomena such as empathy and altruism, or the workings of different kinds of moral judgments (e.g. deontological, duty-oriented judgments vs. consequentialist, that is, consequence-oriented judgments (Reynolds 2006; Reynolds et al. 2010)), this other area of neuroethics has perhaps raised deeper questions about the foundation of ethics and its connection to how we come to general moral judgments based on duties, consequences and the like. There are important limitations to this area of research, notably the common small sample sizes and limitations in the geographic representation of samples (Racine et al. 2017b). There has also been a tendency to attempt to justify a given ethics theory based on how the brain works – a brain-based ethics (e.g. Casebeer (2003) and Gazzaniga (2005)), although some have cautioned against this push early on (Greene 2003; Racine 2005), and proposed models of moral and ethical judgment (Dubljević and Racine 2014) and practices to avoid this fallacy (Racine et al. 2017b). Even more importantly, what we mean by human morality is plural and therefore attempts to understand the "moral brain" need to take into account that those very studies are based on specific understandings of what morality is constituted to be (Holtzman 2017). From a neuroscience perspective, it may not be possible to develop a unified or objective and universal picture of the neural correlates of moral judgment if environmental pressures have led the neural structures involved to develop in different ways, as reflected in diverse human moralities.

In summary, neuroethics has somewhat addressed the issue of diversity. From the standpoint of the ethics of neuroscience, progress has been made but there is much more work to be done to tackle socio-cultural diversity and reflect the broad spectrum of issues relevant to populations outside of North America and Western Europe. In terms of both the contributors to the field and, perhaps unsurprisingly, the topics addressed, diversity is limited. The mainstream neuroethics agenda centers on issues that affect populations using new neurotechnologies, which is not representative of issues of importance worldwide and alternative ways of thinking about and approaching neurological and psychiatric health. The attention granted to issues of social inequality, and their contribution to neurological and psychiatric health, is minimal and is typically not a fundamental feature of neuroethics scholarship or bioethics, more generally (Barned et al. 2019; Wilson et al. 2019). Being essentially prompted by neurotechnologies – a point of view that one of us has challenged for many years with limited success – neuroethics, or at least neuroethics as an ethics of neuroscience, has difficulties adopting a truly critical and reflective lens (De Vries 2005, 2007). The neuroethical imaginary, to borrow De Vries' language, is strikingly and unimaginatively populated with fantasies of beneficial or gruesome neurotechnology (De Vries 2005). Richer accounts of human flourishing deserve greater attention as genuine and deeper foci of ethical inquiry (Racine 2010; Racine and Sample 2018). From the standpoints of the neuroscience of morality, there are interesting opportunities to rethink the nature of human morality and the relationship of ethics with a number of assumptions about human morality. For example, this branch of work has questioned rationalist ethics (Damasio 1994; Haidt 2001), the status of ethics theories as competing foundations of moral judgment (Dubljević 2014) and the ability to consciously guide moral judgment given extensive implicit biases about gender, race and so forth (Luo et al. 2006; Kelly and Roeddert 2008; FitzGerald and Hurst 2017). Sometimes, overly strong interpretations have heralded this branch of work as a new foundation of ethics (Changeux 1996; Gazzaniga 2005), but more moderate interpretations may be entirely reconcilable with more generous and inclusive accounts of human morality and ethics (Racine and Sample 2018).

4.4. Should neuroethics take cultural diversity into account, and why?

Neuroethics should deeply care about diversity, both socio-cultural and anthropological, namely because ethics concerns the examination of our habits – including the disruptions of habits when we are confronted with situations where they now seem inappropriate – in light of their moral salience, that is, their impact on human welfare and flourishing. In this sense, neuroethics has to start from lived human experience, and this inexorably means that diversity is an unavoidable matter of fact. However, experience is not simply a given. Experiences of the world are the result of processes. They are transactions with the world – they reveal not only what the world is, but also what we seek in the world and how we think about the world (Dewey 1922). Experiences tell us something about ourselves too. In that light, changes in habits (e.g. disruptions in how neurological and psychiatric conditions are diagnosed, treated, understood, etc.) all have a potential impact on numerous aspects of the lives and welfare of patients, families, clinicians and society. Working with cultural diversity in such contexts means that the discussion of neuroethical questions needs to take into account more explicitly the anchoring of moral habits into experience and that human experiences diverge notably because of the impact of human cultures on that experience. For this to happen, neuroethics must be much more attuned to the actual lived experiences of those it seeks to help, including how knowledge of neurological and psychiatric conditions is generated, what it means and what it does. Whether it is patients, researchers, clinicians or research participants, and even the general public, neuroethics needs to start from the actual situations encountered by different stakeholders and develop approaches that take into account the inherent diversity of human experiences that make up moral situations. This means that the dominant focus on neurotechnology in neuroethics (Wolpe 2004) may be rather irrelevant to some populations because they are basically unconcerned and untouched by them.

This deep commitment to understanding and starting from human experience does not mean that the work of ethics stops there. Ethics does not require that any experience of the world and any human practice with respect to neurological and neuropsychiatric conditions must be respected as such. On the contrary, it starts with the understanding that human moralities, as sets of habits, are ways of cultivating experience, of taking out something out of experience that can then be evaluated (e.g. discussed, examined, scrutinized) for their ability to promote human flourishing in human experience. This critical evaluation of moral habits is the task of ethics. Indeed, once we acknowledge and understand human experience and work with its diversity, we must ask what kinds of scenarios involving new treatments, diagnostic procedures, new clinical information and new insights into the working of the brain are to be favored. To do this, processes of deliberation, which include more traditional ethical analysis, as well as more grounded and well-rounded anticipations about projected courses of action, need to be weighted, evaluated, discussed in light of their potential outcomes and impacts on human well-being and flourishing (Racine 2016). In this sense, starting from experience and its diversity does not renounce the need to engage in ethics as the disciplined and structured analysis of our habits in light of their (negative and positive) impact on human well-being and flourishing. Rather, it situates this effort in a broader and richer context of human experience.

This experience-oriented proposal is uncomfortable because it means there is literally no way out of human experience; there is no vantage point from which objectivity can hinge upon. Exercises of ethical analysis and philosophical abstraction allow us to distance ourselves from current and past experiences, and to reflect critically on them. However, they start from experience and will need to return to experience to have an impact on human existence. This is messy because it also implies that principles cannot simply be applied as God-given truths based on an applied ethics' model of bioethics and neuroethics. They must be tested and evaluated in light of their outcomes, not only in terms of actual consequences, but also implications in all matters – the agent, social relationships and so on. In all fairness, we must acknowledge that some of the more powerful discourses around neurological and neuropsychiatric conditions, such as those emerging from North American and Western European neuroscience, need to be examined critically to reflect on their implications. However, as we set out at the beginning of this chapter, we encourage humility such that no one initiates the conversation under the assumption of having everything right.

The implications of this transactional account of human experience, and a pragmatic vote in favor of deliberation as a path to human growth, imply that diversity is not an end in and of itself. Human diversity can mirror the fact that humans have leeway in differentially pursuing their existence – making their own choices about what they prefer, and the experiences they want to seek and cultivate. Nothing guarantees that all forms of behaviors and practices are necessarily desirable in terms of an ethical goal such as human flourishing because not all human habits are equally able to promote human flourishing. For example, many cultures entertain significant stigma against those who have illnesses, particularly psychiatric and neurological conditions. What good do such habits serve? Should they be respected as such? They must certainly be understood for the role they play and the meaning they have to those who hold them, but considering these practices in terms of promoting human flourishing suggests a negative answer. While it may appear that Western social habits are superior in such discussions, nothing could be more uncertain. Consider the impersonal aspects of human health and social systems in Western societies and their impact on mental health. Consider the incredible socio-economic disparities tolerated in the richest Western countries and their impact on ever-growing impoverished populations, notably children in poverty (Chomsky 2012). What good do these habits do for the neurological and mental health of youth (Katsnelson 2015)? Genuinely and radically asking whether our human habits promote human flourishing likely points to a path of mutual learning, not of colonial or imperial universalism.

A pragmatic take on diversity starts from a commitment to the fallibility of our habits with respect to their ability to achieve what we consider to be the best thing to do. In this context, ethics is much less a process of teaching and preaching than one of learning – learning that is premised on the valuable insights that each of us as humans glean from their existence. While it is true our existences are shaped by our cultures, they are not bound to these cultures since we can learn from each other and transform ourselves and our surroundings based upon what we learn. That said, how can those who think that their way of doing (e.g. their traditional way of understanding and healing; their technologically oriented interventions) is more advanced than those of others that actually learn from others? In order to engage in open dialogue, we need to avoid presumptions of superiority to genuinely open up to what others can bring us. This is why the promotion of intercultural dialogue that fosters open discussion about the merits of our values and interests would seem most important to resist the colonial imposition of a single-minded unidimensional worldview looming over fields like bioethics and neuroethics (Moreno 2004; Chattopadhyay and De Vries 2008). This kind of dialogue is challenging and requires conditions to be in place, such that force does not become the law (Rousseau 1992). This dialogue must be premised on the confidence that we can grow as humans by learning from each other, and that we are not threatened by our differences but potentially enriched by them (Doucet 2005). In this light, when cultural diversity becomes typified and formalized to the point of becoming an essence, it can become an obstacle to genuine dialogue rather than a prompt for open dialogue. Cultures where a sense of superiority has been promoted are challenged by the need to learn from others. On the other hand, cultures which are regularly oppressed will find it difficult to trust a dialogical process where the odds appear against them and openness to change are predetermined to favor the powers that be.

4.5. Conclusion

Neuroethics, like bioethics, is a structured response to a number of moral questions and problems encountered in this case in the understanding and treatment of neurological and neuropsychiatric conditions. It originates from North America and Western Europe, and these origins and its common imaginaries are reflected in its form, in the problems discussed and in the people involved. When distinguishing between anthropological and socio-cultural diversities, we find that neuroethics has been keen to reflect socio-cultural diversity when tackling issues related to new neurotechnology. However, neuroethics has also been ingrained into mainstream liberal frameworks and accounts of diversity. The work on the neuroscience of morality could be a window into deeper anthropological diversity, provided that it is steps away from neuroscience's claim to epistemic supremacy. In response to such challenges, we delineated how a pragmatic approach favors a dialogical approach, which attempts to cultivate the richness of diversity without falling prey to relativism. No culture possesses a totally satisfactory medical and ethical response to all neurological and psychiatric illnesses. Dialogue, and the desire to pull the best habits of thought and action from different cultures, moves away from implicit colonialism encountered in some forms of neuroethics and bioethics (Moreno 2004). Further, it invites openness to change. This stance is at odds with stringent forms of cultural conservatism that entertain the view of cultures as essences, including some forms of multiculturalism which insist on the respect of the integrity of cultures to the point of making cultural exchanges difficult and even suspect. Pragmatism's commitment to dialogical approaches may correspond to some features of "two-eyed seeing" (*Etuaptmumk*), first advanced by Mi'kmaq Elder Albert Marshall (from the community of Eskasoni, Nova Scotia, Traditional Territory of Mi'kma'ki) (Bartlett et al. 2012, 2015; Marshall et al. 2015). This idea is to:

> learn to see from your one eye with the best or the strengths in the Indigenous knowledges and ways of knowing [...] and learn to see from your other eye with the best or the strengths in the mainstream (Western or Eurocentric) knowledges and ways of knowing [...] but most importantly, learn to see with both these eyes together, for the benefit of all (Elder Albert Marshall, cited in Bartlett (2017, p. 1)).

Some obstacles to a dialogical approach may be due to common narrow understandings of neuroethics as an ethics of neurotechnology or as a neuroscience of morality (Racine and Sample 2018). We contrasted these mainstream ideas to pragmatic neuroethics, which calls for a contextualized form of ethics response and committed deliberation that reflects human diversity (Miller et al. 1996; Racine 2008a, 2008b, 2016). Dialogical ethics could be an approach that allows neuroethics to engage in a mutual learning process whereby alternative and traditional models of neurological and psychiatric disease management and associated ethical practices are not minimized. As Arnett et al. (2009) remind us:

> [D]ialogic ethics begins with one basic prescription – respect whatever is before you and take it seriously. The reality before us is all there is; we must learn from what presents itself, whether wanted or not (p. 91).

In order to move away from a neuroethics, anchored in colonial epistemologies and blindly promoting values such as individualism, it is important to move towards the mutual understanding that can occur in a context of tensions between values (Létourneau 2018). This would help avoid the imposition of neoliberalism and individualism on cultures where representations of disease and corresponding treatments mobilize collectives such as family, relatives and neighborhoods in alternative treatment modalities. However, from an international perspective, is neuroethics, primarily concerned with scientific advances in neurology or psychiatry, ready to engage in a dialogue with the ethical models of alternative or traditional medicines? It is questions such as these that dialogical ethics will have to address by proposing, for example, international models of ethics that promote mutual learning.

4.6. References

Abiodun, S.J. (2019). Seeing color. A discussion of the implications and applications of race in the field of neuroscience. *Fronts. Hum. Neurosci.*, 13. doi:10.3389/fnhum.2019.00280.

Acemoglu, D., Johnson, S., Robinson, J.A. (2002). Reversal of fortune: Geography and institutions in the making of the modern world income distribution. *Q. J. Econ.*, 117(4), 1231–1294.

Arnett, R.C., Fritz, J.M.H., Bell, L.M. (eds) (2009). Dialogic ethics: Meeting differing grounds of the "good". In *Communication Ethics Literacy: Dialogue and Difference*. Sage Publications, Thousand Oaks, CA.

Barned, C., Racine, E., Lajoie, C. (2019). Addressing the practical implications of intersectionality in clinical medicine: Ethical, embodied and institutional dimensions. *Am. J. Bioeth.*, 19(2), 27–29.

Bartlett, C. (2017). Two-eyed seeing: Elder Albert's Marshall's guiding principle for inter-cultural collaboration [Online]. Available at: http://www.integrativescience.ca/uploads/files/Two-Eyed%20Seeing-AMarshall-Thinkers%20Lodge2017(1).pdf [Accessed 11 May 2021].

Bartlett, C., Marshall, M., Marshall, A. (2012). Two-eyed seeing and other lessons learned within a co-learning journey of bringing together indigenous and mainstream knowledges and ways of knowing. *J. Environ. Stud. Sci.*, 2, 331–340.

Bartlett, C., Marshall, M., Marshall, A., Iwama, M. (2015). Integrative science and two-eyed seeing: Enriching the discussion framework for healthy communities. In *Ecosystems, Society and Health: Pathways through Diversity, Convergence and Integration*, Hallstrom, L.K., Guehlstorf, N., Parkes, M. (eds). McGill-Queen's University Press, Montreal.

Casebeer, W. (2003). Moral cognition and its neural constituents. *Nat. Neurosci.*, 4(10), 841–846.

Changeux, J-P. (1996). Le point de vue d'un neurobiologiste sur les fondements de l'éthique. In *Cerveau et psychisme humains : quelle éthique ?* Huber, G. (ed.). John Libbey Eurotext, Paris.

Chattopadhyay, S. and De Vries, R. (2008). Bioethical concerns are global, bioethics is Western. *Eubios J. Asian Int. Bioeth.*, *EJAIB*, 18(4), 106–109.

Chomsky, N. (2012). *Occupy*. Penguin Press, London.

Choudhury, S. and Slaby, J. (eds) (2011). *Critical Neuroscience: A Handbook of the Social and Cultural Contexts of Neuroscience*. Wiley-Blackwell, Hoboken, NJ.

Churchland, P.S. (1986). *Neurophilosophy: Toward a Unified Science of the Mind-Brain.* MIT Press, Cambridge, MA.

Churchland, P.M. (1995). *The Engine of Reason, the Seat of the Soul: A Philosophical Journey into the Brain.* MIT Press, Cambridge, MA.

Churchland, P.S. (2012). *Braintrust.* Princeton University Press, NJ.

Cohan, A. and Howlett, C.F. (2017). John Dewey and his evolving perceptions of race issues in American democracy. *Fac. Works: Educ.*, 17, 16–22.

Damasio, A.R. (1994). *Descartes' Error.* G.P. Putnam Sons, New York.

De Vries, R. (2005). Framing neuroethics: A sociological assessment of the neuroethical imagination. *Am. J. Bioeth.*, 5(2), 25–27.

De Vries, R. (2007). Who will guard the guardians of neuroscience? Firing the neuroethical imagination. *EMBO Rep.*, 8(S1), S65–S69.

Dei, G.S.D. and Atweneboah, N. (2014). The African scholar in the Western Academy. *J. Black Stud.*, 45(3), 167–179.

Dewey, J. (1922). *Human Nature and Conduct: An Introduction to Social Psychology.* H. Holt and Company, New York.

Doucet, H. (2005). Imagining a neuroethics which would go further than genetics. *Am. J. Bioeth.*, 5(2), 29–31.

Dubljević, V. and Racine, E. (2014). The ADC of moral judgment: Opening the black box of moral intuitions with heuristics about agents, deeds, and consequences. *AJOB Neurosci.*, 5(4), 3–20.

Dubljević, V., Saigle, V., Racine, E. (2014). The rising tide of tDCS in the media and academic literature. *Neuron*, 82(4), 731–736.

FitzGerald, C. and Hurst, S. (2017). Implicit bias in healthcare professionals: A systematic review. *BMC Med. Ethics*, 18(1), 19. doi: 10.1186/s12910-017-0179-8.

Fox, R.C. and Swazey, J.P. (2005). Examining American bioethics: Its problems and prospects. *Camb. Q. Healthc. Ethics*, 14(4), 361–373.

Gazzaniga, M.S. (2005). *The Ethical Brain.* Dana Press, New York and Washington.

Gerstle, G. (1993). The limits of American universalism. *Am. Q.*, 45(2), 230–236.

Gilbert, F., Viana, J., Ineichen, C. (2018). Deflating the "DBS causes personality changes" bubble. *Neuroeth*, 19, 1–17.

Greene, J.D. (2003). From neural "is" to moral "ought": What are the moral implications of neuroscientific moral psychology? *Nat. Rev. Neurosci.*, 4(10), 847–850.

Greene, J.D. (2011). Social neuroscience and the soul's last stand. In *Social Neuroscience: Toward Understanding the Underpinnings of the Social Mind, Social Cognition and Social Neuroscience*, Todorov, A., Fiske, S., Prentice, D. (eds). Oxford Academic, New York.

Hagner, M. (2001). Cultivating the cortex in German neuroanatomy. *Sci. Context*, 14(4), 541–563.

Haidt, J. (2001). The emotional dog and its rational tail: A social intuitionist approach to moral judgment. *Psych. Rev.*, 108(4), 814–834.

Harari, Y.N. (2016). *Sapiens: A Brief History of Humankind*. Penguin Random House, New York.

Hedgecoe, A. (2004). Critical bioethics: Beyond the social science critique of applied ethics. *Bioethics*, 18(2), 120–143.

Hedgecoe, A. (2010). Bioethics and the reinforcement of socio-technical expectations. *Soc. Stud. Sci.*, 40(2), 163–186.

Holtzman, G.S. (2017). Neuromoral diversity: Individual, gender, and cultural differences in the ethical brain. *Front. Hum. Neuro.*, 11, 501. doi: 10.3389/fnhum.2017.00501.

Jotterand, F. and Dubljević, V. (eds) (2016). *Cognitive Enhancement: Ethical and Policy Implications in International Perspectives*. Oxford University Press, Oxford.

Jotterand, F. and Ienca, M. (2017). The biopolitics of neuroethics. In *Debates about Neuroethics: Perspectives on Its Development, Focus, and Future*, Racine, E. and Aspler, J. (eds). Springer, Cham.

Katsnelson, A. (2015). The neuroscience of poverty. *Proc. Natl. Acad. Sci.*, 112(51), 15530–15532.

Kaya, H.O. and Seleti, Y.N. (2013). African indigenous knowledge systems and relevance of higher education in South Africa. *Int. Educ. J.: Comp. Perspect.*, 12(1), 30–44.

Kelly, D. and Roeddert, E. (2008). Racial cognition and the ethics of implicit bias. *Philos. Compass*, 3(3), 522–540.

Kokwaro, G. and Kariuki, S. (2001). Medical research in Africa: Problems and some solutions. *Malawi Med. J.*, 13(3), 40.

Kymlicka, W. (1995). *Multicultural Citizenship: A Liberal Theory of Minority Rights*. Clarendon Press, Oxford.

Lanre-Abass, B. (2010). Racism and its presuppositions: Towards a pragmatic ethics of social change. *Hum. Aff.*, 20(4), 364–375.

Leefmann, J., Levallois, C., Hildt, E. (2016). Neuroethics 1995–2012. A bibliometric analysis of the guiding themes of an emerging research field. *Front. Hum. Neurosci.*, 10, 336. doi: 10.3389/fnhum.2016.00336.

Létourneau, A. (2018). Differing versions of dialogic aptitude. Bakhtin, Dewey, and Habermas. In *Dialogic Ethics*, Arnett, R.C. and Cooren, F. (eds). John Benjamins Publishing, Amsterdam.

Lombera, S. and Illes, J. (2009). The international dimensions of neuroethics. *Dev. World Bioeth.*, 9(2), 57–64.

Luo, Q., Nakic, M., Wheatley, T., Richell, R., Martin, A., Blair, R.J. (2006). The neural basis of implicit moral attitude-an IAT study using event-related fMRI. *Neuroimage*, 30(4), 1449–1457.

Marshall, A., Marshall, M., Bartlett, C. (2015). Two-eyed seeing in medicine. In *Determinants of Indigenous Peoples' Health in Canada; Beyond the Social*, Greenwood, M., De Leeuw, S., Lindsay, N.M. (eds). Canadian Scholar Press, Toronto.

Meltzer, M. (1993). *Slavery: A World History*, revised edition. DaCapo Press, Cambridge, MA.

Messikomer, C.M., Fox, R.R., Swazey, J.P. (2001). The presence and influence of religion in American bioethics. *Perspect. Biol. Med.*, 44(4), 485–508.

Mill, J.S. (1956). *On Liberty*. Bobbs-Merrill Company, Indianapolis, IN [Original published 1859].

Miller, F.G., Fins, J.J., Bacchetta, M.D. (1996). Clinical pragmatism: John Dewey and clinical ethics. *J. Contemp. Health Law Policy*, 13(9), 27–51.

Moreno, J.D. (2004). Bioethics imperialism. *ASBH Exchange*, 7(3), 2.

Ortega, F. (2009). The cerebral subject and the challenge of neurodiversity. *Biosocieties*, 4(4), 425–445.

Pekarsky, D. (1990). Dewey's conception of growth reconsidered. *Educ. Theory*, 40(3), 283–294.

Pitts-Taylor, V. (2010). The plastic brain: Neoliberalism and the neuronal self. *Health*, 14(6), 635–652.

Racine, E. (2005). Pourquoi et comment doit-on tenir compte des neurosciences en éthique ? Esquisse d'une approche neurophilosophique émergentiste et interdisciplinaire. *Laval théologique et philosophique*, 61(1), 77–105.

Racine, E. (2008a). Interdisciplinary approaches for a pragmatic neuroethics. *Am. J. Bioeth.*, 8(1), 52–53.

Racine, E. (2008b). Which naturalism for bioethics? A defense of moderate (pragmatic) naturalism. *Bioeth.*, 22(2), 92–100.

Racine, E. (2010). *Pragmatic Neuroethics: Improving Treatment and Understanding of the Mind-Brain*. MIT Press, Cambridge, MA.

Racine, E. (2016). Can moral problems of everyday clinical practice ever be resolved? A proposal for integrative pragmatist approaches. In *Ethics in Child Health: Principles and Cases in Neurodisability*, Rosenbaum, P.L., Ronen, G.M., Racine, E., Johannesen, J., Bernard, D. (eds). Mac Keith Press, London.

Racine, E. and Forlini, C. (2010). Cognitive enhancement, lifestyle choice or misuse of prescription drugs? Ethics blind spots in current debates. *Neuroeth.*, 3(1), 1–4.

Racine, E. and Sample, M. (2018). Two problematic foundations of neuroethics and pragmatist reconstructions. *Camb. Q. Healthc. Ethics*, 27(4), 566–577.

Racine, E. and Zimmerman, E. (2012). Pragmatic neuroethics and neuroscience's potential to radically change ethics. *The Neuroscientific Turn: Transdisciplinarity in the Age of the Brain*, 135–151.

Racine, E., Bar-Ilan, O., Illes, J. (2005). fMRI in the public eye. *Nat. Rev. Neurosci.*, 6(2), 159–164.

Racine, E., Bar-Ilan, O., Illes, J. (2006). Brain imaging: A decade of coverage in the print media. *Sci. Commun.*, 28(1), 122–142.

Racine, E., Waldman, S., Palmour, N., Risse, D., Illes, J. (2007). Currents of hope: Neurostimulation techniques in US and UK print media. *Camb. Q. Healthc. Ethics*, 16(3), 314–318.

Racine, E., Waldman, S., Rosenberg, J., Illes, J. (2010). Contemporary neuroscience in the media. *Soc. Sci. Med.*, 71(4), 725–733.

Racine, E., Aspler, J., Forlini, C., Chandler, J.A. (2017a). Contextualized autonomy and liberalism: Broadening the lenses on complementary and alternative medicines in preclinical Alzheimer's disease. *Kennedy Inst. Ethics J.*, 27(1), 1–41.

Racine, E., Dubljević, V., Jox, R.J., Baertschi, B., Christensen, J.F., Farisco, M., Jotterand, F., Kahane, G., Müller, S. (2017b). Can neuroscience contribute to practical ethics? A critical review and discussion of the methodological and translational challenges of the neuroscience of ethics. *Bioethics*, 31(5), 328–337.

Racine, E., Nguyen, V., Saigle, V., Dubljević, V. (2017c). Media portrayal of a landmark neuroscience experiment on free will. *Sci. Eng. Ethics*, 23(4), 989–1007.

Reynolds, S.J. (2006). Moral awareness and ethical predispositions: Investigating the role of individual differences in the recognition of moral issues. *J. Appl. Psychol.*, 91(1), 233–243.

Reynolds, S.J., Leavitt, K., Decelles, K.A. (2010). Automatic ethics: The effects of implicit assumptions and contextual cues on moral behavior. *J. Appl. Psychol.*, 95(4), 752–760.

Rollins, O. (2021). Towards an antiracist (neuro)science. *Nat. Hum. Behav.*, 5(5), 540–541.

Rommelfanger, K.S., Jeong, S.J., Montojo, C., Zirlinger, M. (2019). Neuroethics: Think global. *Neuron*, 101(3), 363–364.

Rousseau, J.-J. (1992). *Du contrat social.* Groupe Flammarion, Paris.

Sabelo, J.N.-G. (2013). Coloniality of power in postcolonial Africa. *Myths of Decolonization*. CODESRIA (Council for the Development of Social Science Research in Africa), Dakar.

Schaub, J-F. (2017). The formation of racial categories. *Politika* [Online]. Available: https://www.politika.io/en/notice/the-formation-of-racial-categories#1 [Accessed 30 April 2021].

Shevell, M.I. (1999). Neurosciences in the third Reich: From ivory tower to death camps. *Can. J. Neurol. Sci.*, 26(2), 132–138.

Solymosi, T. and Shook, J. (2013). Neuropragmatism: A neurophilosophical manifesto, *Eur. J. Pragmat. Ame. Philos.*, 1. doi: 10.4000/ejpap.671.

Sullivan, S. (2017). On the harms of epistemic injustice: Pragmatism and transactional epistemology. In *The Routledge Handbook of Epistemic Injustice*, Kidd, I.J., Medina, J., Pohlhaus Jr., G., (eds). Routledge, London.

Turner, L. (2003). The tyranny of "genethics". *Nat. Biotechnol.*, 21(11), 1282.

Turner, L. (2004). Bioethic$ Inc. *Nat. Biotechnol.*, 22(8), 947–948.

Vidal, F. (2009). Brainhood, anthropological figure of modernity. *Hist. Hum. Sci.*, 22(1), 5–36.

Vidal, F. and Ortega, F. (2017). *Being Brains: Making the Cerebral Subject.* Fordham University Press, New York.

Wilson, Y., White, A., Jefferson, A., Danis, M. (2019). Intersectionality in clinical medicine: The need for a conceptual framework. *Am. J. Bioeth.*, 19(2), 8–19.

Wolpe, P.R. (2004). Neuroethics. In *The Encyclopedia of Bioethics*, Post, S.G. (ed.). MacMillan, New York.

World Health Organization (2001). *Mental Health: New Understanding, New Hope.* The World Health Report 2001. World Health Organization, Geneva.

World Health Organization (2006). *Neurological Disorders: Public Health Challenges.* World Health Organization, Geneva.

Zizzo, N., Bell, E., Racine, E. (2016). What is everyday ethics? A review and a proposal for an integrative concept. *J. Clin. Ethics*, 27(2), 117–128.

Zizzo, N., Bell, E., Racine, E. (2017). What are the focal points in bioethics literature? Examining the discussions about everyday ethics in Parkinson's disease. *Clin. Ethics*, 12(1), 19–23.

5

Neurofeminism in BCI and BBI Ethics as a Prelude to Political Neuroethics

Mai IBRAHIM and Veljko DUBLJEVIĆ
*Department of Philosophy and Religious Studies,
North Carolina State University, Raleigh, USA*

Recent years have witnessed incredible progress in the fields of neuroscience and neurotechnology, which has affected society as a whole. This chapter discusses the existing sexist and androcentric biases within neuroscientific research. We highlight the fundamental role of neurofeminists in challenging the prevalent neurosexism. We take a feminist materialist approach to explore the dynamic human–technology relationship. By focusing on brain-to-brain interfaces (BBIs) as a case study, we investigate the techno-socio-cultural entanglement of the different agential actors. We employ Karen Barad's theory of agential realism to re-conceptualize the brain as a co-constitutive active actor, as opposed to an independent one, and introduce BBIs as a hybrid phenomenon. In addition, we discuss the political perspective in neuroethics to address potential sexist (and racist) problems that may ensue upon the deployment of BBIs. This chapter concludes by emphasizing the importance of explicitly tackling the socio-political worldviews underscoring research and development, which help to provide sound ethical guidance and ground the discussion in scientific facts.

5.1. Introduction

Neuroscience is a rapidly developing field that is redefining what it means to be a human today. There is an increasing tendency to appeal to the authority of neuroscience to better understand how humans should act (Hoffman and Bluhm 2016). The general findings and applications of neuroscience (e.g. neuroplasticity) have affected society as a whole (Schmitz 2016). Advancements in neurotechnologies motivate the theorizing about socio-political clashes that need to be addressed in neuroethics scholarship, and neurosexism is a relevant case here.

Neuroscience also inspired neurosexism, a belief that Fine (2010) explains as the fundamental fixed differences between male and female brains, leading to the supposition that men are intellectually superior, resulting in damaging stereotypes in society. Such biased and gendered categorizations that devalue women are often perceived to be hardwired (Fine 2008). Hardwired traits are problematic and create controlling images, as they suggest that gender differences are permanent, innate and genetically determined (Bluhm et al. 2012).

Feminist research has much to offer to the field of neuroscience in general and in response to neurosexism, in particular. Feminists have long examined the gender roles and social relations in neuroscientific research. According to feminist science studies, sex is neither solely a physical nor a material construction, but is "deeply interwoven with social and cultural constructions" (Fausto-Sterling (2012), as cited in Schmitz and Höppner (2014)). Given the prevalent idealization of the masculine figure, feminist neuroscientists have sought to examine the implications of neuroscientific research on gender. Neurofeminism, in contrast to neurosexism, reassesses such gendered assumptions and underscores the entanglement between the "development of biological matter and social influences" (Schmitz and Höppner 2014, pp. 1–2). Moreover, neurotechnologies traverse the boundaries between biology and technology. Accordingly, to overcome gendered stereotypes, feminists seek to understand the relationship between gender and the brain by embracing the concept of brain plasticity, which transcends biological determinism (Schmitz and Höppner, 2014). Neuroplasticity posits that "brain structures and brain functions are not […] determined by evolution or remain unchanged during a life span" (Schmitz and Höppner, 2014, p. 4). In other words, the brain is not completely fixed; it is malleable and can respond to the environment (Bluhm et al. 2012).

The purpose of this chapter is twofold. First, we take a critical approach to explore the prevalent sexist attitudes in neuroscience and discuss the neurofeminist response. We adopt a materialist feminist approach to discuss the dynamic human–technology relationship. We focus on BBIs which represent a useful test-case to investigate the techno-socio-cultural inseparability by looking at brains, human bodies, neuroimaging, computers and stimulation devices. We employ Barad's (2007) onto-epistemological framework of agential realism to understand the agency of the various actors which intra-act to create BBIs as a hybrid becoming. Our second purpose in this chapter is to discuss the political perspective in neuroethics so as to address potential sexist and racist problems that may surface upon the widespread deployment of brain–computer interfaces and BBIs. Up until now, the political perspective in neuroethics has been construed (and defended) in terms of a Rawlsian framework of justice (see Dubljević (2019a)), which may alienate scholars from the traditions of critical theory and feminism (see Dubljević (2019b)). More specifically, Mannette (2022) agrees that there is a need for a "political neuroethics", but asserts that a Marxian analysis of "assigned sexual, familial, and economic ideological considerations" (Mannette 2022, pp. 38–39) is necessary to offer an expanded account. Thus, at a minimum, a "feminist political neuroethics" needs to be sketched.

This feminist-inspired exploration of BBIs will be discussed in four sections. First, we introduce BBIs as a particular development of brain–computer interfaces (BCIs). In the second section, we shed light on some of the existing sexist and androcentric biases within neuroscientific research by discussing neurosexism and highlighting how it necessitates a feminist intervention. Third, we build on Barad's agential realism to discuss the mutual shaping of the brain, bodies, technology and society when intra-actively creating BBI as a hybrid phenomenon. Finally, we connect the neurofeminist scholarship with the recently proposed political perspective in neuroethics (see Dubljević et al. (2021)). Such work is a much-needed addition to the field and highlights potential social concerns, as well as questions that neuroethics scholarship should address.

5.2. Brain-to-brain interfaces

Recent technological advances in neuroscience have set the stage for novel ways to cognitively enhance and augment humans. While the notion that brains can connect and directly communicate with a computer or another brain may feel like an episode of a science fiction show, it is not. Today, brains integrate technologies via their connectivity to machines. One such technology that is of increasing interest to researchers, engineers and clinicians is brain–computer interface (BCI) (Burwell et al. 2017). By coupling human brains with computers, the purpose of BCI is to enhance or restore human performance (Glannon 2014). BCI interprets brain activity through invasive or non-invasive monitoring devices (Shih et al. 2012). In addition, BCIs are used for medical purposes to provide alternative ways for patients to interact more fully with the world, as well as for non-medical uses such as gamers (Coin et al. 2020). Researchers have highlighted that BCI demonstrates the possibility of "decoding motor, visual, and even conceptual information from neural activity via a range of recording techniques" (Rao et al. 2014). A specific development of BCI is brain-to-brain interface (BBI) (Coin et al. 2020). While BCI enhances the human's capacity to handle cognitive tasks (Maksimenko et al. 2019), BBI allows for direct communication between human brains. BBI combines BCIs with computer-to-brain interfaces (CBIs) as well as multi-BBIs (Coin et al. 2020) to improve the human-to-human interaction with the aid of neuroplasticity. BBIs allow certain content to be extracted "from the neural signals of a 'sender' brain, digitizes it, and delivers it to a 'receiver' brain" (Jiang et al. 2019, p. 1).

Brain-to-brain interfacing (BBI) has the potential to allow for a network of brains to detect signals in the neural system from one brain and transmit these signals to another brain linked to the same network (Hildt 2019). While this appears to be a figment of the imagination, it is technically feasible, although it must be noted that the research is still in the "emerging" stage. Crucially, research funding for clinical and military uses of BBI is available. For instance, the National Institutes of Health (NIH) actively fund research on assistive devices that could allow paralyzed patients limited but direct brain communication (see Hildt (2019)). Similarly, the Defense Advanced Research Projects Agency (DARPA) is actively funding cutting-edge brain–computer interface and BBI research that could enhance human capacities (see Trimper et al. (2014)).

BBI systems transfer neural signals from one brain to a computer-based program where they are translated and executed, usually with the use of neurostimulation devices (Dubljević 2015), effectively transferring commands to another brain (Coin et al. 2020). The first "proof of principle" study was conducted on networks of animal brains. In an early experiment, animals exchanged information and served as a foundation for social interaction by using computing devices. This study by Pais-Viera et al. (2013) was the first to demonstrate a real-time transfer of information from one subject to another through the utilization of BBI. The research community was impressed with the results, which "indicate that animal brain dyads or even brain networks could allow animal groups to synchronize their behaviors following neuronal-based cues" (Pais-Viera et al. 2013, p. 6).

It was not long before a successful human study followed (Rao et al. 2014). However, a more exciting development was the "brain net" study by Jiang et al.

(2019), in which three human brains were communicating and influencing each other through the use of transcranial magnetic stimulation (TMS), in order to defeat a common computer game. BrainNet works by having two "senders" and one "receiver", with the receiver making a decision based on the senders' information. Even though there are current limitations to the technology, most notably the low bandwidth ultimately reducing communication to binary responses, future developments are striving towards a hive connected mind where information is shared by all through a linked neural network. In the words of neuroengineers at the forefront of this endeavor: "BrainNet allows Receivers to learn to trust the Sender who is more reliable, in this case, based solely on the information transmitted directly to their brains. Our results raise the possibility of future brain-to-brain interfaces that enable cooperative problem solving by humans using a 'social network' of connected brains" (Jiang et al. 2019).

Unsurprisingly, neuroethicists have identified numerous prospective applications of BBIs, such as gaming, human enhancement, user state monitoring and encryption. They have voiced a number of ethical and safety concerns, including, but not limited to: safety, agency, shared control, accountability, privacy and identity (see Trimper et al. (2014); Hildt (2019)). The consensus position is that multi-person BBIs further complicate ethical issues. In what follows, we argue that neuroethics scholarship needs to adequately incorporate a distinctively political perspective (Dubljević et al. 2021, see also Dubljević (2019a)) in order to appreciate the full extent of ethical and social issues raised by BBIs. Although research is underway to explore how to use BBIs to "deliver more complex information such as semantic concepts" (Jiang et al. 2019), we do not have to imagine future developments to appreciate how BBIs may further entrench existing biases and odious positions such as sexism and racism. Namely, even with limited, binary communication, "like" or "dislike" responses in the emerging multi-brain network may enable receivers to block or reduce certain types of senders. The use of social networks has rapidly facilitated sexist and racist content in the past (see Misselhorn 2018), and there is no reason to doubt that this will also be the case with BBIs. Due to reasons of space, and a lack of neuroethics scholarship on that particular topic, we will not discuss neuroethics of race but merely issue a call for a more relevant scholarship. However, there is an established body of neuroethics scholarship on neurosexism, to which we now turn.

5.3. Neurosexism

There is a widespread belief in the general public that men and women have irreconcilable differences that are difficult to overcome. Typically, men are portrayed as uncaring, logical and incapable of multitasking, while women are portrayed as emotional, empathetic and good listeners (Fine 2008). These are some of the gendered and sexist beliefs that have never really disappeared and have, in fact, began to resurface, couched in (neuro)scientific terms. The supposition that we can "sex" a brain, that is, to understand a brain as "male" or "female" and attribute certain characteristics and abilities to one sex or the other, has permeated our intellectual history. Female brains have been persistently described as undersized, underdeveloped and generally defective, resulting in the understanding of women as inferior and vulnerable (Rippon 2019, p. 10). In his article *Hegemonic Masculinity on the Mound*, Nick Trujillo offers an important insight into (neuro)sexism by exploring media representations of baseball pitcher Nolan Ryan, which underscores the extent of hegemonic masculinity in sports. Trujillo argues that media have functioned hegemonically by embodying Ryan as an "archetypical male athletic hero" (Trujillo 1991, p. 290).

The sexist practices and understandings in neuroscience are what Cordelia Fine labels "neurosexism" (Rippon 2019, 98). The term neurosexism describes "the phenomenon of using neuroscientific practices and results to promote sexist conclusions" (Hoffman and Bluhm 2016, p. 716). The investigation of male/female differences has an "unsavory past", which led to unjustified scientific claims that further proliferated gender stereotypes (Fine 2012, p. 369).

Neurosexism can take multiple forms that can support gender stereotypes and often exaggerate gender differences. One of the forms of neurosexism is exaggerating or even fabricating brain differences between men and women. This assumes that brain differences depend on a gender binary, which incorrectly posits that gender categories are mutually exclusive "essences" (Hoffman and Bluhm 2016). Not surprisingly, the categorization of genders is not as clear-cut as many may think, as the male–female binary remains contestable. Similarly, neurosexism holds that there is a categorical difference between male and female brains (Hoffman and Bluhm 2016, p. 718). Nevertheless, this remains a mistaken assumption, as differences in the brain concern brain size rather than sex since "male and female brains are [...] far more similar than they are different" (Fine 2010, p. 171). Furthermore, it is also pertinent to highlight that the phenotypic variation between individual brains renders the gender-based classification of brains unrealistic. A phenotype is understood as the observable and dynamic expression of an individual's genotype, which is influenced by both the genes and the environmental factors (Wojczynski and Tiwari 2008). In the past two decades, "genomic analyses have identified thousands of genetic variants that contribute statistically to variation in complex phenotypes" (as cited in Rosenberg et al. (2019, p. 27)). Such variation in phenotype underscores the fact that ascribing brain differences to gender is a misconception.

Remarkably, social pressures in neuroscience research seem to be tailored to highlight the differences, and not the similarities, between genders (Fine 2010, p. 171). For instance, gender stereotypes may motivate research questions or unconscious bias in the interpretation of data. Thus, there seems to be a myriad of ways in which neuroscience can accentuate the differences. Neuroscientific research and its various interpretations in publications and media outlets often promote sexist views engendering neurosexism (Hoffman and Bluhm 2016, p. 716). For instance, in 2004, Victoria Brescoll and Marienne LaFrance explored over 250 articles from multiple U.S. newspapers which discussed whether a given sex/gender was innate or acquired. Their research concluded that the ideological stance of the newspapers impacted the publication of scientific research. Conservative newspapers, for example, had the predisposition to ascribe sex differences to biological factors (Brescoll and LaFrance (2004), as cited in Jordan-Young and Rumiati (2012)). More often than not, neuroscientific studies that fail to find categorical differences between genders do not get published, whereas other studies that get published are misinterpreted either by researchers themselves or by the media (Hoffman and Bluhm 2016, p. 719). In the same vein, Fine (2008) maintains that some books draw on neuroscientific research that claims that the differences between sexes are hardwired. Brizendine's (2006) book *The Female Brain* is one example that draws on scientific research to further advance views on brains being gendered and having different structures. Brizendine rejects the presence of a unisex brain and insists that males' and females' brains are different by nature. Girls are wired as girls, and boys are wired as boys. She discusses how a three-year-old girl was presented with unisex toys such as a firetruck as opposed to a doll. Rejecting the role of the environment and the socialization process in shaping the brain and accentuating the differences between males and females, Brizendine's notes that the little girl started cuddling the firetruck wrapped in a blanket. This is to say, the girl's brain dictated her behavior.

Another form of neurosexism posits that the brain differences between men and women are permanent and that gender/sex traits are unalterable (Hoffman and Bluhm 2016, pp. 721). In other words, neurosexism contributes to the belief in "hardwiring" (Rippon 2019, pp. 288). The term hardwiring has enjoyed popularity in reference to sex differences and has become associated with traits of innateness, essentialism and immanence (Grossi 2017, p. 1050). Hardwiring explains the development of the human mind by emphasizing the permanence of the correlation between sex and gender, on the one hand, and the function or structure of the brain, on the other hand, and thus overlooks the effect of the social world in creating the sex/gender differences (Bluhm et al. 2012, p. 115). In this context, hardwiring understands the differences between male and female brains as fixed, and consequently results in reinforcing stereotypes (Rippon 2019, p. 288). This is the product of scientific explanations that defer gender differences to biological materiality (Bluhm et al. 2012, p. 235). For instance, some studies suggested that exposure to hormones during the fetal development phase resulted in a permanent yet dissimilar development pattern for male and female brains. Sociobiology, an influential and controversial position (see Kitcher (1985)), aggravated the situation by highlighting that the basis for hormonal sex differences was genetic (Bluhm et al. 2012, pp. 232–235). Such deterministic perspectives resulted in feelings of cynicism towards neuroscientific understandings of sex differences (Bluhm et al. 2012, p. 231).

Gender discrimination is very much alive, and bias still underpins much of the neuroscience of sex difference. Many neuroscientific studies on sex differences seem to only exacerbate the gender gap (Bluhm et al. 2012, p. 105). The deeply embedded beliefs in stereotypes still result in researchers designing studies and interpreting data in terms of male–female characteristics (Rippon 2019, p. 11). Imaging technologies such as functional magnetic resonance imaging (fMRI), which were used to investigate sex differences, are a case in point. The results of critical reviews demonstrated that imaging research was biased and reinforced gender stereotypes, resulting in further accusations of neurosexism (Fine 2012, p. 370). In the same vein, in developing the empathizing–systemizing hypothesis, Baron-Cohen (2003) emphasized that gender inequalities are not strictly due to culture or the socialization process. According to this view, biology, as well as essential differences between men and women, play a significant role (p. 92).

Neurofeminists have been challenging the prevalent neurosexism and the belief that biology is destiny (Rippon 2019, 13–17). A central approach they employ to challenge and deconstruct neurosexism is through the emphasis on neuroplasticity, a concept which underscores the idea that the brain is changeable and malleable in response to environmental pressures (Bluhm et al. 2012, p. 35). The theory of brain plasticity was first proposed by James (1890). James noted that "organic matter, especially nervous tissue, seems endowed with a very extraordinary degree of plasticity" (p. 64). That is, the brain is endowed with an extraordinary degree of plasticity and can be influenced. It was not until 1948 that Polish neuroscientist Jerzy Konorski coined the term "neuroplasticity", suggesting a theory by which "neurons which have been activated by closeness of an active neural circuit, change and incorporate themselves into that circuit" (Konorski (1948), as cited in Demarin et al. (2014)).

Neuroplasticity today questions the essentialist-gendered constructions, as well as the reductionist biological traits, and underscores the "influences of experiences and social interactions on the structural and functional developments in the brain"

(Schmitz 2016, p. 142). Insufficient attention to the theory of neuroplasticity has resulted in a bias towards understanding gender differences as fixed and reinforcing gender essentialism (Fine 2012, p. 370).

Empirical data shows that the brain responds to the environment, which affects its structure, such as its size and weight (Bluhm et al. 2012, p. 35). The brain's neural circuits develop and are changed by experience (Fine 2012, p. 397). The "gendered life experiences and social constructions of gender (such as leisure activities, educational interests, poverty and status) have material effects on the body, including the brain" (Fine 2012, p. 397). For instance, the performance of females in mathematics can differ across different social contexts where sometimes performance is improved when tests downplay the stereotype that males are better at math. This is to say that the social context impacts gendered behavior (Fine 2012, p. 397).

While the concept of neuroplasticity is useful in overcoming neurosexism and reframing sexist bias in neuroscience research, it may be interpreted to imply that the brain is a passive organ by indicating that social experiences impact the brain's structure (Schmitz 2016, p. 143). In other words, neuroplasticity may be interpreted as another form of essentialism, if it reifies the cultural influences on behavior by shaping the neurobiology of the brain, which is only a passive recipient. This understanding results in framing the brain as a "passive reactor to the attribution of gendered (and intersected) significations" (Schmitz 2016, p. 143). To counter this (mistaken) view, and demonstrate that the brain is an active actor, we take a new materialist approach to re-evaluate the meaning of brain plasticity by applying Barad's onto-epistemological framework of agential realism to BBI. We reconceptualize brain plasticity as an intra-active phenomenon that results from the entanglement of multiple agential forces.

5.4. Agential realism

Materialist feminists hold that there are "mutually implicating relationships that exist in the dynamics that constitute the discursive and material world" (Trites, Roberta Seelinger 2017). Materialist feminists explore the intrinsic relationship between the natural, the human and the nonhuman, with a particular focus on the question of agency. They explore the "interaction of culture, history, discourse, technology, biology, and the 'environment', without privileging any one of these elements" (Alaimo and Hekman 2008, p. 7).

New materialism arose as a reaction to the representationalist and constructivist movements that overlooked the material and in turn, resulted in postulating dualisms between what was observed and the external reality (Ferrando 2019). New materialism holds that biology is culturally mediated as much as culture is materially constructed. It denounces conventional thinking which perceives matter as inert, passive and static, and proposes to understand it as active and agentic (Ferrando 2019). Barad (2007) introduces an "ontoepistemological" framework, which she labels "agential realism", to demonstrate how humans and nonhumans create the world together. Agential realism criticizes representationalism, the idea that "representations and the objects (subjects, events or states of affairs) they purport to represent are independent of one another" (Barad 2007, p. 28). Representationalism assumes the existence of two independent entities: representations and the entities to be represented. The notion of separation is at the heart of representationalism, and thus, conflicts with feminist epistemologies (Anderson 2020) that denounce dualisms and separations, and work to reinforce connections and entanglements. Accordingly, Barad shifts the questions of descriptions and reality to matters of practices and actions, hence moving from representationalism to performativity.

In Barad's (2007, 2012) agential realism, matter is not a fixed substance, but a dynamic and an intra-active agent. Matter's dynamism is about the engagement with a new understanding of the world where bodies are not simply situated in the world but rather, bodies and the environment are intra-actively co-constituted. Agency in this account is not a capacity that someone or something has. Rather, agency is understood as a matter of intra-acting. Intra-action is also a new way to think about causality. Causality is not interactional, but intra-actional. In contrast with the concept of interaction, which presumes the prior existence of independent entities or relata, intra-action presents a shift in that it signifies "the mutual constitution of objects and agencies of observation within phenomena" (Barad 2007, p. 197). Thus, similar to Haraway's (1991) seminal "Cyborg Manifesto", Barad (2007) challenges dichotomies by incorporating the human and nonhuman, material and discursive, natural and cultural, and stresses that subjects, objects and agencies of observation are entangled and intra-act with one another. It is only through agential intra-actions that the properties and boundaries of the components of any given phenomena become meaningful (Barad 2007).

BBIs allow for direct communication between two brains. Brain–computer interfaces (BCIs), which constitute BBIs, detect neural signals in a brain and translate them to computer commands. The CBI then delivers these commands to another brain (Hildt 2019). This understanding of BBIs presumes linearity since the neural signals are transferred from one brain to a computer program, and then to another brain (Coin et al. 2020). A new materialist lens exposes the narrowness of this understanding of causation, positing that it is only when we think about causation in "linear terms that essentialism can be seen as the inevitable outcome of an attempt to think about the agency of matter or biology" (Frost 2011, p. 71). Therefore, new materialists aim to transform the understanding from "a framework within which the agency of bodies and material objects is understood largely as an effect of power – a unidirectional account of agency – to a framework within which, for example, culture and biology have reciprocal agentive effects upon one another" (Frost 2011, p. 71). This underscores the view that causation is not linear, but is rather conceived as "complex, recursive and multi-linear" (Frost 2011, p. 71).

A second challenge posed by this simple understanding of BBI is that of inertness, where the human brains are perceived as passive actors throughout the transmission process from one brain to another. New materialists challenge the idea that matter is passive and inert, emphasizing that it is active and dynamic (Alaimo and Hekman 2008, p. 75 and p. 135). Barad develops the notion of performativity, which is a "materialist, naturalist, and posthumanist elaboration that allows matter its due as an active participant in the world's becoming" (Alaimo and Hekman 2008, p. 122).

BBIs can be interpreted to demonstrate the power of agential realism. Agential realism (Barad 2007) permits the understanding of BBI as a material-discursive practice where actors are always already entangled. The relationship between the actors materializes through the continuous intra-actions, and thus challenges the biological determinisms that are used to legitimize gendered norms (Schmitz 2016, p. 141). In other words, BBIs are better conceptualized as intra-active phenomena that are constituted out of the intra-actions of multiple forces, that is, the intra-action of "the discursive, technological, and material", which produce a new configuration that has agency (Alaimo and Hekman 2008, p. 106). Although communication using BBIs seems linear, this is not the case. In fact, BBIs demonstrate the effect of mutual communication and information exchange on biological formations (Schmitz 2016, p. 151). BBI "rests on two pillars: the capacity to read (or 'decode') useful information from neural activity and the capacity to write (or 'encode') digital information back into neural activity" (Rao et al. 2014, p. 1). Therefore, BBI is a phenomenon that is constituted as a result of the reciprocal exchange of information between the brain, body, technologies and the environment.

The BBI experiment conducted by Rao et al. (2014) emphasizes this point. In this experiment, the BBI communication channel was built using EEG to record brain signals and TMS to stimulate the brain. A non-invasive BBI was applied to participants who were given the roles of senders and receivers in playing a computer game. The brain activity from the sender was recorded using EEG, and the resulting signal was used to control a cursor movement. When the cursor hit the assigned target on the screen, the sender's computer sent a signal over the Internet to the receiver's computer. The sender's and receiver's computers communicated using hypertext transfer protocol (HTTP). When the receiver's computer received a command, a magnetic pulse was sent to the receiver's brain, producing a movement of their right hand placed over a touchpad, triggering a click on the touchpad and resulting in the execution of the request initiated by the sender (Rao et al. 2014, p. 2).

This experiment supports the view that BBIs are an entangled network where the multiple actors intra-act together to create the BBI phenomenon. The components of the computer hardware and software intra-act with the brain and body via the screens, the EEG cap, the servers, and touchpad (Rao et al. 2014, pp. 2–4). That is, there is an intertwinement between the materiality of the participants' brains and bodies, as well as the inputs from the computer and the creation of certain outputs, which are stimulated through the TMS technology and translated through the EEG. This is also accompanied by the materiality of the touch and the movement of the cursor and the touchpad. This entanglement emphasizes that "the communicative intra-actions between brain, bodies and technical apparatuses are not conceptualized only in one direction" (Schmitz 2016, p. 148). In this way, BBIs can be framed through the dynamics of intra-activity, which disrupts the conventional understanding of linearity.

Similarly, in conceiving BBIs through the framework of agential realism, the brain is understood to be shaped by a multiplicity of external forces rather than an independent actor. The brain here is posited as a co-constituted and dynamic actor that is intra-acting with all others. In their experiment, Rao et al. stress that "imagery is central to the demonstration of brain-to-brain communication, as the movement intention that is initially imagined in the sender's brain is remotely executed by the receiver's brain" (Rao et al. 2014, p. 2). This demonstrates that the brain is not shaped solely by external forces, but is an agential force itself that initiates actions by imagining and intending certain movements, which results in sending and executing activities by the sender's and receiver's brain, respectively. The brain–body relationship cannot be exclusively characterized by auto-poetic dynamics; rather "the dynamics of brain codes, communication, and information patterns are always open to environmental entanglements" (Schmitz 2016, p. 149). While brain plasticity stresses how the brain and body integrate sensory inputs through learning, agential changes "always intra-act with unconscious and conscious experiences, information from the environment, meaning-making and discursive settings" (Schmitz 2016, p. 149). In this respect, the brain is not an isolated material force, but is the result of the "agential convergence that constitutes behavior and communication" (Schmitz 2016, p. 150).

5.5. Political perspective in neuroethics

Given the widespread neurosexism and essentializing tendencies in neuroscience-inspired cultural debates, we cannot overlook the likelihood of BBIs further intensifying the existing sexism and bias. To counter this possibility, we need to increase the socio-political sensibilities of neuroethicists to address the issues that will be raised by BBIs. As indicated earlier, there is a scarcity of neuroethics scholarship on BBIs due to the research being in its infancy. Therefore, our purpose here is to proactively discuss potential ethical concerns and issue a call for more relevant scholarship. Additionally, we hope to open the door of political neuroethics scholarship to traditions of thought (e.g. critical theory, feminism) that are at the same time less represented and more tuned into social conflicts.

Racine (2010) has identified three perspectives on neuroethics which aim to understand the implications of neurotechnology: namely, knowledge-driven perspective, which emphasizes research; technology-driven perspective, which defines neuroethics based on the technologies it examines, and healthcare-driven perspective, which posits that neuroethics is a subfield of bioethics. Dubljević et al. (2021) introduced a fourth political perspective in neuroethics scholarship, which accounts for the "interplay between the behavioral as well as the brain sciences and the socio-political system" (Dubljević et al. 2021, p. 3). This perspective takes into account the political and social regulations that are required and has the potential to address the challenges that result from introducing new technologies (Dubljević et al. 2021, p. 5). We build on this understanding by applying the political perspective to explore the potential implications and challenges that may stem from the application of BBIs. At the same time, we correct the (up until now) overtly liberal and Rawlsian tone of political neuroethics by incorporating insights from neurofeminism and agential realism.

Brain–computer and BBIs do not come into existence in a vacuum. Rather, they are realized in "social, political, and economic settings, discourses and power relations" (Schmitz 2016, p. 152): taxpayers provide funds, which are directed by federal agencies (civilian and military), which are awarded to researchers who conduct experiments on animals and humans and publish their results, which are then taken up by the media and interpreted by the general public. This raises important questions such as who has the power to decide on the content, as well as plan the communication and information processes underlying these multidirectional relationships (Schmitz 2016, p. 152). Therefore, a main objective of the socio-political perspective is to keep the public and policymakers informed and offer ethical guidance (Dubljević et al. 2021, p. 9). BBI is simultaneously a social as well as a political technology that is influenced by the publics' social attitudes and perspectives, as well as science policy. Agential realism, as discussed above, is only one way of conceptualizing the pervasiveness of BBI technologies that impact the human condition. As such, a political neuroethics perspective should be taken up by multiple neuroethicists to provide answers to a set of questions based on divergent (including liberal, socialist and conservative) political views. The first question that such political neuroethics scholarship should answer is:

What are the most legitimate public policies for regulating the development and use of BBIs by healthy adults in a reasonably just, though not perfect, democratic society?

Following Dubljević et al. (2021), we posit that political neuroethics will have to answer four additional questions to ensure that the discussion of BBI technology is realistic and not merely hype:

1) What are the criteria for assessing the relevance of cases of BBI technology to be discussed?

2) What are the relevant policy options for the targeted regulation of BBIs (e.g. of research, manufacture, use)?

3) What are the relevant external considerations for policy options (e.g. international treaties)?

4) What are the foreseeable future challenges that public policy may have to grapple with (e.g. proliferation of racism or sexism via BBIs)?

As a way of grounding the discussion of the neuroethics of BBI, we build on Dubljević et al. (2021) to propose several criteria for the relevance of specific cases. The first is the ability to increase the number of senders and responders in the system to something approximating a social group, even if communication is merely reduced to binary responses. Another criterion is the potential for the uptake of the BBI technology in society. Dubljević et al. (2021) highlight five approaches that need to be considered in regard to public policy: mandatory use, encouraged use, laissez-faire, discouraged use and prohibition. Even though policy considerations (on the national or international level) for the manufacture and use may be premature, given that BBI development is still in its early stages, the regulation of BBI research on humans may need to be considered, and the full range of policy options should be weighed. The potential future challenges of BBI-exacerbated racism and sexism are something that neuroethicists and public policies need to consider before such neurotechnologies are introduced in society.

It has to be noted that questions 3 and 4 cannot be answered from a single perspective. In fact, *liberal*, *socialist* and *conservative* political neuroethics will provide different responses. For instance, classical liberal viewpoints may not be impressed by international treaties, whereas conservative viewpoints may de-emphasize sexism as a problem in BBI technology (and beyond). Rather than lamenting the fact that disagreements about political premises will emerge, neuroethics would serve society best if such disagreements were made explicit and offered to public scrutiny in a democratic dialogue.

5.6. Conclusion

In this chapter, we have taken a neurofeminist approach to discuss BBIs as one of the latest developments in the field of neurotechnology. We applied Karen Barad's agential realism as a framework to BBIs for two reasons. The first reason was to highlight the potentially powerful impact and provide a better understanding of BBIs on the overall human condition that stresses the entanglement of multiple actors, as well as challenges the human–technology dichotomy. Second, agential realism provided an alternate viewpoint to the conventional understanding of the notions of causality and agency, as well as to the (up until now) liberal interpretation of political neuroethics. By employing the concept of intra-actions, we provided a new understanding of BBIs that is not understood as linear, where a single cause precedes one or more effects, but rather as an intertwinement of various actors that do not interact but rather intra-act. Furthermore, and in contrast to the present understanding of neuroplasticity, we reconceptualized the brain not as an independent entity, but as an entangled actor in the hybrid becoming of the BBI phenomenon.

BBI technology is still in the early stages of research, and therefore discussing the prospects of social and ethical challenges during research is urgent. By underscoring the pervasiveness of neurosexism in neuroscience and acknowledging the possibility of BBIs further exacerbating the existing sexist bias, this chapter highlighted the importance of a socio-political perspective that aims to provide ethical guidance and ground the discussion in scientific fact, as opposed to science fiction. This perspective embraces the diversity of political views and hopes to avoid overt conflict by fostering reasonable disagreement.

5.7. References

Alaimo, S. and Hekman, S.J. (2008). *Material Feminisms*. Indiana University Press, Bloomington, IN.

Anderson, E. (2020). Feminist epistemology and philosophy of science. In *The Stanford Encyclopedia of Philosophy* (Spring 2020 Edition), Zalta, E.N. (ed.) [Online]. Available at: https://plato.stanford.edu/entries/feminism-epistemology/ [Accessed 17 May 2021].

Barad, K. (2007). *Meeting the Universe Halfway*. Duke University Press, Durham, NC.

Barad, K. (2012). Matter feels, converses, suffers, desires, yearns and remembers: Interview with Barad. In *New Materialism: Interviews and Cartographies*, Dolphijn, R. and Van der Tuin, I. (eds). Open Humanities Press, Ann Arbor, MI [Online]. Available at: http://open–humanitiespress.org/books/download/Dolphijn-van-der-Tuin_2013_New-Materialism.pdf.

Baron-Cohen, S. (2003). *The Essential Difference: The Truth About the Male and Female Brain*. Basic Books, New York.

Bluhm, R., Jaap Jacobson, A., Lene Maibom, H. (eds) (2012). *Neurofeminism: Issues at the Intersection of Feminist Theory and Cognitive Science*. Palgrave Macmillan, New York.

Brescoll, V. and LaFrance, M. (2004). The correlates and consequences of newspaper reports of research on sex differences. *Psychological Science*, 15, 515–520.

Brizendine, L. (2006). *The Female Brain*. Broadway Books, New York.

Burwell, S., Sample, M., Racine, E. (2017). Ethical aspects of brain computer interfaces: A scoping review. *BMC Medical Ethics*, 18, 60

Coin, A., Mulder, M., Dubljević, V. (2020). Ethical aspects of BCI technology: What is the state of the art? *Philosophies*, 5(4), 31. doi: 10.3390/philosophies5040031.

Demarin, V., Morović, S., Béné, R. (2014). Neuroplasticity. *Periodicum Biologorum*, 116, 209–211

Dubljević, V. (2015). Neurostimulation devices for cognitive enhancement: Toward a comprehensive regulatory framework. *Neuroethics*, 8(2), 115–126.

Dubljević, V. (2019a). *Neuroethics, Justice and Autonomy: Public Reason in the Cognitive Enhancement Debate*. Springer, Heidelberg.

Dubljević, V. (2019b). Public reason and reasonable conceptions of justice. *Sofia Philosophical Review*, 12(1), 24–46.

Dubljević, V., Trettenbach, K., Ranisch, R. (2021). The socio-political roles of neuroethics and the case of klotho. *AJOB-Neuroscience*, 13(1), 10–22. doi: 10.1080/21507740.2021.1896597.

Fausto-Sterling, A. (2012). *Sex/Gender: Biology in a Social World*. Routledge, New York.

Ferrando, F. (ed.) (2019). From new materialisms to object-oriented ontology. In *Philosophical Posthumanism*. Bloomsbury Academic, London.

Fine, C. (2008). Will working mothers' brains explode? The popular new genre of neurosexism. *Neuroethics*, 1(1), 69–72.

Fine, C. (2010). *The Real Science Behind Sex Differences*. Oxford, Icon Books.

Fine, C. (2012). Is there neurosexism in functional neuroimaging investigations of sex differences? *Neuroethics*, 6(2), 369–409. doi: 10.1007/s12152-012-9169-1.

Frost, S. (2011). The implications of the new materialisms for feminist epistemology. In *Feminist Epistemology and Philosophy of Science*, Grasswick, H.E. (ed.). Springer, New York & London.

Glannon, W. (2014). Ethical issues with brain-computer interfaces. *Frontiers in Systems Neuroscience*, 8, 136.

Grossi, G. (2017). Hardwiring: Innateness in the age of the brain. *Biology & Philosophy*, 32(6), 1047–1082. doi: 10.1007/s10539-017-9591-1.

Haraway, D. (1991). A cyborg manifesto: Science, technology, and socialist-feminism in the late twentieth century. In *Simians, Cyborgs and Women: The Reinvention of Nature*. Routledge, New York.

Hildt, E. (2019). Multi-person brain-to-brain interfaces: Ethical issues. *Frontiers in Neuroscience*, 13(1177) [Online]. Available at: www.frontiersin.org/articles/10.3389/fnins.2019.01177/full.

Hoffman, G.A. and Bluhm, R. (2016). Neurosexism and neurofeminism. *Philosophy Compass*, 11(11), 716–729. doi: 10.1111/phc3.12357.

James, W. (1890). *The Principles of Psychology*. Holt, New York.

Jiang, L., Stocco, A., Losey, D.M., Abernethy, J.A., Prat, C.S., Rao, R.P.N. (2019). BrainNet: A multi-person brain-to-brain interface for direct collaboration between brains. *Scientific Reports*, 9(6115). doi: 10.1038/s41598-019-41895-7.

Jordan-Young, R. and Rumiati, R.I. (2012). Hardwired for sexism? Approaches to sex/gender in neuroscience. *Neuroethics*, 5(3), 305–315. doi: 10.1007/s12152-011-9134-4.

Kitcher, P. (1985). *Vaulting Ambition: Sociobiology and the Quest for Human Nature*. MIT Press, Cambridge, MA.

Konorski, J. (1948). *Conditioned Reflexes and Neuron Organization*. Cambridge University Press, New York.

Maksimenko, V., Hramov, A., Runnova, A., Pisarchik, A. (2019). Brain-to-brain interface increases efficiency of human-human interaction. *7th International Winter Conference on Brain-Computer Interface (BCI)*. doi: 10.1109/iww-bci.2019.8737316.

Mannette, R. (2022). Between *neurodiscourse* and *ideology*: Expanding on the socio-political dimension in neuroethics. *AJOB – Neuroscience*, 13(1), 38–40.

Misselhorn, C. (2018). Artificial morality: Concepts, issues and challenges. *Society*, 55, 161–169.

Pais-Vieira, M., Lebedev, M., Kunicki, C., Wang, J., Nicolelis, M.A.L. (2013). A brain-to-brain interface for real-time sharing of sensorimotor information. *Scientific Reports*, 3(1319). doi: 10.1038/srep01319.

Racine, E. (2010). *Pragmatic Neuroethics: Improving Treatment and Understanding of the Mind-Brain*. MIT Press, London.

Rao, R.P.N., Stocco, A., Bryan, M., Sarma, D., Youngquist, T.M., Wu, J., Prat, C.S. (2014). A direct brain-to-brain interface in humans. *PLoS ONE*, 9(11), e111332. doi: 10.1371/journal.pone.0111332.

Rippon, G. (2019). *The Gendered Brain: The New Neuroscience that Shatters the Myth of the Female Brain*. Vintage, London.

Rosenberg, N.A., Edge, M.D., Pritchard, J.K., Feldman, M.W. (2019). Interpreting polygenic scores, polygenic adaptation, and human phenotypic differences. *Evolution, Medicine, and Public Health*, (1), 26–34. doi: 10.1093/emph/eoy036.

Schmitz, S. (2016). The communicative phenomenon of brain-computer interfaces. In *Mattering: Feminism, Science, and Materialism*, Pitts-Taylor, V. (ed.). New York University Press, New York and London.

Schmitz, S. and Höppner, G. (2014). Neurofeminism and feminist neurosciences: A critical review of contemporary brain research. *Frontiers in Human Neuroscience*, 8. doi: 10.3389/fnhum.2014.00546.

Shih, J., Krusienski, D.J., Wolpaw, J.R. (2012). Brain-computer interfaces in medicine. *Mayo Clinic Proceedings*, 87(3), 268–279.

Trimper, J.B., Wolpe, P.R., Rommelfanger, K.S. (2014). When "I" becomes "We": Ethical implications of emerging brain-to-brain interfacing technologies. *Frontiers in Neuroengineering*, 7(4). doi: 10.3389/fneng.2014.00004.

Trites, R.S. (2017). Material meminism, adolescent "becoming", and Libba Bray's beauty queens. *Tulsa Studies in Women's Literature*, 36(2), 379–400. Project MUSE. doi: 10.1353/tsw.2017.0027.

Trujillo, N. (1991). Hegemonic masculinity on the mound: Media representations of Nolan Ryan and American Sports Culture. *Critical Studies in Mass Communication*, 8, 290–308. doi: 10.1080/15295039109366799.

Wojczynski, M.K. and Tiwari, H.K. (2008). Definition of phenotype. *Advances in Genetics*, 75–105. doi: 10.1016/S0065-2660(07)00404-X.

6

Neuroethics as an Anthropological Project

Fabrice JOTTERAND[1,2]
[1] Medical College of Wisconsin, Milwaukee, USA
[2] University of Basel, Switzerland

6.1. Introduction

Whether neuroethics should be more speculative and technology-driven or clinically oriented in scope is a question that has been part of an ongoing debate in neuroethics literature (Racine 2010; Fins 2017; Racine and Aspler 2017). In this chapter, I argue that such a debate ought to be supplemented by a framework that characterizes neuroethics as a project aiming at the analysis of the anthropological implications of neuroscientific and neurotechnological advances. Specifically, I contend that neuroethics should promote upstream philosophical anthropology analysis as neurotechnologies have the capacity to alter, control and manipulate human beings within social and clinical contexts.

My analysis is structured according to four main sections. The first section outlines the nature of neuroethics. It reveals a dichotomy in its orientations that impedes, to a certain extent, a more fundamental explorations of the impact of neurotechnologies on what it means to be human beyond mere pragmatic and utilitarian considerations. The next section focuses on why and how the brain, as the organ of the mind, has a special status and thus some form of neuroexceptionalism must characterize neuroethics. In the last part of my analysis, I specifically address how the brain and the mind have become objects of potential manipulation and control, not only in their structural dimensions but also in terms of output and behavior, including cognitive, emotional and volitional aspects. For this reason, I argue for the protection of the brain *and* the mind under what I call *identity integrity*. My analysis leads to the conclusion that in light of the pace of neurotechnological development and the increasing blurring line between the clinical implementations and social applications, neuroethics must incorporate more explicitly philosophical anthropology as its core concern.

6.2. The nature of neuroethics

6.2.1. *Neuroethics as a second-order discipline*

Neuroscientist and philosopher Adina Roskies has provided an influential distinction about the nature of neuroethics. She divides it into the ethics of neuroscience and the neuroscience of ethics (Roskies 2002). The former refers to the analysis of the ethical implications associated with the use of neurotechnologies (e.g. brain–computer interfaces, neuroimaging technologies, brain stimulation techniques) in the clinical environment and the broader social context. The latter focuses on subject matters related to personhood such as the neurobiology of morality, free will, the nature of consciousness, personal identity[1], the relation between the mind and the brain, etc. Since its inception, neuroethics research has drastically increased in both domains, whether providing ethical guidance for brain research or for the development and implementation of neurotechnologies. Both approaches have relied on home disciplines with well-established methods of investigation and have provided useful insights. The issues investigated in neuroethics, however, have already been deliberated in academic circles in the past and therefore neuroethical issues do not depend on new theoretical frameworks for their analysis. As pointed out by Jon Leefmann and colleagues in a 2016 literature review of neuroethics scholarship and research between 1995 and 2012, the fact that "after more than one decade there still is no dominant agenda for the future of neuroethics research…calls for more reflection about the theoretical underpinnings and prospects to establish neuroethics as a marked-off research field distinct from neuroscience and the diverge branches of bioethics" (Leefmann et al. 2016).

---

1. In this chapter, personal identity or the notion of the self will be defined as follows: "[it] is composed of two dimensions that are interrelated: psychological and biological aspects of human agency. Psychological features of human experience correspond to mental capacities or character traits, such as reason or rationality, language, moral agency, and so on. A human being, however, does not experience the world as if the brain is detached from the material realities of the body. The body, as a biological organism, is the physical medium through which time and space are experienced and interpreted" (Jotterand 2018, 415).

This lack of theoretical underpinnings reflects, to a certain extent, competing views about the nature of neuroethics, its historical origin, its academic significance and its practice. Beyond this theoretical plurality, what demarcates previous analyses – that is, prior to the advent of neuroethics as a field of investigation in its own rights – from current ones is the particular attention paid to their practical implications as the deployment of actual neurotechnologies (as opposed to imaginary scenarios involving hypothetical technologies) could possibly change our understanding of what it means to be a human being. More importantly in the context of understanding the nature of neuroethics, the failure to provide distinct theoretical foundations raises the question as to the degree in which neuroethics is, as a field of research, distinct from bioethics and neuroscience. In short, is neuroethics a discipline in its own rights, with its own methodologies, set of core texts, educational requirements, and standards for neuroethical practice?

The same questions and concerns have been raised regarding bioethics, and consequently a comparison with bioethics may be useful. Albert Jonsen, commenting on the nature of bioethics as a discipline, observed that bioethics does not have any methodology or master theory and thus is a "demi-discipline", that is, "only half of bioethics counts as an ordinary academic discipline... the other half... is the public discourse" (Jonsen 1998, 346). The dual nature of bioethics is also a feature of neuroethics. On the one hand, neuroethics uses traditional methods of investigation such as philosophical inquiry, ethical analysis, as well as scientific evidence, to name a few, to determine and address conundrums associated with advances in neuroscience and neurotechnologies. The product of these explorations is ideally communicated to the public via various means such as academic publications, conferences, social media, op-eds, newspapers articles, etc. As I have noted elsewhere, neuroethics can then be considered a second-order discipline that focuses on analyzing the ethical and policy implications of neuroscientific knowledge to guide brain research and address neurological disorders. But contrary to disciplines like philosophy, law or mathematics, it lacks a well-defined scholarly canon. Individuals practicing in a second-order discipline depend on works, standards of competency and pedagogical criteria of their home discipline(s) (Jotterand and Ienca 2017).

6.2.2. *Neuroethics, neurology and brain research*

To a certain extent, the above analysis might imply that the boundaries between bioethics and neuroethics are rather blurry and that the role of *neuro*-ethicists and *bio*-ethicists is somewhat similar since both disciplines share common methodologies and theoretical frameworks. For instance, in the early phase of the development of neuroethics, neurologist Ron Cranford saw a distinct role of "neuroethicists" in clinical ethics consultations, especially in the process of clarifying issues related to clinical neurology and clinical ethics (Cranford 1989), a role very similar to traditional bioethicists. Echoing Cranford, neurologist and neuroethicist James L. Bernat likewise sees a special role of neuroethicists in the clinical setting related to clinical cases of patients with severe brain damage injuries or neurological conditions. In his influential book *Ethical Issues in Neurology*, he states that there are:

[m]any issues provoking clinical ethics consultations in the general hospital surround the prognosis of patients who are severely brain damaged or paralyzed, including patients who are brain dead, those in a coma after a cardiac arrest or in a persistent vegetative state and those suffering from dementia, the effects of a stroke and locked-in syndrome. Neurologists and neurosurgeons can combine their specialized knowledge and experience with these states with their knowledge of clinical ethics to provide a unique combination of clinical-ethical advice (Bernat 2008, 124).

While Cranford and Bernat attribute a distinct role of neuroethics in the clinical setting, they do not regard the brain as a special organ that deserves a separate consideration compared to other organs in the body. On their account, the distinction between neuroethics and clinical ethics, as two distinct domains of ethical inquiry, is simply unwarranted because it is unclear what the overall utility of neuroethics per se is in the practice of medicine. Neuroethics only makes sense in relation to clinical neurology. This vision of neuroethics, however, has not become the dominant view and consequently "redirected the trajectory of neuroethics in a way that has made it more speculative and less relevant to real patients and the needs of the clinic" (Fins 2017, 59).

At a deeper level, though, the question remains regarding the object of neuroethical investigation. Neuroethics specifically focuses on the brain, neurological disorders, and neurotechnologies, and therefore is better equipped than bioethics or moral philosophy because neuroethical inquiry works at the intersection of actual clinical conundrums and hypothetical scenarios related to the dual use neurotechnologies. Hence, neuroethics assumes a neuroessentialist posture (i.e. the idea that human behavior, mental states, personal identity and sense of self can be reduced to neurobiology; see Jotterand 2016, 42–56) in ways that bioethical inquiry does not. William Safire is often credited for coining the term neuroethics at a meeting organized in 2002 by the Dana Foundation entitled *Neuroethics: Mapping the Field*. During the event, he defined neuroethics as "the examination of what is right and wrong, good and bad about the treatment of, perfection of, or unwelcome invasion of and worrisome manipulation of the human brain" (Marcus 2004, 5). It is precisely the manipulation of the human brain through technological means, in the clinical milieu or in the broader social context, that distinguishes neuroethics from bioethics.

Two orientations of neuroethics are emerging from the above analysis. The first one is pragmatic, healthcare-driven, and does not consider the brain as an organ with a special status. Neuroethics is a sub-specialization of clinical ethics qua bioethics. The aim is not primarily to anticipate the potential implications of advances in neuroscience and neurotechnologies but rather to examine how these advances impact the reality of clinicians and patients in the context of care. The contrasting view sees the role of neuroethics engaged in ethical inquiry with an emphasis on anticipation. On this account, neuroethics examines the potential future trajectories of brain research and advances in neurotechnologies to anticipate their potential in ethical, clinical and social implications (Vidal and Piperberg 2017). This approach is more speculative in nature, techno-driven, and deems the brain as an "exceptional" organ because of its importance on questions of personhood, personal identity, and our understanding of the mind and its relation to the brain.

In this chapter, I question the relevance of these two divergent views about the nature of neuroethics. Considering the increasing use of neurotechnologies in the clinical context and social setting (i.e. consumer neurotechnologies), there are justified reasons to move the discussion in a different direction to anticipate future trajectories of these technologies. The current broad paradigm focuses on pragmatic and utilitarian concerns. It emphasizes the instrumentality of knowledge whereby knowledge becomes only useful to organize human reality. Thus, notions of truth and values can merely be tested pragmatically, not theoretically. In addition, concerns of human identity are reduced to morphological freedom, the right to choose and individuals' autonomy. The body is not considered as constitutive of our identity but as a property that can be manipulated, controlled and altered at will through technological means. The question is no longer "am I my body?" but "do I have a body?" (Agacinski 2019). On this account, people have the right to choose their "self-directed technological transformations", including brain functions and structures that determine personal self, cognitive capacities, etc. (More 2013). The lack of a deeper reflection on what the notions of embodiment mean for human identity marginalizes philosophical anthropology in ways that are detrimental to the development of ethical frameworks which sustain laws and policies regulating neurotechnologies. The power of technology, either to order the world and serve human ends or to reshape human identity, is reorienting human beings from a position of orderers of the world to objects of control and manipulation. French philosopher Jean d'Ormesson summarizes our situation well. He rightly observes that:

Science and technology have taken precedence over nature, over power, over poetry, over philosophy and over religion. This is the heart of the matter. They turned our life upside down (Ormesson 2013, author's translation)[2].

The incursion of technology in our everyday life, in politics, in our intellectual life demands greater attention at the core of the brain research, and neurotechnological development through a redefined neuroethics project that places philosophical anthropology at the front and center.

6.3. Neuroethics as an anthropological project

What I mean by an anthropological project is the philosophical examination of the implications of innovative neuroscientific knowledge and neurotechnologies regarding the nature of human beings, the human condition and what it means to be human in a technological age. It is in this way that neuroethics is not bioethics applied to brain research. Neuroethics is concerned with the ethics of neuroscience research and the clinical implications of neurotechnologies but at a more fundamental level, neuroethics should be concerned with how new knowledge challenges our understanding of our own humanity in its embodied, psychological and social dimensions. As Martha J. Farah remarks:

> [t]here is more to neuroethics than classic bioethics applied to neuroscience. New ethical issues are arising as neuroscience gives us unprecedented ways to understand the human mind and to predict, influence, and even control it. These issues lead us beyond the boundaries of bioethics into the philosophy of mind, psychology, theology, law, and neuroscience itself… (Farah 2010, 2).

Beneath the neuroethics umbrella, there is nothing new under the sun since various disciplines already have a long tradition of scholarly investigation on topics pertaining to the mind, its relation to the brain, the nature of mental states and consciousness, and personal identity. But in a very new way, the scope of neuroethical investigation escapes the gaze of bioethics, as pointed out by Farah. Advances in neurotechnologies allow the collecting of brain data for the development of a novel model of explanation of brain functions and structures leading to cutting-edge research and novel models of explanation. For instance,

---

2. "La science et la technique ont pris le pas sur la nature, sur le pouvoir, sur la poésie, sur la philosophie et sur la religion. Voilà le cœur de l'affaire. Elles ont bouleversé notre vie."

researchers using brain–computer interfaces (BCIs) can now record brain data associated with a specific activity in a mouse and use the same data to stimulate the brain of the same mouse and "recreate" the activity at will (Yuste et al. 2021). At this point, such research is limited to animal models but some researchers speculate that human mental states (thoughts, memories, imaginations, decisions, emotions, etc.) as the product of neural activity could be likewise mined, hence potentially affording "the real possibility of human thoughts being decoded or manipulated using technology" (Yuste et al. 2021, 155).

The potential ethical issues arising from this type of research, if it moves into human experimentation, are not simply related to traditional questions encountered in the bioethics or clinical literature. They go to the core of what it means to be human, how we conceptualize personhood and human agency. Such concerns have not been at the forefront of bioethical inquiry. Neuroethics, with its dual foci on neurology/brain science and emerging technologies, is well positioned to provide the intellectual environment to engage in a project that examines the impact of neuroscientific knowledge and neurotechnologies on conceptualization of human identity. While Farah does not include neurotechnologies as a source of potential transformation (mistakenly in my view), she appreciates how neuroscientific knowledge challenges some of our profound beliefs concerning the nature of human beings. She points out that "some of the most profound ethical challenges from neuroscience come not from new technologies but from new understandings. Neuroscience is calling our age-old understanding of the human person into question" (Farah 2010, 8).

I posit that the way neuroscience and neurotechnologies challenge our understanding of human personhood should be addressed according to the following three criteria: impact, interface and identity.

Firstly, impact. As previously stated, neuroscientific knowledge and neurotechnologies must consider their impact on human beings, their brain and mind in the clinical environment and the social context. The availability of neurotechnologies to treat patients from neurological disorders and brain injuries should not hide the fact that consumer neurotechnologies are becoming likewise accessible in the market. Putting in place the right regulations and policies is a complex process that must bridge clinical aspirations and societal requests. However, before making these neurotechnologies available, it is primordial to carefully consider how they could impact patients and individuals qua consumers in their humanity, brain and sense of self from an ontological and moral perspective.

Secondly, interface. The second anthropological issue that neuroethics should continuously address with more depth is the degree to which human beings should interface with various neurotechnologies, whether at the level of embodiment or the mind. In the clinical setting, neurotechnologies are used for the treatment of mental disorders such as schizophrenia, Parkinson's disease, or Alzheimer's disease. In the social setting, there are various types of neurotechnologies and devices used for cognitive enhancement, well-being, social interaction, companionship, and gaming. The exact boundaries of such human–machine, brain–computer interactions are not clearly understood and determined. In addition, the moral and legal status of some of these devices (social robots or companionship robots, for instance) remains a subject of contention as it could be argued that some interactive robots might deserve a moral and legal status (Jotterand 2018). We might think of the robot Sophia "who" was granted citizenship in Saudi Arabia, a country that limits the freedom of women and violates human rights on a regular basis.

The third and final criterion is identity. The issue here is the extent to which the cognitive, affective, and volitional attributes of the human mind should be altered, controlled, or manipulated. In some instances, neurotechnologies can be used in the clinical context to manage condition like Parkinson's disease, Alzheimer's disease or mental impairments but such neurointerventions can have side effects including changes in personality (Jotterand and Giordano 2011; Jotterand et al. 2015; Jotterand 2019; Jotterand and Bosco 2020). Various neurotechnologies are also suggested for their implementation in the broader social context as an alleged potential means to "enhance" moral capacities or cognitive abilities (Douglas 2008; Persson and Savulescu 2016; Baccarini and Malatesti 2017; for a critical appraisal, see Jotterand 2011, 2016; Jotterand and Levin 2017; Jotterand 2022).

Because the use of many neurotechnologies is not limited to the clinical context but expand to various other applications, there is a high likelihood that these technologies will transform our way of life, how we relate to technological devices, and potentially our understanding of what it means to be human. These three dimensions (impact, interface and identity) acknowledge more fundamental questions about human identity and the human condition regardless of whether in the clinical or social setting.

6.4. Protecting the brain and the mind

6.4.1. *Neuroessentialism and neuroexceptionalism*

Earlier in the chapter, I outlined the reason why I hold a neuroexceptionalist view when it comes to the nature of neuroethics. My conceptualization of personhood, however, is not limited to the brain and therefore it does not embrace neuroessentialism. In other words, the brain constitutes an important organ for the mind to emerge, but human identity encompasses biological (the whole body), psychological (the mind) and social (how we engage with others and the surrounding world) dimensions necessary to develop a sense of self and personal identity. Thus, the claim that "we are our brains" (Reiner 2011) or that our mental states, notions of self, personal identity, and demeanor can be reducible to neurobiology should be rejected. I do acknowledge that brain chemical imbalance and traumatic insults to the brain impact behavior, mood, personal identity and can lead to deviant conduct. But human behavior encompasses more than the firing of synapses in the brain. Neural activity is necessary but not sufficient in the development of our sense of self. The mind, as an emerging property of the brain, provides the content of what characterizes personal identity. It justifies, through the ability to reason, normative claims about moral beliefs, desires, passions, etc. Neural pathways constantly adapt according to environmental factors, what a person reads and learns, and how we interact with other human beings. Neuroplasticity is the ability of the brain to adapt to these external factors.

6.4.2. *Identity integrity: protecting brain and mind*

The concerns related to the disruption of personal identity or the potential mining of brain data unbeknownst to individuals have been addressed in various ways at a national level (although not specifically dealing with neurotechnologies; the Fourth Amendment of the United States Bill of Rights prohibits unreasonable searches; in Chile the Senate's Constitutional Reform Bill and the Neuroprotection Bill of Law approved the inclusion of neurorights; see below) and international level (e.g. the Declaration of Human Rights; the European Convention for the Protection of Human Rights and Fundamental Freedoms). There has also been a push, as an extension of existing human rights, to create neurorights (Goering and Yuste 2016; Ienca and Andorno 2017; Goering et al. 2021). These rights include a) the right to personal identity, b) the right to free-will, c) the right to mental privacy, d) the right to equal access to mental augmentation and e) the right to protections from algorithmic bias (Neurorights Initiative).

The right to personal identity is particularly relevant to our discussion. It states that "boundaries must be developed to prohibit technology from disrupting the sense of self. When Neurotechnology connects individuals with digital networks, it could blur the line between a person's consciousness and external technological inputs". The protection and preservation of our sense of self through the right to personal identity is undoubtedly an important approach. However, implicit to the right to mental integrity is a focus on the datafied brain, that is, detectable neural activity, while omitting the mind. Here, the distinction between the brain and mind is key. On the one hand, the brain is the locus of neural activity and provides the necessary neuro-architecture for the mind to emerge. On the other, the mind is what generates and expresses, through the ability to reason and communicate, the content of our identity. Considering that current brain–computer technologies are moving in the direction of "reading and writing" on the brain, potentially in human, the right to mental integrity should be replaced by the right to identity integrity (for a full analysis on the concept of identity integrity, see Chapter 8, Jotterand 2022). Not only our mental capacities should be protected (neural activity) but also the content of our mind. It is in that sense that the brain has a special status (neuroexceptionalism). It should not be subject to technological manipulations regarding its capacities, nor should the content of the mind be altered without a careful consideration of how it could impact an individual sense of self and personal identity. We should resist the temptation to perceive human beings, their bodies, their brain, and their minds as entities of infinite resources that can be mined for scientific and technological hubris.

6.5. Concluding remarks

The purpose of this chapter has not been to question the very nature of neuroethics or its relevance. Rather, my goal has been to stress the importance of philosophical anthropology in neuroethics as I see a failure on the part of the field to consider it more thoroughly, whether in the clinical context or the social milieu. Medicine is already struggling with the question as to whether some neurotechnologies (e.g. AI neurotechnologies) will re-humanize or de-humanize its practice. The increasing availability of consumer technologies is likely to follow the same trend. It is my contention that neuroethics can play an important role in making sure our technological gourmandize, to use Herve Chneiweiss' expression, does not lead to fundamental changes in our understanding of what it means to be human in ways that degrade our humanity.

6.6. References

Agacinski, S. (2019). *L'Homme désincarné : du corps charnel au corps fabriqué*. Gallimard, Paris.

Baccarini, E. and Luca, M. (2017). The moral bioenhancement of psychopaths. *Journal of Medical Ethics*, 43(10), 697–701. DOI: 10.1136/medethics-2016-103537.

Bernat, J.L. (2008). *Ethical Issues in Neurology*, 3rd edition. Wolters Kluwer Lippincott Wiliams & Wilkins, Philadelphia, PA [Online]. Available at: https://www.wolterskluwer.com/en/solutions/ovid/4602.

Cranford, R.E. (1989). The neurologist as ethics consultant and as a member of the institutional ethics committee. The neuroethicist. *Neurologic Clinics*, 7(4), 697–713.

Douglas, T. (2008). Moral enhancement. *Journal of Applied Philosophy*, 25(3), 228–245. DOI: 10.1111/j.1468-5930.2008.00412.x.

Farah, M.J. (2010). Neuroethics: An overview. In *Neuroethics: An Introduction with Readings*, Martha, J. (ed.). MIT Press, Cambridge, MA [Online]. Available at: https://mitpress.mit.edu/books/neuroethics.

Fins, J.J. (2017). Toward a pragmatic neuroethics in theory and practice. In *Debates About Neuroethics: Perspectives on Its Development, Focus, and Future*, Racine, E. and Aspler, J. (eds). Springer International Publishing, Cham. DOI: 10.1007/978-3-319-54651-3_4.

Goering, S. and Rafael, Y. (2016). On the necessity of ethical guidelines for novel neurotechnologies. *Cell*, 167(4), 882–885. DOI: 10.1016/j.cell.2016.10.029.

Goering, S., Eran, K., Sullivan, L.S., Wexler, A., Arcas, B.A.Y., Bi, G., Carmena, J.M., Fins, J.J., Friesen, P., Gallant, J. et al. (2021). Recommendations for responsible development and application of neurotechnologies. *Neuroethics*. DOI: 10.1007/s12152-021-09468-6.

Ienca, M. and Roberto, A. (2017). Towards new human rights in the age of neuroscience and neurotechnology. *Life Sciences, Society and Policy*, 13(1), 5. DOI: 10.1186/s40504-017-0050-1.

Jonsen, A.R. (1998). *The Birth of Bioethics*. Oxford University Press, New York.

Jotterand, F. (2011). Virtue engineering' and moral agency: Will post-humans still need the virtues? *AJOB Neuroscience*, 2(4), 3–9. DOI: 10.1080/21507740.2011.611124.

Jotterand, F. (2016). Moral enhancement, neuroessentialism, and moral content. In *Cognitive Enhancement: Ethical and Policy Implications in International Perspectives*, Fabrice, J. and Veljko, D. (eds). Oxford University Press, Oxford and New York.

Jotterand, F. (2018). The boundaries of legal personhood. In *Human, Transhuman, Posthuman: Emerging Technologies and the Boundaries of Homo Sapiens*, Bess, M. and Walsh, P.D. (eds). Macmillan Reference USA, Farmington Hills, MI.

Jotterand, F. (2019). Personal identity, neuroprosthetics, and Alzheimer's disease. In *Intelligent Assistive Technologies for Dementia*, Fabrice, J., Marcello, I., Tenzin, W., Bernice, E. (eds). Oxford University Press, Oxford and New York [Online]. Available at: https://oxfordmedicine.com/view/10.1093/med/9780190459802.001.0001/med-9780190459802-chapter-11.

Jotterand, F. (2022). *The Unfit Brain and the Limits of Moral Bioenhancement*, 1st edition. Palgrave Macmillan, Singapore.

Jotterand, F. and Clara, B. (2020). Keeping the "Human in the Loop" in the age of artificial intelligence. *Science and Engineering Ethics*, 26(5), 2455–2460. DOI: 10.1007/s11948-020-00241-1.

Jotterand, F. and James, G. (2011). Transcranial magnetic stimulation, deep brain stimulation and personal identity: Ethical questions, and neuroethical approaches for medical practice. *International Review of Psychiatry (Abingdon, England)*, 23(5), 476–485. https://doi.org/10.3109/09540261.2011.616189.

Jotterand, F. and Marcello, I. (2017). The biopolitics of neuroethics. *Debates About Neuroethics*, 125, 247–261. DOI: 10.1007/978-3-319-54651-3_17.

Jotterand, F. and Susan, B.L. (2017). Moral deficits, Moral motivation and the feasibility of moral bioenhancement. *Topoi*, April, 1–9. DOI: 10.1007/s11245-017-9472-x.

Jotterand, F., McCurdy, J., Elger, B.S. (2015). Cognitive enhancers and mental impairment: Emerging ethical issues. In *Rosenberg's Molecular and Genetic Basis of Neurological and Psychiatric Disease*, 5th edition, Rosenberg, R.N. and Pascual, J.M. (eds). Academic Press, Boston, MA. DOI: 10.1016/B978-0-12-410529-4.00011-5.

Leefmann, J., Clement, L., Elisabeth, H. (2016). Neuroethics 1995–2012. A bibliometric analysis of the guiding themes of an emerging research field. *Frontiers in Human Neuroscience*, 10. DOI: 10.3389/fnhum.2016.00336.

Marcus, S.J. (ed.) (2004). *Neuroethics: Mapping the Field*, 1st edition. Dana Press, New York.

d'Ormesson, J. (2013). *Un jour je m'en irai sans en avoir tout dit*. Robert Laffont, Paris.

Persson, I. and Julian, S. (2016). Moral bioenhancement, freedom and reason. *Neuroethics*, 9(3), 263–68. DOI: 10.1007/s12152-016-9268-5.

Racine, E. (2010). *Pragmatic Neuroethics: Improving Treatment and Understanding of the Mind-Brain*, Illustrated edition. MIT Press, Cambridge, MA.

Racine, E. and John, A. (eds) (2017). *Debates About Neuroethics: Perspectives on its Development, Focus, and Future*. Advances in Neuroethics. Springer International Publishing, Cham. DOI: 10.1007/978-3-319-54651-3.

Reiner, P.B. (2011). The rise of neuroessentialism. SSRN Scholarly Paper ID 3128444. Social Science Research Network, Rochester, New York [Online]. Available at: https://papers.ssrn.com/abstract=3128444.

Roskies, A. (2002). Neuroethics for the New Millenium. *Neuron*, 35(1), 21–23. DOI: 10.1016/S0896-6273(02)00763-8.

Vidal, F. and Michelle, P. (2017). Born free: The theory and practice of neuroethical exceptionalism. In *Debates About Neuroethics: Perspectives on its Development, Focus, and Future*, Eric, R. and John, A. (eds). Springer International Publishing, Cham. DOI: 10.1007/978-3-319-54651-3_5.

Yuste, R., Genser, J., Hermann, S. (2021). It's time for neuro-rights: New human rights for the age of neurotechnology. *Horizons: Journal of International Relations and Sustainable Development*, 18, 154–165.

PART 2

Cultural Influences on Neuroethics

7
Neuroethics and Culture

Arleen SALLES[1,2]

[1] Center for Research Ethics and Bioethics, Uppsala University, Sweden
[2] Neuroetica Buenos Aires (NEBA) BsAs, Argentina.

7.1. Introduction

Neuroethics is a multi- and interdisciplinary field that focuses on the ethical issues raised by rapidly developing neuroscience and attempts to find new answers to age-old philosophical questions[1].

Multiple explanations of its nature, methodology, content and goals have been offered (Roskies 2002; Racine 2017). Here, I follow a useful distinction among three main methodological approaches: neurobioethics, empirical neuroethics and conceptual neuroethics (Evers et al. 2017a; Farisco et al. 2018a). According to this distinction, "neurobioethics" is primarily normative. It applies ethical theory and reasoning to practical issues arising from neuroscientific research and its clinical applications, and to issues raised by public communication of neuroscientific findings and their impact. "Empirical neuroethics" is descriptive and occasionally explanatory: it uses empirical data to inform theoretical (e.g. what is moral reasoning) and practical issues (e.g. who is really a moral agent) (Northoff 2009). Finally, "conceptual neuroethics" is primarily theoretical and analytical focusing not just on conceptual clarification of key notions but also on how neuroscientific knowledge is constructed, the plausibility and legitimacy of scientific and

---

1. Multidisciplinary because it draws from several disciplines and interdisciplinary because it is intended to synthesize their insights into a coherent whole.

philosophical interpretations, and their impact on clinical and social contexts (Salles and Evers 2017; Farisco et al. 2018a).

In the last decade, neuroethics has been expanding across geographical borders. Its potential role in enhancing neuroscientific research is formally acknowledged by existing international brain initiatives such as the European Human Brain Project and US BRAIN among others (Rommelfanger et al. 2019); it is a recognized line of research in several developed countries and is beginning to be discussed in developing countries that make use of advanced neuroscientific techniques and tools and conduct neuroscientific research.

The neuroethical approach that has become prominent is neurobioethics. Neurobioethics, however, has been permeated by Western philosophical traditions and has often taken for granted the portability of such traditions across geographical borders. Its expansion, on the one hand, and the cultural and social pluralism of present societies, on the other, have led to calls for a more explicit recognition of how culture shapes people's encounter with and experience of the outcomes of neuroscience (Herrera-Ferra 2018)[2]. In practice, this resulted in some engagement with the question of how to accommodate cultural diversity. Beyond this, however, the topic of culture has not attracted significant neuroethical interest.

In this chapter, I have two main objectives, one mostly empirical and the other more theoretical. The mostly empirical objective is to provide an outline (limited, of course) of how neuroethics has been grappling with culture, that is, how it is addressing what is considered to be the challenge of cultural diversity. The more theoretical one is to propose that neuroethics can play a different and potentially useful role in the discussion of some key conceptual issues regarding the brain and culture. This chapter is accordingly divided into two main parts. In the first, I provide an overview of different neuroethics approaches to accommodating culture. Here, I note that the discussion has mostly been kept at a neurobioethical (practical) level. In this sense, it mirrors the bioethical treatment of cultural differences. In the second part, I argue that when kept at this level, neuroethics does not yield a new understanding of the issues involved and, therefore, is unlikely to contribute significantly to advancing the discussion. However, I proceed to suggest that neuroethics can still play a modest but important role in the discussion by engaging more deeply with some of the conceptual issues raised by recent work on culture and the brain.

---

2. I am aware that the issue of culture is not exhausted by regional considerations. There are many sociocultural contexts, such as religious and disciplinary cultures, with different languages and conceptions. In this paper, I concentrate on cultural variation across geographical regions.

7.2. Neuroethics and the challenge of cultural diversity

Culture is a multifarious concept, with a number of categories (religion, gender, and class, among others) often considered to be under its remit. Divergent and not always fully compatible characterizations of the concept have been offered (Jahoda 2012). Despite the lack of a broadly accepted definition, there is agreement in that culture comprises a dynamic set of beliefs or a world vision that intentionally or unintentionally and directly or indirectly impacts the behavior and way of life of those associated with it. Recently, the topic of how to approach what appears to be the challenge of cultural diversity when addressing the ethical and social issues raised by brain research and emerging neurotechnologies has been attracting some attention within neuroethics. Below, I present an overview of how the field has generally approached this issue.

7.2.1. International neuroethics

An opening to the recognition of cultural diversity lies in calls for increasing "international diversity and representation in neuroethics" (Lombera and Illes 2009a; Racine 2017). The underlying idea is that even if neuroethics is dominated by some (developed and typically wealthy) countries, examination of the social, legal and ethical implications of brain research is relevant to all, which is shown by the fact that those issues are raised in a significant number of countries well beyond the developed world[3]. An international neuroethics, then, does not necessarily focus on how culture impacts the examination of normative concerns, but rather on how to include more international stakeholders, particularly from poor and developing countries, so that they can participate in the discussion. Such participation might lead to sharing concerns and to developing strategies to address them (Illes et al. 2005; Matshabane 2021). Whether, the extent to which, and in what sense international participation will lead to actual cross-cultural engagement, furthering understanding of how different cultures and histories shape the identification and approach to specific topics of concern is generally left unattended. Nor is there any reflection on whether and which cultural differences are ethically relevant and why. Instead, an international neuroethics approach calls for recognizing that "the usual suspects" do not exhaust the discussion on the ethical, legal and social issues raised by neuroscience and that more countries, both developed and developing, should be given a voice.

---

3. This type of approach seems evident at the intersection of neuroethics and law, for example, Spranger (2012), Stein and Singh (2020), Chandler et al. (2021), Dubljevic and Bottenberg (2021).

7.2.2. *Culturally aware/engaged neuroethics*

While internationalization does not quite equate to greater cultural engagement, it might further awareness of the existence of cultural differences and, ideally, reflection on their implications. Indeed, recent years have seen several articles that illustrate how a repertoire of local, social and cultural concerns might influence people's beliefs on the brain and mental health as well as their attitude towards the use of neuroscience and neurotechnology for the diagnosis and treatment of neuropsychiatric disorders, and even the academic neuroethical discussion (e.g. on the relevance, usefulness and permissibility of neuro-enhancement) (Sakura 2012; Bonete 2013; Salles 2014, 2018; Fukushi et al. 2018; Herrera-Ferrá et al. 2019). These articles reinforce the idea that cultural differences do not come in one size. Cultural differences may express themselves in the articulation and use of different ethical frameworks or in how a same ethical framework is applied in different cultural realities. Moreover, they may emerge even before the normative ethical discussion, for example, in diverse metaphysical and epistemic understandings of the issues involved (e.g. what is consciousness, what is the brain and how is it perceived) and different valorative concerns (why is the brain valuable?) (Chien-Chang Wu and Fukushi 2012; Tsomo 2012a). Unfortunately, however, within mainstream neuroethics, these insights often remain unexplored. While some efforts have been taken in the direction of the integration of cultural perspectives when discussing specific topics (Specker Sullivan 2019), they continue to be the exception rather than the rule. Often, cultural awareness in neuroethics appears to entail mere acknowledgment of cultural diversity from a disengaged standpoint, without attention to exactly what is at stake and why.

This has led some neuroethicists to suggest the need for a cross-cultural methodology that aims at true cultural engagement (Global Neuroethics Summit Delegates 2018). A cross-cultural approach is intended to go beyond awareness: it consists of a two-way flow of information that apprehends and seeks to understand the specificity of people's experiences (including our own) and enriches all (for more on this topic, see Rommelfanger and Sullivan, this volume). The idea is that without a cross-cultural lens, neuroethics will remain culturally limited[4].

---

4. While desirable, however, cross-cultural conversations are not without problems. How do we carry them out without missing conceptual and linguistic subtleties? And how do we overcome the practical issues such as accessibility of the relevant evidence, or the different structures for which different societies might have to address issues? For a discussion of some, see Rommelfanger and Specker Sullivan in this volume.

7.2.3. *Towards a global neuroethics*

A cross-cultural approach is intended to take cultural diversity as a starting point and to build intercultural understanding. It is often hoped that it will help to do two things: 1. identify commonalities, thus going beyond particular national concerns, and 2. harmonize (when possible) some ethical standards in order to build a type of intercultural moral consensus. In that sense, for some it is a condition for a "global neuroethics".

The term "global neuroethics" is often preferred precisely due to its more inclusive implications as compared to "international neuroethics" that still suggest national distinctions[5]. The term is used to convey the idea of transcending territorial borders even if the issues raised by neuroscientific research are actually experienced within specific, local contexts. Thus, a global neuroethics generally attempts to identify and articulate rules and norms that reflect common values and a common general vision (Chen and Quirion 2011) or at least articulate a convergent ethical framework that leads to some type of consensus (Kellmeyer et al. 2019). Interestingly, the extent to which global approaches actually undertake prior cross-cultural reflection is not too clear. To illustrate, the search for some type of global consensus is evident in recent proposals that invite a fresh rethinking of human rights and potentially the articulation and establishment of a set of neurorights (Goering and Yuste 2016; Ienca and Andorno 2017; Yuste et al. 2017; Goering et al. 2021; Ienca 2021) and in the development of a modest "cosmopolitan neuroethics" that commits itself to defending "a coherent set of moral priorities" that can be translated into midlevel principles (Shook and Giordano 2014, 2016). Despite their differences, these proposals have some things in common: (1) they have clear normative objectives, insofar as they present a set of specific considerations (which are based on international legislation or on a philosophical understanding of what humans are) that they think, should be used to unify the current ethically fragmented treatment of questions and problems raised by neuroscience and its applications, (2) they take those considerations to be important for upgrading international policies and legislations for the governance of brain research and emerging neurotechnologies and (3) they appear to be top down. They are culturally aware (taking as a starting point the existence of different cultures and traditions) but it is not clear to what degree they are the result of a cross-cultural dialogue and engagement with cultural differences (even if the expectation is that the proposals themselves would foster such a dialogue) nor do they attempt to focus on what the application of the proposed frameworks in different cultures would entail.

---

5. I thank Karen Rommelfanger for this point.

7.3. Can neuroethics contribute to the discussion?

The discussion around the best way to address and accommodate cultural differences is not new[6]. As is well known, the general topic of how to handle the "challenge" of cultural multiplicity and the potential diversity of values when confronting concrete ethical issues has been addressed in ethics and moral philosophy for millennia, with heated debates revolving around the denial or acceptance of universally valid values and principles. More recently, the discussion over cultural diversity and the desirability of global approaches has attracted attention within the domain of bioethics, a field that has shown significant concern with its own development, its underlying philosophical Western assumptions, and the effects and implications of importing them across borders (Campbell 1999b; Marshall and Koenig 2004; Holm and Williams-Jones 2006a; Myser 2011; Chattopadhyay and De Vries 2013).

Given the above, can neuroethics contribute to the discussion over culture and diversity? I would like to advance that, in general terms, to the extent that neuroethics scholarship on culture is kept mainly at the normative-prescriptive (neurobioethical) level, there are reasons to be skeptical about its potential for yielding a particularly novel approach or strategy for solving this longstanding ethical discussion. Indeed, calls for internationalization, cultural engagement and for attending to the global landscape are important, but the issues raised are not unlike those that have been and continue to be addressed within the field of bioethics. However, this does not mean that neuroethics should be seen as trivial regarding the topic of culture. In what follows, I propose that neuroethics can take a productive direction by using its conceptual resources to scrutinize neuroscientific studies that focus on "the brain on culture" and thus hopefully contribute to a better framing of the normative discussion.

7.3.1. *The brain on culture*

The view that people's self-perception and interpretative frameworks are importantly shaped by the social and the cultural is not new. Now, a growing body of scientific studies sheds further light on how cultural and social dimensions are engaged at different levels of biological functions. One of those levels concerns the brain: recent studies give support to the view that the human brain's structural and functional architecture is shaped by culture and that cultural contexts themselves are

---

6. See, for example, the latest issue of *Global Bioethics* 33:1 (2022) which focuses on what a global bioethics does and must do in the context of a global pandemic.

biologically grounded (Ames and Fiske 2010; Zhou and Cacioppo 2010; Wexler 2011).

A number of empirical findings in epigenetics have triggered a richer understanding of the interaction between the genome and diverse environments and of how environmental influences can have a lasting impact on gene expression at different levels of biological structures and systems. Epigenetic mechanisms have also been found to play a key role in the development of the central nervous system and the brain (Sweatt 2013; Dias et al. 2015). This has led to a significant transformation of how the brain is scientifically understood (Rubin 2009). Indeed, rather than as a simple, mechanistic input–output processing device with precise connectivity that reads external signals, the brain is now thought of as a highly malleable and plastic organ deeply shaped by diverse environments through life. There is evidence that while genetic factors are responsible for invariant features, the brain's architecture emerges from a dynamic selection interplay between its species-specific genetic make-up and its many environments (Changeux 1985; Edelman 1987) which impact the growth of new neurons in developing and adult brains (Eriksson et al. 1998; Yao et al. 2016) and the process of synapse formation (Edelman 1987; Changeux 2017). In humans, the malleability of the brain is particularly significant due to the increased number of neurons (as compared to other brains) and the lengthy period of brain development which takes place over the course of a 25-year period after birth (Wexler 2011). This makes it possible for different environments to leave their imprint on the connectivity of the developing brain. There is significant evidence that synaptogenesis is the result of a selection process shaped by changing environments and adaptation, and constrained by positive or negative reinforcement from repeated external stimuli. Moreover, there is evidence that epigenetic mechanisms operate in adult brain functions as well (Sweatt 2013). In the case of humans, the environmental cues that shape the process of synapse formation are not just physical (nutrition, stress, etc.) but also cultural. This has led some thinkers to bring the issue of culture and the brain to the forefront, to speculate about the extent to which lines between the biological and the cultural are blurring and to consider the potential relevance of epigenetic findings to social policy and ethics education (Evers and Changeux 2016; D'Ambrosio and Colage 2017; Evers 2020b).

More specific attention to the interaction between biological, social and cultural factors becomes apparent in social neuroscience in general, and the interest in unveiling the neural underpinnings of cultural variations is evident in the emerging field of cultural neuroscience in particular. In its attempt to better understand human diversity, cultural neuroscience calls for explaining "how theoretical and empirical approaches across distinct fields within the social and natural sciences may further an understanding of how cultural and genetic factors influence the human mind, brain and behavior not only across a lifespan, but also within the situation and across evolutionary timescales" (Chiao et al. 2013).

The scientific goal of cultural neuroscience is to study how neural structures may serve similar functions across cultures and at the same time be the basis for cultural experience-dependent malleability. It attempts "to extract experimentally or by other empirical methods anthropological universals and cultural specifics" (Bao and Poppel 2012). Research in cultural neuroscience integrates insights from anthropology, psychology, social cognitive neuroscience and genetics. It uses tools such as transcultural neuroimaging to understand the culture's influence on brain structure and function (Han and Northoff 2008) and has provided evidence of culture-specific patterns of brain activity underlying not just low-level perceptual and attentional processes (which depend on physical stimuli) but also high-level social cognition and representations (which require mental stimuli) such as perspective taking and the sense of self (Wu and Keysar 2007; Kitayama and Park 2010). It has been claimed that the field can have important implications, for example, provide a richer understanding of the causes of health disparities in different populations (Chiao et al. 2013), advance psychiatric science by furthering the understanding of how cultural, socioeconomic and geographic factors help explain psychopathologies (Choudhury and Kirmayer 2009), contribute in the long-standing controversy over nature and nurture (Kitayama and Park 2010; Kim and Sasaki 2014), and enrich our grasp of how the mind and the self result from the interaction of culture and the brain (Seligman et al. 2016).

7.3.2. Culture and the brain: some challenges

Studies in epigenetics and in cultural neuroscience present potential opportunities. Nonetheless, it has been pointed out that they are not without challenges. Here, I only highlight some of the already identified conceptual issues that would benefit from more attention, starting with epigenetics.

The elusive nature of terms such as epigenetics or environment (Landecker and Panofsky 2013; Meloni and Testa 2014), and the lack of clarity on what epigenetic research is intended to comprise is a central critique leveled at the field of epigenetics in general. A point frequently made is that while the flexibility of the notion of epigenetics might have helped the rise of the field, the impact of the epigenetic discourse in a number of areas warrants awareness of hyperbolic expectations regarding what the scientific findings show and a careful evaluation of what the implications of the research really are (Meloni 2014; Meloni and Testa 2014). Another general concern is that even if epigenetics attempts to decentralize the gene and its sovereignty, it is not clear whether it truly moves beyond reductionist views of the interaction of biology and the environment. A tendency to reductionism, it is claimed, can become evident in epistemically different ways: in attempts to biologize the social and cultural, in trying to give separate and different roles to genes and the environment, and even in reducing nature and nurture to an integrated blurred nature nurture ontogenetic niche (Meloni and Testa 2014).

Importantly, conceptual precision regarding epigenetic findings and what they mean seems key given their practical implications. Indeed, epigenetic findings can be overinterpreted or misinterpreted creating unrealistic expectations (e.g. regarding the extent to which humans have a capacity for self-improvement and neurological control (Pickersgill 2013; Meloni 2014)) or unfounded worries (e.g. that people are determined by their environments). Moreover, the fact that these findings are often used to ground normative claims at ethical and political levels warrants more careful attention to what the operative notions mean and the extent to which they can be used to drive societal change.

Regarding cultural neuroscience, as a discipline that focuses on how culture shapes the brain, understanding culture appears to be a key requisite. However, even when many studies provide a definition of culture, it has been argued that the field should engage more critically with its conception of culture, for the evaluation of empirical findings on cultural differences and their implications will depend, to a great extent, on the theoretical framework used to understand culture, its salient dimensions and how they are measured. In particular, there are worries about the potential reductionistic implications of seeking culture at the neural level (Vidal and Ortega 2017), about the political, racial and ethnic biases implicit in the questions addressed by relevant studies in cultural neuroscience research (Martinez Mateo et al. 2012), about the field's tendency to reduce culture to discrete categories (ethnic or national identity, for example) that typically leads to sidelining internal diversity and the reification of cultural identities (Seligman et al. 2016), and the potential for reproducing problematic stereotypes (Choudhury and Kirmayer 2009).

A related issue that has been highlighted has to do with how cultural differences are measured and interpreted and their generalizability given both the brain's plasticity and malleability (Hyde et al. 2015) and the fact that cultural neuroscience is itself embedded in specific cultural and historical contexts. This explanation calls for a more nuanced concept of culture and cultural differences (Gutchess and Goh 2013), for careful examination of how to interpret the interaction between culture and the developing brain and for increased awareness of the social, political and cultural situatedness of science itself (Vidal and Ortega 2017).

7.3.3. *Is there a role for neuroethics?*

Despite the recognized need to consider the challenges outlined above and the attention they are receiving among some scholars in the social sciences (Choudhury and Kirmayer 2009; Seligman and Brown 2010; Landecker and Panofsky 2013; Meloni 2014; Meloni and Testa 2014; Jablonka 2016; Seligman et al. 2016), sustained attention from neuroethics is lacking. This, I think, is partly due to the tendency of neuroethics to conceive itself chiefly as "neurobioethics", a fundamentally practical approach that focuses mainly on the normative discussion and can generally do without a careful conceptual analysis.

The problem with this tendency is that it presents an impoverished understanding of the field. Neuroethics is not and should not be exhausted by the normative management of the issues raised by neuroscience (Roskies 2002; Evers, Salles, Farisco 2017). As noted earlier in this chapter, in addition to the normative and descriptive neuroethics approaches, it is also possible to identify a conceptual approach that seeks to increase the conceptual soundness of the research and of the reflection on its potential and actual practical implications[7]. A conceptual neuroethics approach takes, as a starting point, the conviction that conceptual analysis plays a central role in dealing with the multiple issues raised by neuroscience but, beyond this, it emphasizes the need to develop and use a critical methodology for effectively integrating scientific (e.g. neuroscience) and philosophical (e.g. ethics) interpretations in order to inform the ethical analysis itself (Evers et al. 2017b; Salles et al. 2017; Farisco et al. 2018). Thus, a conceptual approach's main focus is not on the identification of the ethical issues raised by neuroscientific findings and applications or the use of ethical theory to solve them but rather on exploring what the scientific findings are, how they are framed and interpreted and then how they can be potentially used in the normative discussion. In the case of culture, prior to the normative discussion, it is relevant to ask questions such as how is the neuroscientific knowledge about culture and the brain constructed? How are the findings about culture and the brain interpreted? What are the theoretical and potentially practical limits of those suggested interpretations? What criteria should be used for assessing the relevance of the findings about culture and the brain?

---

7. The claim that a conceptual neuroethics approach attempts to increase the conceptual soundness of the research should not be interpreted as entailing that neuroscience is by definition conceptually deficient: neuroscience includes a significant conceptual component. However, as a relatively young discipline, such a conceptual component is not as developed as it is in other more developed scientific disciplines (Farisco et al. 2018a).

In fact, a conceptual neuroethics analysis may be key in order to fully unpack the ethical applied issues raised by value diversity and brain research[8]. Consider, for example, how studies in epigenetics are lately mentioned as relevant for explaining a number of ethical and social ills (such as prevalent poverty, marginalization and inequality) and for possibly grounding specific public policies and strategies to alleviate them. Careful conceptual analysis of the key scientific and philosophical notions, of underlying theoretical assumptions, of scientific, philosophical and political interpretations (e.g. of the relation between brain and environment) and of their theoretical and practical advantages and risks has the potential for enabling a more robust practical discussion of challenges and promises. Indeed, it can be argued that such reflective process is a precondition for designing and implementing the relevant practical strategies to solve the issues.

Conceptual analysis should also play a role when engaging in the discussion over how to manage cultural difference and the desirability or undesirability of the different options (cultural awareness, cultural engagement, globality). Indeed, such a discussion requires reflection on the translatability (or lack thereof) of the concepts and values at stake, but importantly it appears blind without attention to the neuroscientific research on culture, its promises and its limitations. While I am very sympathetic to the idea that we must avoid neurologization of the social and cultural, I am equally wary of undertaking the normative discussion of how to handle ethical and social issues in a culturally sensitive way without some reflection on how culture impacts the brain, when and why this cultural impact is ethically relevant, and what it means.

7.4. Concluding remarks and the way forward

Lately, neuroethicists have been calling for more engagement with culture. In practice, this has generally resulted in more awareness of value diversity and a focus on what it means to do global, international or cross-cultural neuroethics and on promoting a better understanding of how to deal with different cultural practices, beliefs and concerns. Without minimizing the importance of this discussion, I think that it has been both limited and limiting for neuroethics. Indeed, if these are the

---

8. The claim that a conceptual neuroethics approach can play such a role is not a call for displacing the social sciences which, as noted before, have identified important issues raised by neuroscientific research on culture. Rather, the claim is that a conceptual neuroethics can collaborate with both, the social sciences and neuroscience, by asking new questions, and attempting to refine and develop the kind of conceptual framework needed to better analyze the research's assumptions, results and interpretations and inform the ethical discussion without giving primacy to biological or social explanations of the phenomena in question.

main issues to be addressed regarding culture, I am reluctant to claim that neuroethics can offer a unique perspective not least because it is unlikely that a unique perspective on how to deal with cultural diversity is forthcoming. However, I do not think that the neuroethical discussion needs are exhausted by this. There are still big questions to ask regarding culture and the brain, about the culture-related issues that the science attempts to answer and why, about how findings on culture are interpreted, what they mean and how they can be applied. A conceptual neuroethics approach can be instrumental in examining these issues and thus enhancing the theoretical framework used to conduct research in neuroscience and culture. Moreover, such conceptual work can have implications on how the neurobioethical discussion is carried out. Indeed, at present, it seems that the mainstream normative discussion on how to handle cultural diversity in neuroethics and the resulting proposals tend to trigger one of the following three (generally unreflective) attitudes: (a) minimization or denial of the role of culture in shaping how issues raised by neuroscience are understood and addressed, (b) mere passive acknowledgment that there are cultural differences without really engaging with whether and how they shape the identification and treatment of issues and (c) criticism of (a) and (b) that fails to offer constructive strategies to move forward. Interestingly, what these attitudes have in common is that usually they do not problematize what it means for the brain and culture to shape each other nor do they engage with how that happens, whether and how it is relevant, and why. To the extent we continue to succumb to any of those attitudes, we risk not advancing in the normative discussion either. By engaging with issues such as culture, the assumptions behind neuroscience's findings, interpretations and meanings, a conceptual neuroethics approach can contribute to a way forward.

7.5. Acknowledgments

Special thanks to Michele Farisco, Karen Rommelfanger and Kathinka Evers for their very useful comments to a previous version of this chapter. Thanks also to the participants of the CRB seminars for stimulating discussion on some points developed in this chapter. This project/research was funded by the European Union's Horizon 2020 Framework Program for Research and Innovation under the specific grant agreement no. [945539] (Human Brain Project SGA3).

7.6. References

Ames, D.L. and Fiske, S.T. (2010). Cultural neuroscience. *Asian Journal of Social Psychology*, 13, 72–82.

Bao, Y. and Poppel, E. (2012). Anthropological universals and cultural specifics: Conceptual and methodological challenges in cultural neuroscience. *Neurosci Biobehav Rev*, 36, 2143–2146.

Bonete, E. (2013). Neuroethics in Spain: Neurological determinism or moral freedom? *Neuroethics*, 225–232.

Campbell, A.V. (1999). Presidential address: Global bioethics – dream or nightmare? *Bioethics*, 13, 183–190.

Chandler, J.A., Cabrera, L.Y., Doshi, P., Fecteau, S., Fins, J.J., Guinjoan, S., Hamani, C., Herrera-Ferrá, K., Honey, C.M., Illes, J. et al. (2021). International legal approaches to neurosurgery for psychiatric disorders. *Frontiers in Human Neuroscience*, 14.

Changeux, J.P. (1985). *Neuronal Man*. Princeton University Press, Princeton, NJ.

Changeux, J.P. (2017). Climbing brain levels of organisation from genes to consciousness. *Trends Cogn Sci*, 21, 168–181.

Chattopadhyay, S. and De Vries, R. (2013). Respect for cultural diversity in bioethics is an ethical imperative. *Med Health Care Philos*, 16, 639–645.

Chen, D. and Quirion, R. (2011). From the internationalization to the globlization of neuroethics: Some perspectives and challenges. In *Oxford Handbook of Neuroethics*, Illes, J. and Sahakian, B.J. (eds). Oxford University Press, Oxford.

Chiao, J.Y., Cheon, B.K., Pornpattanangkul, N., Mrazek, A.J., Blizinsky, K.D. (2013). Cultural neuroscience: Progress and promise. *Psychol Inq*, 24, 1–19.

Chien-Chang Wu, K. and Fukushi, T. (2012). Neuroethics in Taiwan: Could there be a Confucian solution? *East Asian Science, Technology and Society: An International Journal*, 6, 321–334.

Choudhury, S. and Kirmayer, L.J. (2009). Cultural neuroscience and psychopathology: Prospects for cultural psychiatry. In *Progress in Brain Research*, Chiao, J.Y. (ed.). Elsevier, Amsterdam.

D'Ambrosio, P. and Colage, I. (2017). Extending epigenesis: From phenotypic plasticity to the bio-cultural feedback. *Biol Philos*, 32, 705–728.

Dias, B.G., Maddox, S., Klengel, T., Ressler, K.J. (2015). Epigenetic mechanisms underlying learning and the inheritance of learned behaviors. *Trends Neurosci*, 38, 96–107.

Dubljevic, V. and Bottenberg, F. (2021). *Living with Dementia: Neuroethical Issues and International Perspectives*. Springer, Cham.

Edelman, G. (1987). *Neural Darwinism: The Theory of Neuronal Group Selection*. Basic Books, New York.

Emerging Issues Task Force International Neuroethics Society (2019). Neuroethics at 15: The current and future environment for neuroethics. *AJOB Neuroscience*, 10(3), 104–110.

Eriksson, P.S., Perfilieva, E., Bjork-Eriksson, T., Alborn, A.M., Nordborg, C., Peterson, D.A., Gage, F.H. (1998). Neurogenesis in the adult human hippocampus. *Nat Med.*, 4, 1313–1317.

Evers, K. (2020). The culture bound brain: Epigenetic proaction revisited. *Theoria*. DOI: 10.1111/theo.12264.

Evers, K. and Changeux, J.P. (2016). Proactive epigenesis and ethical innovation: A neuronal hypothesis for the genesis of ethical rules. *EMBO Rep*, 17, 1361–1364.

Evers, K., Salles, A., Farisco, M. (2017). Theoretical framing for neuroethics: The need for a conceptual aproach. In *Debates About Neuroethics*, Racine, E. and Aspler, J. (eds). Springer, Cham.

Farisco, M., Salles, A., Evers, K. (2018). Neuroethics: A conceptual approach. *Camb Q Healthc Ethics*, 27, 717–727.

Fukushi, T., Isobe, T., Nakazawa, E., Takimoto, Y., Akabayashi, A., Specker Sullinca, L., Sakura, O. (2018). Neuroethics in Japan. In *The Routledge Handbook of Neuroethics*, Johnson, L.S.M. and Rommelfanger, K.S. (eds). Routledge, New York.

Global Neuroethics Summit Delegates (2018). Neuroethics questions to guide ehtical research in the international brain initiatives. *Neuron*, 100(1), 19–36.

Goering, S. and Yuste, R. (2016). On the necessity of ethical guidelines for novel neurotechnologies. *Cell*, 167, 882–885.

Goering, S., Klein, E., Specker Sullivan, L., Wexler, A., Aguera, Y.A.B., Bi, G., Carmena, J.M., Fins, J.J., Friesen, P., Gallant, J. et al. (2021). Recommendations for responsible development and application of neurotechnologies. *Neuroethics*, 1–22.

Gutchess, A. and Goh, J.O.S. (2013). Refining concepts and uncovering biological mechanisms for cultural neuroscience. *Psychological Inquiry*, 24, 31–36.

Han, S. and Northoff, G. (2008). Culture-sensitive neural substrates of human cognition: A transcultural neuroimaging approach. *Nat Rev Neurosci*, 9, 646–654.

Herrera-Ferrá, K., Salles, A., Cabrera, L. (2018). Global neuroethics and cultural diversity: Some challenges to consider. *The Neuroethics Blog* [Online]. Available at: http://www.theneuroethicsblog.com/2018/04/global-neuroethics-and-cultural.html.

Herrera-Ferrá, K., Zavala, G.S., Nicolini Sancehz, H., Pinedo Rivas, H. (2019). Neuretica en Mexico: Reflexiones medicas, legles y socioculturales. *ScienceDirect BIOETHICS Update*, 5, 89–106.

Holm, S. and Williams-Jones, B. (2006). Global bioethics – myth or reality? *BMC Med Ethics*, 7, E10.

Hyde, L., Tompson, S., Creswell, J.D., Falk, E.B. (2015). Cultural neuroscience: New directions as the field matures. *Cult Brain*, 3, 75–92.

Ienca, M. (2021). On neurorights. *Front Hum Neurosci*, 15, 701258.

Ienca, M. and Andorno, R. (2017). Towards new human rights in the age of neuroscience and neurotechnology. *Life Sci Soc Policy*, 13, 5.

Illes, J., Blakemore, C., Hansson, M.G., Hensch, T.K., Leshner, A., Maestre, G., Magistretti, P., Quirion, R., Strata, P. (2005). International perspectives on engaging the public in neuroethics. *Nat Rev Neurosci*, 6, 977–982.

Jablonka, E. (2016). Cultural epigenetics. *Sociol Rev Monogr*, 64, 42–60.

Jahoda, G. (2012). Critical reflections on some recent definitions of "culture". *Cul Psychol*, 18.

Kim, H.S. and Sasaki, J.Y. (2014). Cultural neuroscience: Biology of the mind in cultural contexts. *Annu Rev Psychol*, 65, 487–514.

Kitayama, S. and Park, J. (2010). Cultural neuroscience of the self: Understanding the social grounding of the brain. *Soc Cogn Affect Neurosci*, 5, 111–129.

Landecker, H. and Panofsky, A. (2013). From social structure to gene regulation, and back: A critical introduction to environmental epigenetics for sociology. *Annu Rev Sociol*, 333–357.

Lombera, S. and Illes, J. (2009). The international dimensions of neuroethics. *Dev World Bioeth*, 9, 57–64.

Marshall, P. and Koenig, B. (2004). Accounting for culture in a globalized bioethics. *J Law Med Ethics*, 32, 252–266, 191.

Martinez Mateo, M., Cabanis, M., Echeverria Loebell, N., Krach, S. (2012). Concerns about cultural neurosciences: A critical analysis. *Neuroscience and Behavioural Reviews*, 36, 152–161.

Matshabane, O.P. (2021). Promoting diversity and inclusion in neuroscience and neuroethics. *EBioMedicine*, 67, 103359.

Meloni, M. (2014). The social brain meets the reactive genome: Neuroscience, epigenetics and the new social biology. *Front Hum Neurosci*, 8, 309.

Meloni, M. and Testa, G. (2014). Scrutinizing the epigenetics revolution. *Biosocieties*, 9, 431–456.

Myser, C. (2011). *Bioethics Around the World*. Oxford University Press, New York.

Northoff, G. (2009). What is neuroethics? Empirical and theoretical neuroethics. *Curr Opin Psychiatry*, 22, 565–569.

Pickersgill, M. (2013). The social life of the brain: Neuroscience in society. *Curr Sociol*, 61, 322–340.

Racine, E.A.J. (2017). *Debates about Neuroethics: Perpectives on its Development, Focus, and Future*. Springer International Publishing, Dordrecht.

Rommelfanger, K.S., Jeong, S.J., Montojo, C., Zirlinger, M. (2019). Neuroethics: Think global. *Neuron*, 101, 363–364.

Rubin, B. (2009). Changing brains: The emergence of the field of adult neurogenesis. *Biosocieties*, 4, 407–424.

Sakura, O. (2012). A view from the far east: Neuroethics in Japan, Taiwan, and South Korea. East Asian Science. *Technology and Society: An International Journal*, 6, 297–301.

Salles, A. (2014). Neuroethics in a "psy" world. *The Case of Argentina. Camb Q Healthc Ethics.*, 23, 297–307.

Salles, A. (2018). Neuroethics in context: The development of the discipline in Argentina. In *The Routledge Handbook of Neuroethics*, Johnson, L.S.M. and Rommelfanger, K.S. (ed.). Routledge, New York.

Salles, A. and Evers, K. (2017). Social neuroscience and Neuroethics: A fruitful synergy. *Social Neuroscience and Social Science: The Missing Link*, Ibanez, A. Sedeno, L., Garcia, A. (eds). Springer, Cham.

Seligman, R. and Brown, R.A. (2010). Theory and method at the intersection of anthropology and cultural neuroscience. *Soc Cogn Affect Neurosci*, 5, 130–137.

Seligman, R., Choudhury, S., Kirmayer, L. (2016). Locating culture in the brain and in the world: From social categories to the ecology of mind. In *The Oxford Handbook of Cultural Neuroscience*, Chiao, J.Y., Li, S.C., Seligman, R., Turner, R. (eds). Oxford University Press, Oxford.

Shook, J.R. and Giordano, J. (2014). A principled and cosmopolitan neuroethics: Considerations for international relevance. *Philos Ethics Humanit Med*, 9, 1.

Shook, J.R. and Giordano, J. (2016). Neuroethics beyond Normal. *Camb Q Healthc Ethics*, 25, 121–140.

Specker Sullivan, L. (2019). Insight and the no-self in deep brain stimulation. *Bioethics*, 33(4), 487–494.

Spranger, T.M. (2012). *International Neurolaw: A Comparative Analysis*. Springer, Berlin.

Stein, D. and Singh, I. (2020). *Global Mental Health and Neuroethics*, 1st edition. Elsevier Academic Press, Amsterdam.

Sweatt, J.D. (2013). The emerging field of neuroepigenetics. *Neuron*, 80, 624–632.

Tsomo, K.L. (2012). Compassion, ethics, and neuroscience: Neuroethics through Buddhist eyes. *Sci Eng Ethics*, 18, 529–537.

Vidal, F. and Ortega, F. (2017). *Being Brains: Making the Cerebral Subject*. Fordham University Press, New York.

Wexler, B. (2011). Neuroplasticity, culture, and society. In *Handbook of Neuroethics*, Illes, J. and Sahakian, B. (ed.). Oxford University Press, Oxford.

Wu, S. and Keysar, B. (2007). The effect of culture on perspective taking. *Psychol Sci*, 18, 600–606.

Yao, B., Christian, K.M., He, C., Jin, P., Ming, G.L., Song, H. (2016). Epigenetic mechanisms in neurogenesis. *Nat Rev Neurosci*, 17, 537–549.

Yuste, R., Goering, S., Arcas, B.A.Y., Bi, G., Carmena, J.M., Carter, A., Fins, J.J., Friesen, P., Gallant, J., Huggins, J.E. et al. (2017). Four ethical priorities for neurotechnologies and AI. *Nature*, 551, 159–163.

Zhou, H. and Cacioppo, J. (2010). Culture and the brain: Opportunities and obstacles. *Asian Journal of Social Psychology*, 13, 59–71.

8

Globalization of Neuroethics: Rethinking the Brain and Mind "Global Market"

Karen HERRERA-FERRÁ
Asociación Mexicana de Neuroética, Mexico City, Mexico

8.1. Introduction

Globalization is a word that is commonly and widely used in a variety of domains; however, we may not always be aware of its importance and impact on the multiple dimensions of human life.

Globalization began as an alternative solution to complement the poor or lack of resources to satisfy certain needs in a specific location and hence, forced to search, produce and/or move products from one place to another (World Economic Forum 2019). For this process to take place it needs "a market", which is made up of producers and consumers of resources which can be local or foreign, and in globalization the producers/consumers are referred as the "global market" (International Monetary Fund 2002). In other words, the globalization process began as an economic system that aimed to increase trade in goods and services around the world (Turner and Holton 2011; Cambridge 2021). It is important to remember that when referring to *the world*, each country[1], as an independent agent, should be

---

1. A geographical area.

expected to have its own context[2] and culture[3]. Therefore, context(s) and culture(s) will shape specific needs, which in turn are related to different local factors relevant to globalization such as income, global integration, human capital and workforce skills, the facilitation of resource reallocation, etc. (The World Bank 2020a, 2020b).

For instance, in theory, a high-income country will achieve a rapid satisfaction of basic needs and will perceive the need to produce/consume more complex goods which are expected to facilitate and improve human life, such as advanced science and technology for medicine, society, public safety, etc. Moreover, greater production and acquisition of advanced products and services will require interrelation with other disciplines such as law, in order to develop proper public policies and regulations for the safe and sound use of these goods (The World Bank 2020a).

Thus, globalization will not only have an impact on economy but will also affect other areas on human life. In this sense, it would be naïve to think of globalization as only or mainly an economic system that improves material and/or service provisioning, because globalization is much more than a movement of resources but is rather an expanding and intertwined multi-domain interaction.

This expansion also entails a significant and profound interconnectedness between people, lifestyles, ideas, beliefs, cultures, religions, etc. (Beckford 2003). This unique human interconnectivity has resulted, among many other things, in emergent ideologies or movements related to the consciousness of this globalization (Beckford 2003) and therefore could lead either to social idealization or disapproval. One good example is the *#MeToo* movement which although had receptivity and a positive effect in many parts of the world, did not achieve a global consensus because of the repressive or punitive consequences in specific cultural contexts (e.g. Mexico). Moreover, it has been described that the global availability and accessibility of novel goods and ideas have led to legal, social, cultural and political reactions and debates regarding imperialism or world order, arising dichotomic postures such as synergism versus nationalism, inclusion versus discrimination, multiculturalism versus cultural identity, power versus vulnerability, equity versus inequity, etc. (Turner and Holton 2011).

These reactions and debates should make us aware that "global markets", although similar in many ways, may differ not only in contexts and cultures, but also

---

2. This includes economic, legal, social, ethical, health and environmental systems (i.e. the ecosystem), etc., circumstantial variables (e.g. poverty, political instability, violence, etc.) and individual factors (age, gender, race, religion, health, migratory status, etc.).
3. Set of beliefs, customs, attitudes, rituals, habits, etc. that characterizes a specific group.

in cognitive processes such as perceptions and meanings. Thus, the multi-effect of globalization on humanity will constantly spark a variety of ethical, legal, social, cultural and political issues and concerns.

In this light, globalization should be understood as a complex dynamic, interactive and multidirectional process that embraces and has an effect – and has an affect – on many domains of human life (i.e. economy, medicine, law, arts, humanities, ethics, science, public safety, society, culture, etc.).

Thus, the multidomain convergence, the plausible and real effect on human life, and the global and collective reflection – and perceptions – of benefits, burdens and risks that result from the globalization of goods and ideas, calls for further consideration and proactive integration of diverse ideologies, cognitive processes and ethical systems. Specially regarding the impact of scientific and technological goods on sensitive topics such as the meaning of being human, human dignity and human rights, although many of them have been addressed in a variety of international treaties such as the United Nations Universal Declaration of Human Rights (United Nations 1948), the Universal Declaration on the Human Genome and Human Rights (UNESCO 1998) and the Universal Declaration on Bioethics and Human Rights (UNESCO 2005), among others. Nevertheless, and despite these efforts, the potential and emergent concerns related to the demanding and escalated production and use at a global scale of more complex goods are not contemplated or fully addressed within these international treaties for proper protection, as is the case of neuroscience, neurotechnology and some forms of artificial intelligence (Ienca and Andorno 2017; OECD 2017; International Brain Initiative 2018a; NeuroRights Initiative 2020). Furthermore, these sophisticated tools and techniques for the brain have granted data about the plausible correlational biological basis of the most sensitive topics related to the mind and human essence. Some of these topics include consciousness, emotions, behaviors, cognitions, identity, decision-making, the self, free will, autonomy, empathy and morality, among others (International Brain Initiative 2018a). As such, neuroethics as a discipline and set of practices dedicated to the study and analysis of these issues and concerns has been recognized as an instrumental reflective discipline towards the development and use of neuroscience, neurotechnology and some forms of artificial intelligence (Giordano 2015; OECD 2017; International Brain Initiative 2018a, 2018b).

However, the perceived benefits, risks and burdens, as well as specific ethical, legal, social and cultural issues and concerns currently addressed by neuroethics, might differ between countries, contexts and cultures and hence, globalization of neuroethics concerns as a "one-size-fits-all product in and for the global market" requires further reflection.

8.2. Neuroethics within the global market: the "normality" problem

Although it may seem that because developing and non-developed countries[4] have not yet satisfied basic needs for their entire population such as food, clean water, shelter, etc. (which is reflected in the poverty index), they will not have the required financial means to be positioned as "buyers" or "consumers" of advanced and emergent tools and techniques for the brain; however, this does not reflect current realities.

For instance, advanced, novel, innovative and sophisticated neuroscience and neurotechnology mainly produced by governmental projects or initiatives in some high-income countries such as the United States, countries from the European Union, Japan, Australia, China, South Korea and Canada (i.e. "developers") are currently used in many parts of Latin America[5] in areas such as mental health (i.e. neuromodulatory techniques such as Deep Brain Stimulation, transcranial electrical stimulation, transcranial magnetic stimulation (Herrera-Ferrá et al. 2019, 2020; Ramirez-Zamora et al. 2020)). Moreover, the use of these tools and techniques for the brain is taking place not only without proper legal and ethical regulation but also with poor or lack of neuroethical, legal, social and cultural reflections, which could affect the safe and sound use of neuroscience and neurotechnology, at least in these specific contexts (e.g. Latin America) (Herrera-Ferrá 2019, 2021; Herrera-Ferrá et al. 2020).

This begs the question as to whether or not, in neuroethics (as in many other fields), a "normality" has been assumed for and within the global market (i.e. "the perception and concerns of those who rule will be the perception and concerns of all"), and therefore, it would seem unnecessary to include alternative perceptions in the global discourse.

Furthermore, the dichotomous category of developer countries as a cluster versus consumer countries as "the other" cluster, might undermine neuroethical concerns of

---

4. Some of the ways to classify countries are by income (e.g. low- and middle-income countries, high-income countries, emerging economies, etc.). However, the terms developed, developing and non-developed when used in this chapter will refer to the level of development of the context (i.e. economic, legal, social, health systems, etc.; poverty, political instability, violence, etc.).

5. And other parts of the world, however as the author lives in Mexico, emphasis will be placed on the Hispanic world. Nevertheless, I strongly encourage professionals related to the brain area from under-represented countries and cultures, to share neuroethical concerns related to their context and culture.

each country – whether developer or consumer – as an independent agent, with its own context, culture, needs, interpretations and meanings.

As such, I believe that for the field of neuroethics to be globalized it is necessary to consider that contextual and cultural anthropological variables, which have an impact on neural synapsis (Causadias et al. 2017; Ferdinand et al. 2019), shape fundamental concepts (i.e. cognitive diversity) related to the meaning of being human. In this light, my work has focused on advocating that cultural cognitive diversity will result not only in alternative perspectives and meanings on the use of neuroscience, neurotechnology and some forms of artificial intelligence, but also on the perceived importance, perceived need and perceived relevance of neuroethics *as field or discipline* (Herrera-Ferrá et al. 2019; Herrera-Ferrá 2020, 2021).

Certainly, the inclusion of other perceptions, values and meanings is not new and has been described in relevant areas that precede the globalization of neuroethics, including global mental health (the perception and understanding of concepts such as mental well-being, mental health, Western medicine, etc. (World Health Organization 2013; Herrera-Ferrá 2020)) and human rights in ethical globalization. To illustrate, at the Yale Center for the Study of Globalization at Yale University, during a debate on the issue of human rights and ethics raised by globalization, former Mexican President Ernesto Zedillo asked former President of Ireland and former United Nations High Commissioner for Human Rights Mary Robinson (who founded Realizing Rights: The Ethical Globalization Initiative (Global Health Workforce Alliance 2002):

> Why speak of ethical globalization? … Do you see globalization as a threat or as an opportunity to strengthen, enforce human rights throughout the world? … Do you see globalization as an instrument that could accelerate the establishment of the values and practices of human rights in the world or do you see in globalization something that inherently could weaken the prevalence of human rights in the world? (Yale 2003).

These questions reiterate the issues I pose in this chapter (a) the non-universal perception and meaning of a concept, (b) the need to include cognitive diversity for a responsible globalization (of neuroethics as a discipline), and (c) the relevance and opportunity to build a responsible, responsive and proactive bridge between countries, contexts and cultures.

In this sense, being mindful and inclusive of other ways of thinking, and being proactive in building collaborations, could strengthen and complement current efforts to anticipate potential risks and benefits in the development and use of neuroscience, neurotechnology and some forms of artificial intelligence. Moreover, these efforts would be diligent with international initiatives aims such as "[c]atalyzing and advancing neuroscience research through international collaboration and knowledge sharing, uniting diverse ambitions to expand scientific possibility, and disseminating discoveries for the benefit of humanity" (International Brain Initiative 2018a); "addressing the social and ethical implications of neuroscience as a global endeavor" (International Brain Initiative 2018b), and the need for "international cooperation to anticipate and address ethical, legal and social issues raised by emerging neurotechnologies" to "enrich the regulatory and policy discourse around neurotechnology" (OECD 2017).

On the contrary, it is important to highlight that efforts to address neuroethical, legal, social, cultural issues and concerns raised by brain sciences are not limited to governmental initiatives, projects and/or organisms. They also include civil associations that have been fostering the awareness of such concerns such as the International Neuroethics Society (INS) (International Neuroethics Society 2021), the Neuroethics Network (Neuroethics Network 2021), the Italian Society of Neuroethics (SINe) (Società Italiana di Neuroetica 2021) and the Mexican Association of Neuroethics (AMNE) (Asociación Mexicana de Neuroética 2021), among others. These efforts demonstrate the importance, interest and intention in the divulgation and development of the field of neuroethics as a global and multicultural endeavor.

It should be noted that the proactive integration of Mexico (which does not have a multi-million dollar brain project and is primarily a "consumer" country), has been well received and has led to international collaborations not only to address common concerns, but also because the AMNE has focused on the analysis of specific issues related to context and culture, which have not been reported in the academic literature of "producer" countries (García-López et al. 2019; Herrera-Ferrá et al. 2019; Herrera-Ferrá et al. 2020; Ramirez-Zamora et al. 2020). Some of these issues include (1) perceptions on neuroenhancement and on the use of neuromodulatory techniques (Cabrera and Herrera-Ferrá 2019; Herrera-Ferrá et al. 2019, 2020; Ramirez-Zamora et al. 2020), (2) the need to develop regulatory frameworks on the development and use of neuroscience, neurotechnology and some forms of artificial intelligence (Herrera-Ferrá 2021), (3) contextual considerations on human rights for the brain and mind (2021) and (4) the use of neuroscience and neurotechnology in criminal law (García-López et al. 2019; Mercurio et al. 2020).

In other words, although neuroethical concerns of developers and consumers indeed converge, they also present alternative perspectives, and therefore, there is a neuroethical gap that needs to be considered and addressed.

8.2.1. *The neuroethical gap: cultural – beliefs and perspectives of the brain and mind*

In this chapter, I have highlighted some concerns on the dichotomous category of developer countries as one cluster and of consumer countries as "the other" cluster, which undermines *global cognitive diversity*[6].

However, the importance of including the cognitive diversity of current and potential consumers, as well as their specific neuroethical, legal, social and cultural concerns, have not yet been adequately addressed. To illustrate, I would like to highlight some relevant information regarding the current state of the field of neuroethics in the world.

Firstly, neuroscience, neurotechnology and some forms of artificial intelligence are mainly being developed by rich and industrialized countries according to their specific needs, social, legal, ethical and cultural systems and expectations. As such, it is implicitly understood that the neuroethical, legal, social and cultural issues (NELSCI) and concerns that have been depicted, analyzed and addressed are (and will continue to be) diligent with these specific countries (contexts and cultures) (International Brain Initiative 2018b).

Nevertheless, these countries represent 32% of the world's population (OECD 2021a) – and cultural cognitive diversity – and therefore, only a third of the global needs, perspectives and meanings on the development and use of neuroscience, neurotechnology and some forms of artificial intelligence are known. Moreover, the NELSCI of developers, as well as *the importance and value of neuroethics as a discipline*, might not be the same for current and potential consumers, who represent 68% of the global population (OECD 2021a).

Consequently, moving the field of neuroethics as a good or service (since it serves for a better development and use of neuroscience, neurotechnology and some forms of artificial intelligence) in a multi-transnational way to a global market,

---

6. The globalization of the field of neuroethics will require pertinent and prudent considerations, inclusion and collaboration of cultures from consumer countries and from cultures and minority groups within developer countries such as indigenous people, migrants and refugees.

means one will face ethnographic and cognitive-diverse gaps. These gaps include various conceptualizations related to the brain and mind, such as mental health and mental well-being, emotions, cognitions, behaviors, consciousness, the self, free will, autonomy, personal identity, empathy, morality and decision-making, among others (Herrera-Ferrá et al. 2019, 2020). In addition, these conceptualizations related to the brain and mind are also considered characteristic traits of the human person[7] and their essence. Therefore, these traits underlie fundamental and unique culturally shaped perceptions, interpretations and meanings, and as a result, the given value, importance and sacredness of these traits, may vary among contexts and cultures (Schlicht et al. 2009; Herrera-Ferrá et al. 2019). This is important because these psychological processes related to the brain, mind and human essence, shape receptive or resistant attitudes regarding biological explanations as well as on the use of scientific and technological resources on, for example, mental health (Saxena et al. 2007; Herrera-Ferrá 2020) and on the importance and need for neuroethics as a field. As such, alternative perspectives – within and between countries – are expected regarding the impact of neuroscientific and neurotechnological tools and techniques not only in the brain and mind, but also in the human person. Moreover, socio-cultural reactions to the replication of some of the characteristic traits of the person in the form of artificial intelligence might also differ. Simply put, since each country, context and culture bring different perceptions and challenges, that which is *global* within the brain and mind should not be limited to being mindful of "other ways of thinking or acting". Rather, it should aim at a proactive effort to try to understand and sympathize with alternative contextually and culturally shaped *cognitions* such as perspectives, expectations, needs, explanations and narratives regarding ourselves, others and the world; and in this case, the perceived dimensional impact of neuroscience, neurotechnology and some forms of artificial intelligence in the human person (i.e. NELSCI), and on the importance of neuroethics as a field.

In this light, I would like to share the experience of moving neuroethics to a consumer country through narrative. Narrative, although subjective, is an instrument for exploring the interpretative system in a context and/or culture and helps to place self-experiences, selfhood and identity at the core and within a context and/or culture and is therefore key for ethnographic explanatory purposes (Vassilieva 2016).

8.3. Neuroethics in a consumer country: a narrative from Mexico

Mexico is an independent country with its own context and culture, which is actively integrated in the global market. These variables have been relevant

---

7. I am completely aware of the importance on the debate of other moral statuses as well as dignity; nevertheless, this is out of the scope of this chapter.

regarding shaping the current state of brain sciences and neuroethics. To illustrate, I will briefly give a general framework.

The country is geographically located in the south area of North America – being geopolitically strategic for facilitated accessibility to advanced neuroscience and neurotechnology produced in the United States of America.

Mexico has a population of 125.3 million people (OECD 2021a), is one of the 15 largest economies in the world and the second largest economy in Latin America (The World Bank 2021). The poverty headcount ratio[8] is 41.9% (OECD 2021b) and gross domestic spending on general research and development is 0.30%[9] (compared to 2.82% in the USA (OECD 2021b)).

Resource allocation for *brain research* is estimated[10] to be of 0.3% of the total budget for federal research (Cámara de Diputados 2020) while resource allocation for *brain health* is only 2% of the federal budget for health in general (Senado de la República 2019). Moreover, there is not only a mental health treatment gap[11] but also a mental health, neuroscientific and neurotechnological regulatory gap[12] (Herrera-Ferrá et al. 2019). Both gaps highlight that past and current governmental authorities have not yet ranked brain-related issues as a priority that needs immediate attention. This is the case with reports of other non-high income countries where investment in mental health is not perceived as an emergency, resulting in separate and disparate resource allocation between mental health and health in general (Herrman and Swartz 2007; Herrera-Ferrá 2020). Also, it makes sense that if the development and use of neuroscience and neurotechnology for mental health is not a priority, neither are the development of legal regulatory frameworks or the inclusion of a reflective discipline for the ethical use of neuroscience and neurotechnology (i.e. neuroethics).

---

8. "National poverty headcount ratio is the percentage of the population living below the national poverty line(s)" (OECD 2020). "National poverty lines [...] usually reflect the line below which a person's minimum nutritional, clothing, and shelter needs cannot be met in that country" (The World Bank 2021).
9. This data is from 2018, before the Covid-19 pandemic and before the current administration, which further lowered the budget for research and development.
10. The annual budget reported for the national institutes dedicated only to mental health, neurology and neurosurgery.
11. Discrepancy between the level of mental health treatment that is required and the actual level that is provided.
12. Discrepancy between the level of regulatory frameworks that are required and the actual level available.

Another issue is that a regulation is intended to solve a social problem; however, if society does not perceive mental health and/or the use of brain sciences as a problem, or as a problem that needs to be regulated, this will reinforce authorities' apathy (Herrera-Ferrá 2020, 2021). For instance, socio-cultural perceptions and attitudes towards a biological basis of mental health, mental well-being and more specifically, of human traits related to the mind and human essence such as emotions, cognitions, behaviors, empathy, morality, etc., are still resistant as this may not be completely compatible with the intertwined pre-Hispanic and Catholic spiritual view of these human traits. This may be due to the misperception that a biological basis "neutralizes" the rest of the human dimensions (i.e. social, psychological, philosophical, etc.). For example, there is a significant amount of the population (more than 70%) who believe that alternative medicine including shamanic[13] methods are underestimated by science (Instituto Nacional de Estadística y Geografía 2017). Thus, it is not uncommon that mental health patients seek advice from spiritual leaders (being that there is also a strong Catholic presence)[14], use herbalism or other pseudoscientific methods before approaching a mental health facility or professional.

Interestingly, the use of neuroscience and neurotechnology are not only focused on high burden neuropsychiatric disorders, but also there has been increasing attention on exploring the potential use of neuroscience in criminal law, which has even resulted in the development of a PhD program in Neurolaw and Forensic Psychopathology at the National Institute on Criminal Sciences (Instituto Nacional de Ciencias Penales 2021).

As such and being that brain sciences are mainly focused on mental health and violence, neuroethical concerns are primarily – but not exclusively – proactively addressed by "ground battle" professionals dedicated to these areas: mental health professionals and lawyers.

I think it is important to underscore that the slow development of neuroethics *as a discipline or field* in consumer countries such as Mexico does not necessarily mean that there are not *neuroethical concerns*. Neuroethical concerns related to the plausible effect on the use of neuroscience and neurotechnology in the human essence are not new, especially in diverse areas of medicine such as anesthesiology, neurology, neurosurgery and psychiatry, and more recently under the umbrella of "neurolaw" and "neurorights". For instance, in medicine, many neuroethical concerns might be labeled as moral or medical dilemmas and are expected to be addressed through the ontological duties of medicine and/or bioethics. This could be

---

13. Rituals performed by healers based on the spiritual dimension.
14. 70% of the population is Catholic (Instituto Nacional de Estadística y Geografía 2020).

due to the possible perception that neuroethics, as a field, only addresses concerns related to advanced, innovative, novel and sophisticated neuroscience and neurotechnology developed, available and applicable in geographies with costly brain projects, and thus, the discipline may be perceived unnecessary, impractical or not valuable for low and middle-income countries' immediate priorities.

As a result, neuroethics *as a field* in Mexico is a continued work in progress: (a) it is currently being approached by small groups of scholars, (b) is still not officially part of the national agenda (although is mentioned in the National Guide of Hospital Committees of the National Commission of Bioethics (Comisión Nacional de Bioética 2015), and efforts have been made to promote awareness of the importance of neuroethics), so there is no specific governmental budget for the field, and (c) is mainly a self-funded extra-curricular activity, which results in poor and slow academic productivity.

Accordingly, current efforts in Mexico position neuroethical concerns and neuroethics as a field within two different scopes: literature reviews and practical research.

Literature reviews are published in peer-reviewed journal publications and books predominantly in Spanish and from journals and editorials from Latin America; as for many, *English is not an accessible language* and therefore in some cases, not knowing English means poor or no accessibility to academic literature related to neuroethics (and other philosophical or scientific areas). These publications in Spanish mainly rely on international reviews explaining the field of neuroethics including history, concepts and topics related to the development and use of neuroscience, neurotechnology and some forms of artificial intelligence from brain projects and initiatives from high-income countries. This approach is mainly informative and educational and does not contemplate local NELSCI.

The second scope is mostly analytic and argumentative, which is grounded on local and contextual problem-solving, being the most prominent on neuromodulation (Herrera-Ferrá et al. 2020), neuroenhancement (Cabrera and Herrera-Ferrá 2019) and neurolaw (García-López et al. 2019), among others.

In this light, AMNE's efforts intend to address national and international ethical, legal, social and cultural issues and concerns, raised by emerging neurotechnologies to foster ethical and responsible use, which is adaptable to specific contexts and cultures – with special emphasis in Iberoamerica – to strengthen, develop and complement global efforts (Asociación Mexicana de Neuroética 2021) in line with international endeavors (OECD 2017; International Brain Initiative 2018b).

However, current efforts in neuroethics as a field are not part of any neuroscientific, medical or law program, but rather are based on bioethics' departments and programs, and therefore expose some challenges to fully embrace and accept the value of bioethics in general, and of neuroethics in specific. For instance, the field of bioethics is not always perceived as a secular reflective discipline since it was first introduced under a Catholic scope. Furthermore, the hospital committees of bioethics registered in the country from 2013 to 2019 are 1023 out of 2885 private hospitals and 1395 public hospitals (Instituto Nacional de Estadística y Geografía 2019), and many of them have non-compliance issues (Valdez-Martínez et al. 2015). Moreover, as institutional research ethical committees are limited, inconsistent and slow (Herrera-Ferrá et al. 2020), there is a general view of ethics as radical and restrictive, resulting in limited receptiveness either in the clinical, research and policy-making areas. Thus, bioethics as the main platform to introduce neuroethics as the reflective discipline of neuroscience and specific brain-mind issues does not face a smooth transnational receptiveness.

Thus, as a first step, the development of neuroethics *as a field* in a consumer country such as Mexico (and other parts of Latin America) aims to be a prudent, viable, valuable and adaptable discipline that can be perceived as meaningful, practical and pragmatical for ground battle concerns. In other words, neuroethics in Mexico needs to be developed to provide reflections to better approach immediate domestic problem-solving in mental health and criminal law with available and accessible neuroscience and neurotechnology.

In sum, this narrative reflects that the transnational development of neuroethics confronts relevant challenges related to the context and culture: poverty, governmental apathy, socio-cultural resistance and poor or lack of financial resources for education and research.

8.4. Conclusion and future directions

The first thing I would like to emphasize is that neuroethical concerns in consumer countries have been and are constantly and continuously present. Nevertheless, since these concerns are not always recognized as a problem "categorized" as neuroethics, this may influence the value given to neuroethics as a discipline, as well as the need to develop it. On the contrary, it is important to acknowledge that global perception on the development and use of neuroscience, neurotechnology, some forms of artificial intelligence and on the respective emergent neuroethical, legal, social and cultural concerns, are shaped according to specific contexts. And although many of these perceptions are similar around the world, we have indeed found alternative perspectives in common problems such as neuromodulation (in Mexico) and neuroenhancement (in the Hispanic world).

Moreover, the limited resources for brain sciences to attend the needs in countries and regions such as Mexico and Latin America focus on contextually urgent issues such as high burden neuropsychiatric disorders and violence. This causes neuroethical concerns to be generated from "the battlefield", so the main need is aimed at these scenarios. However, considering contexts does not mean that there is a relativism in neuroethics as a discipline, but probably neither a universality or normality on the needs, availability of resources, and on the understanding, meaning and value of sensitive topics for the human condition, such as the mind nested in the brain, and of the mind as a fundamental trait of being *human*. That is, it must be considered that neuroethics as a discipline in countries that are mostly consumers is presented (and developed) as an adaptable, practical, prudent and pertinent tool to solve immediate problems in practical and real scenarios, at least in medicine and law (i.e. bottom to top).

Therefore, I believe that future directions towards positioning neuroethics as a field at a global level should focus on national and international efforts to (a) be mindful of the significant transnational and multidirectional interconnectivity of the diverse human dimensions that result from globalization, (b) avoid assuming universality in key concepts of the human mind and essence and (c) be inclusive of other perspectives, narratives and neuroethical concerns on the use of neuroscience, neurotechnology and some forms of artificial intelligence.

Furthermore, there should be multilateral proactivity in building collaborations, as this is fundamental if one wants to truly anticipate potential risks, benefits and concerns and thus be able to develop a responsible global neuroethics.

This should be considered as a global work in progress.

8.5. References

Asociación Mexicana de Neuroética (2021). Asociación Mexicana de Neuroética AC [Online]. Available at: https://www.neuroeticamexico.org [Accessed 20 May 2021].

Beckford, J. (2003). *Social Theory & Religion*. Cambridge University Press, Cambridge.

Cabrera, L. and Herrera-Ferrá, K. (2019). Neuroensanchamiento? Concepts and perspectives about neuroenhancement in the Hispanic literature. *J. Cogn. Enhanc.* [Online]. Available at: https://doi.org/10.1007/s41465-019-00131-w.

Cámara de Diputados (2020). Evolución de los Recursos Federales Aprobados paar la Ciencia y el Desarrollo 2012–2021 [Online]. Available at: https://www.cefp.gob.mx/publicaciones/nota/2020/notacefp0682020.pdf [Accessed 26 May 2021].

Cambridge English Dicitionary (2021). Globalization [Online]. Available at: https://dictionary.cambridge.org/dictionary/english/globalization [Accessed 22 January 2021].

Causadias, J.M., Telzer, E.H., Lee, R.M. (2017). Culture and biology interplay: An introduction. *Cultural Diversity & Ethnic Minority Psychology*, 23(1), 1–4 [Online]. Available at: https://doi. org/10.1037/cdp0000121.

Comisión Nacional de Bioética (2015). Guía nacional para la integración y el funcionamiento de los comités hospitalarios de bioética [Online]. Available at: http://www.conbioetica-mexico.salud.gob.mx/descargas/pdf/registrocomites/Guia_CHB_Final_Paginada_con_forros.pdf [Accessed 22 May 2021].

Feias, J.M., Telzer, E.H., Lee, R.M. (2017). Culture and biology interplay: An introduction. *Cultural Diversity & Ethnic Minority Psychology*, 23(1), 1–4. doi: 10.1037/cdp0000121.

Ferdinand, N.K., Paulus, M., Schuwerk, T., Kühn-Popp, N. (2019). *Social and Emotional Influences on Human Development: Perspectives from Developmental Neuroscience.* Frontiers Media, Lausanne.

García-López, E., Mercurio, E., Nijdam-Jones, A., Morales, L.A., Rosenfeld, B.M. (2019). Neurolaw in Latin America: Current status and challenges. *Int. J. Forensic Ment. Health*, doi: 10.1080/14999013.2018.1552634.

Giordano, J. (2015). No new neuroscience without neuroethics. BioMed central blog network [Online]. Available at: http://blogs.biomedcentral.com/on-health/2015/07/08/no-new-neuroscience-without-neuroethics [Accessed 3 January 2021].

Global Health Workforce Alliance (2002). Global Health Workforce Alliance [Online]. Available at: https://www.who.int/workforcealliance/members_partners/member_list/realizingrights/en/ [Accessed 3 January 2021].

Herrera-Ferrá, K. (2020). Global mental health and the treatment gap: A human rights and neuroethics concern. In *Global Mental Health and Neuroethics*, Stein D. and Singh, I. (eds). Elseiver, London.

Herrera-Ferrá, K. (2021). Bioculture and the global regulatory gap in neuroscience, neurotechnology, and neuroethics. In *Developments in Neuroethics and Bioethics – Regulating Neuroscience: Transnational Legal Challenges*, Hevia, M. (ed.). Academic Press, Cambridge, MA.

Herrera-Ferrá, K., Saruwatari-Zavala, G., Nicolini-Sánchez, H., Pinedo-Rivas, H. (2019). Neuroética en México: Reflexiones médicas, legales y socioculturales. *Bioethics Update* [Online]. Available at: https://doi.org/10.1016/j.bioet.2019.05.001.

Herrera-Ferrá, K., Nicolini, H., Giordano, J. (2020). Professional attitudes toward the use of neuromodulatory technologies in Mexico: Insight for neuroethical considerations of cultural diversity. *CNS Spectrums*, 1–3, doi:10.1017/S1092852920002151.

Herrman, H. and Swartz, L. (2007). Promotion of mental health in poorly resourced countries. *The Lancet* [Online]. Available at: https://doi.org/10.1016/S0140-6736(07)61244-6.

Ienca, M. and Andorno, R. (2017). Towards new human rights in the age of neuroscience and neurotechnology. *Life Sci. Soc. Policy*, 13(5) [Online]. Available at: https://doi.org/10.1186/s40504-017-0050-1.

Instituto Nacional de Ciencias Penales (2021). INACIPE [Online]. Available at: https://inacipe.gob.mx/ofertaConvPosg2021.php [Accessed 1 May 2021].

International Brain Initiative (2018a). We are the International Brain Initiative [Online]. Available at: https://www.internationalbraininitiative.org/ [Accessed 12 December 2020].

International Brain Initiative (2018b). Neuroethics Working Group [Online]. Available at: https://www.internationalbraininitiative.org/neuroethics-working-group [Accessed October 2020].

International Monetary Fund (2002). Globalization: Threat or opportunity? [Online]. Available at: https://www.imf.org/external/np/exr/ib/2000/041200to.htm [Accessed 5 February 2021].

International Neuroethics Society (2021). International Neuroethics Society [Online]. Available at: https://www.neuroethicssociety.org/ [Accessed 1 May 2021].

Intituto Nacional de Estadística y Geografía (2017). Encuesta sobre la percpeción pública de la ciencia y la tecnología (ENPECYT) [Online]. Available at: https://www.inegi.org.mx/programas/enpecyt/2017/ [Accessed 7 November 2020].

Intituto Nacional de Estadística y Geografía (2019). Estadísticas de salud en establecimientos particulares 2019 [Online]. Available at: https://www.inegi.org.mx/contenidos/programas/salud/doc/salud_2019_nota_tecnica.pdf [Accessed 3 January 2021].

Mercurio, E., García-López, E., Morales-Quintero, L.A., Llamas N.E., Marinaro, J.A., Muñoz, J.M. (2020). Adolescent brain development and progressive legal responsibility in the Latin American context. *Front. Psychol.* [Online]. Available at: https://doi.org/10.3389/fpsyg.2020.00627.

Neuroethics Network (2021). Neuroethics Network: Going global and facing the future [Online]. Available at: http://neuroethicsnetwork.com/ [Accessed 19 January 2021].

NeuroRights Initiative (2020). The five NeuroRights [Online]. Available at: https://neurorights-initiative.site.drupaldisttest.cc.columbia.edu/sites/default/files/content/The%20Five%20Ethical%20NeuroRights%20updated%20pdf_0.pdf [Accessed 19 January 2021].

OECD (2017). Neurotechnology and society: Strengthening responsible innovation in brain science. *OECD Science, Technology and Industry Policy Papers*, no. 46 [Online]. Available at: https://doi.org/10.1787/f31e10ab-en.

OECD (2020). OECD data [Online]. Available at: https://data.oecd.org/mexico.htm [Accessed 19 January 2021].

OECD (2021a). Population (indicator) [Online]. Available at: doi: 10.1787/d434f82b-en [Accessed 17 February 2021].

OECD (2021b). Gross domestic spending on R&D (indicator) [Online]. doi: 10.1787/d8b068b4-en [Accessed 6 March 2021].

Prince, M., Patel, V., Saxena, S., Maj, M., Maselko, J.P. (2007). No health without mental. *Lancet*, (370), 859–877.

Ramirez-Zamora, A., Giordano, J., Gunduz, A., Alcantara, J., Cagle, J.N., Cernera, S., Difuntorum, P., Eisinger, R.S., Gomez, J., Long, S. et al. (2020). Proceedings of the Seventh Annual Deep Brain Stimulation Think Tank: Advances in neurophysiology, adaptive DBS, virtual reality, neuroethics and technology. *Front. Hum. Neurosci.*, 14(54), doi: 10.3389/fnhum.2020.00054.

Saxena, S., Thornicroft, G., Knapp, M., Whiteford, H. (2007). Resources for mental health: Scarcity, inequity, and inefficiency. *Lancet*, 370, 878–889.

Schlicht, T., Springer, A., Volz, K.G., Vosgerau, G., Schmidt-Daffy, M., Simon, D., Zinck, A. (2009). Self as cultural construct? An argument for levels of self-representations. *Philosophical Psychology*, 22(6), 687–709 [Online]. Available at: https://doi.org/10.1080/09515080903409929.

Senado de la República (2019). Gaceta del Senado [Online]. Available at: https://www.senado.gob.mx/64/gaceta_del_senado/documento/98907 [Accessed 1 May 2021].

Società Italiana di Neuroetica (2021). SOCIETADINEUROETICA [Online]. Available at: http://societadineuroetica.it/ [Accessed 20 May 2021].

The World Bank (2020a). Fading promise: How to rekindle productivity growth [Online]. Available at: https://pubdocs.worldbank.org/en/778161574888355532/Global-Economic-Prospects-January-2020-Topical-Issue-1.pdf [Accessed 28 March 2020].

The World Bank (2020b). Poverty [Online]. Available at: https://www.worldbank.org/en/topic/poverty/overview [Accessed 28 March 2020].

The World Bank (2021). The World Bank Mexico [Online]. Available at: https://www.worldbank.org/en/country/mexico [Accessed 30 April 2021].

Turner, B.S. and Holton, R.J. (eds) (2011). *The Routledge International Handbook of Globalization Studies.* ProQuest Ebook Central [Online]. Available at: https://ebookcentral.proquest.com.

UNESCO (1998). Universal declaration on the human genome and human rights [Online]. Available at: https://en.unesco.org/themes/ethics-science-and-technology/human-genome-and-human-rights [Accessed 3 January 2021].

UNESCO (2005). Universal declaration on bioethics and human rights [Online]. Available at: https://en.unesco.org/themes/ethics-science-and-technology/bioethics-and-human-rights [Accessed 4 January 2021].

United Nations (1948). Universal Declaration of Human Rights [Online]. Available at: https://www.un.org/en/about-us/universal-declaration-of-human-rights [Accessed 19 Januray 2021].

Valdez-Martínez, E., Mata-Valderrama, G., Bedolla, M., Fajardo-Dolcib, G.E. (2015). Los comités de ética en la experiencia del IMSS: Una instancia en Latinoamérica. *Rev. Med. Inst. Mex. Seguro. Soc.*, 53(4), 490–503.

Vassilieva, J. (2016). *Narrative Psychology: Identity, Transfromation and Ethics.* Palgrave Macmillan, London.

World Economic Forum (2019). A brief history of globalization [Online]. Available at: https://www.weforum.org/agenda/2019/01/how-globalization-4-0-fits-into-the-history-of-globalization/ [Accessed 21 March 2021].

World Health Organization (2013). Traditional medicine strategy 2014–2023 [Online]. Available at: http://apps.who.int/iris/bitstream/handle/10665/92455/9789241506090_eng.pdf?sequence=1 [Accessed 3 January 2021].

Yale University (2003). YaleGlobal Online [Online]. Available at: https://yaleglobal.yale.edu/content/discussion-ethical-globalization [Accessed 3 January 2021].

9

The Dilemma of Cross-Cultural Neuroethics

Laura SPECKER SULLIVAN[1] and Karen S. ROMMELFANGER[2,3]
[1] *Philosophy Department, Fordham University, Bronx, USA*
[2] *Neuroethics and Neurotech Innovation Collaboratory, Center for Ethics Neuroethics Program, Emory University, Atlanta, USA*
[3] *Institute of Neuroethics Think and Do Tank, Atlanta, USA*

9.1. Framing

9.1.1. *Neuroscience and culture?*

Culture is a concept that can be broadly interpreted; any single definition of culture is typically unsatisfying. UNESCO defines culture as "the set of distinctive spiritual, material, intellectual and emotional features of society or a social group, and [...] it encompasses, in addition to art and literature, lifestyles, ways of living together, value systems, traditions and beliefs" (UNESCO 2001). Even more generally, Wong (2006) .describes cultures as sets of commonly recognized values, practices, and ways of making meaning in the world. Lock (2001) clarifies that cultures are not static: any culture can submit to criticism, reflection and transformation. Regardless of any particular definition, science is deeply embedded in culture. Culture impacts which scientific aims are pursued, where scientific practice occurs, how results are disseminated and shared, as well as who is regarded as a scientist. Science is just as much a product of culture as scientific discoveries influence culture.

Arguably, culture is especially relevant for brain science. The brain is special biologically, as the locus of the nervous system, but perhaps even more importantly the brain is special culturally, as the seat of mental life and conscious thought. In 1637, Descartes wrote in the *Discourse on the Method* that "I think therefore I am", identifying it as the only dictum of which we can be certain. Regardless of what Descartes meant by the statement, thinking is surely intertwined with being, as is the mind with the self (Gottlieb 2006). To learn about the science of the brain is in many ways to learn about ourselves, but no individual can be independent from their context. Thus, the brain's cultural meaning cannot be easily divorced from biological findings about the brain nor vice versa. This is one reason why there is controversy over, for example, whether biological information about the brain can be used to determine definitions of death (McMahan 2006) and whether describing addiction as a brain disease can impact criminal responsibility (Morse 2013). Apart from research, brain-derived information is being pursued in the clinical and commercial space to quantify notions of healthy minds or brains, to predict risk for diseases like Alzheimer's or schizophrenia, and even to define mental health resilience in a post-Covid world (Ahlgrim et al. 2019; Türközer and Öngür 2020).

Just as the brain sciences are intimately connected to culture, the ethical questions around developments in the brain sciences are interwoven with culture (Shook and Giordano 2014b). Culture is not a simple set of values that can be used to determine the ethical implications of the brain sciences. Critically considering culture in neuroethics requires acknowledging that for any single cultural concept such as autonomy, there will be traditionalists, skeptics, critics and progressives about its meaning and use to analyze the brain sciences (for a more lengthy description of this idea, see Chapter 7). Due to this complexity, it can be difficult to determine just how to acknowledge the significant role of culture in neuroethical analyses.

This is where cross-cultural methods become helpful: as a way to take culture into account without narrowly defining culture, essentializing any one culture or portraying an entire country or geographic region as monolithic. As we describe below, cross-cultural analyses undertake comparative projects with the aim of bringing into relief cultural differences that only emerge when alternatives are considered. Yet, as we address in the conclusion, cross-cultural neuroethics may paradoxically undermine the project of neuroethics itself, which in its mainstream formulation is arguably premised on the special role of the brain in human understanding. That is, by highlighting just how much of human meaning is determined beyond the circumscribed boundaries of the skull, in broader communal and societal patterns, a truly cross-cultural neuroethics may undermine the special role of the brain, and thus, the premise of neuroethics itself, forcing a choice between the field of neuroethics and genuinely cross-cultural work. In the end, we suggest that "the dilemma of cross-cultural neuroethics" need not condemn the project outright, but that it raises questions for the development of the field.

9.1.2. *State of the art of cross-cultural neuroethics*

As we understand it (and we recognize that other scholars may have different views from our own), "cross-cultural" properly applies to any project which foregrounds the significance of cultural comparison, whether the goal is to identify similarities against a background of differences, or to highlight differences against a background of similarities. A benefit of this definition is that we do not need to agree on an essential set of criteria for a culture; we need only be able to identify where there is a difference between two social groups. Broadly, we think that cross-cultural differences often manifest in terms of beliefs, values and practices: both what they are and how they came to be.

Current cross-cultural work is distinguished by a number of features that we briefly describe here. "Neuroethics" as a neologism and a unique field is still quite young at only 20 years old; we leave it to the chapters in Part 1 of this volume to fill in this history. As a field, neuroethics predominantly flourishes in countries with developed economies, which is not an accident: neuroethicists tend to focus on the ethical significance of developments in neuroscience, neurology and neurotechnology, all of which are pursuits requiring access to significant funding sources and an advanced research infrastructure.

Neuroethics builds on the earlier success of the field of bioethics, which now includes numerous professional societies, dedicated research centers, academic departments and government-sponsored initiatives worldwide. Bioethics itself has grappled with cross-cultural work. A flurry of work in the late 19th to early 20th century questioned whether cross-cultural bioethics research identified entrenched disagreements that threatened the possibility of a global bioethics agenda, although in recent years, interest in this topic seems to have waned (Fox and Swazey 1984; Sakamoto 1999; Campbell 1999a; Engelhardt 2006; Holm and Williams-Jones 2006b; Sullivan 2016). Regardless of its popularity, suffice to say that there is enough work on cross-cultural bioethics to properly constitute a body of literature.

By comparison, cross-cultural work that self-identifies as neuroethics is relatively sparse, including co-author Rommelfanger's Global Neuroethics Summit, co-author Specker Sullivan's work with the Science for Monks program and "environmental neuroethics" work with indigenous groups in Canada, as well as a handful of topical studies, such as on the cultural context of brain death (Yang and Miller 2015; Illes and Lou 2019). There is more cross-cultural work that does not self-identify as neuroethics scholarship, but instead addresses work at the intersection of ethics and neuroscience, such as Margaret Lock's *Twice Dead: Organ Transplants and the Reinvention of Death*, which examines brain death in Japan from a North American perspective, the business literature on decision-making or projects like the Brain Science and Kokoro project funded by the Templeton Foundation (Swanson 2011; Adam et al. 2015). There are also efforts to "translate" neuroethics by mapping onto religious traditions such as Islam and Buddhism, and to use East Asian ideas to challenge assumptions in neuroethics (Moosa 2012; Tsomo 2012b). We also see examples from the AI ethics literature on expanding ethical solutions to include often-ignored relational perspectives, such as Ubuntu and Buddhist perspectives (Mhlambi 2020; Hershock 2021).

We believe there is much room for cross-cultural neuroethics to grow, as we address in more detail in the conclusion. These underexplored areas include, for example, intra-society cross-cultural research, of which work on cochlear implants is a paradigmatic example. As some have argued, there is a proper deaf culture, so any work comparing the deaf with the hearing perspective is cross-cultural, according to our definition (Sparrow 2005). Likewise, research working across cultural lines including race, ethnicity, disability and even geographical regions can highlight where and how neuroethical views differ. There is also immense room for growth in neuroethical research that does not highlight Euro-American ethical concerns (Chattopadhyay and De Vries 2008), and while this is a conceptually difficult space given that neuroethics arose in a Euro-American context, the broad area of concern that the field of neuroethics aims to investigate – at the intersection of the brain sciences and conceptions of the human good – is almost certainly universal.

In the next sections, we identify several benefits and aims of work in cross-cultural neuroethics, and we describe the methods that scholars may choose to reach those aims. Finally, in the conclusion, we reflect on where the field may go from here.

9.2. Benefits and aims of cross-cultural neuroethics

The premise behind cross-cultural neuroethics is aligned with that of many comparative projects: undertaking intellectual analysis across cultural lines is a fruitful way to build intercultural understanding and identify areas of mutual interest, as well as to improve intracultural creativity and recognize and more deeply explore the idiosyncratic features of our own cultural context (McWeeny and Butnor 2014; Van Norden 2019). Thus, the concept of cross-cultural neuroethics includes within it particular conceptions of these aims as valuable goals, as well as methodological ideas about how best to reach them. In the next two sections, we will survey these two sets of ideas in turn: first, the benefits and aims of cross-cultural work in neuroethics, and second, the methods and processes by which individuals undertake this work.

9.2.1. Intercultural understanding

One of the primary instrumental reasons to undertake cross-cultural work is to increase intercultural understanding. A desire to fulfill this aim may even be seen as a prerequisite of the other benefits and aims. Indeed, the mission of the United States' Fulbright program, one of the pre-eminent supporters of cross-cultural work between the United States and the world, sees its paramount mission as increasing "mutual understanding" between countries. The description of the Fulbright U.S. Student Program reads[1]:

> During their grants, Fulbrighters will meet, work, live with and learn from the people of the host country, sharing daily experiences. The program facilitates cultural exchange through direct interaction on an individual basis in the classroom, field, home, and in routine tasks, allowing the grantee to gain an appreciation of others' viewpoints and beliefs, the way they do things, and the way they think. Through engagement in the community, the individual will interact with their hosts on a one-to-one basis in an atmosphere of openness, academic integrity, and intellectual freedom, thereby promoting mutual understanding.

9.2.2. Self-awareness

Just as cross-cultural work can increase understanding of cultures other than our own, cross-cultural research can also allow for greater self-awareness of the distinctive cultural elements of our own work. For instance, much early engagement in cross-cultural bioethics highlighted the special focus on autonomy in Euro-American cultural contexts, as compared with a greater attention to community or group harmony in East Asian contexts, such as Japan, Korea and China (Fan 1997; Akabayashi et al. 1999, Sakamoto, 1999). While a risk of this kind of comparison is cultural essentialism and must be carefully considered, a benefit is the

---

1 Fullbright U.S. Student Program. The Power of International Education, 2021. See: https://us.fulbrightonline.org/about/fulbright-us-student-program.

window into the potential myopia of ethical analyses that focus on autonomy, for example, to the exclusion of other values. The more that neuroethicists engage across cultural boundaries, the better they can identify which values are shared and which are more parochial.

9.2.3. *Mutual interest and cooperation*

Once intercultural understanding and self-awareness are achieved through cross-cultural work, it becomes possible to identify shared goals and interests, should they exist, to enable cooperation in areas of mutual concern. This is the premise behind international research ethics recommendations to honor a principle of positive reciprocity, whereby the researcher and the local community benefits, even after the research is complete (Sofaer 2014; Wenner 2017).

Of course, cross-cultural research need not meet this aim. Some work will achieve intercultural and self-understanding, without being able to identify any areas of agreement or shared concern. This is not a failure of cross-cultural work – it is an achievement to recognize where values differ, even though we cannot move forward on projects of mutual concern.

9.2.4. *Intracultural creativity*

A final benefit of successful cross-cultural work is creativity within a culture. Engagement with alternative belief systems, practices, and ways of life can help researchers to imagine possibilities beyond their current arrangements. Taylor and Rommelfanger (2022) describe that White Western individualistic bias (WWIB) contributed to impoverished technology such as EEGs. A product of cultural racism, EEGs developed in 1920s remain less accurate when reading from scalps of individuals with thick, curly hair as is common in populations of African ancestry, due to optimization for a predominantly White population. The solution was a relatively simple, yet elegant design created by an African American neuroscience student working in an EEG lab (Entienne et al. 2020). Sometimes it is difficult to see, from within our own cultural context, how things could be otherwise.

9.3. Potential forms of cross-cultural neuroethics

There are a number of ways to engage groups from a variety of cultures and societies in neuroethical work. These four potential forms of cross-cultural neuroethics roughly map onto the four categories of benefits and aims delineated above.

9.3.1. *Summits and international meetings*

Among professional neuroethicists, international meetings and summits are increasingly popular means of engaging in cross-cultural work. Co-author Rommelfanger has led the Global Neuroethics Summit for several years. It is a cooperative effort among nations with national and regional brain projects to identify best practices for integrating ethics with cutting edge neuroscience.

The International Neuroethics Society (INS) is another example of a way to engage diverse groups in the sharing of neuroethics research. One challenge here, which we will take up in more detail in the next section, is ensuring that genuine cross-cultural communication and understanding occur, rather than just a myriad of one-way reports.

These types of meetings may be a necessary first step in genuine cross-cultural work, because they create the personal connections that may facilitate such research later on. In addition, it may be easier to achieve intercultural understanding when scholars meet each other in person, than when they only interact through their published work.

9.3.2. *Historical, sociological and anthropological research*

Research in history, sociology and anthropology also often counts as "cross-cultural", whether there is an explicit comparison between two cultural groups or the comparison is between the researcher and the object of research. In ethnographic research, which is a key research method in anthropology, there is considerable discussion about what are described as "insider/outsider" issues – whether or not the researcher is a member of the cultural group being studied, and where both have challenges. As neuroethicists increasingly engage in cross-cultural work, it is essential to consider questions of who can genuinely speak for or represent a community's interests, as we address in more detail in the next section. Nevertheless, to our knowledge, researchers in history, sociology and anthropology do not often self-identify as neuroethicists and so more may need to be done to include these scholars in the field to foster a cross-cultural perspective.

9.3.3. *Community-based participatory research (CBPR)*

Although not commonly deployed in neuroethics research, CBPR is an increasingly popular means of cross-cultural work, especially when the researcher is from a different cultural group than those whose views are being represented. To our knowledge, CBPR is not a common form of neuroethics research (Jull et al. 2017, Musesengwa n.d.). The goal of CBPR is to implement just and equitable research practices across culture and other barriers, such as race and class, in order to balance the relative privilege of the researcher with the empowerment of the local community to identify and achieve their own goals.

An international participatory research framework (IPRF) (Pinto et al. 2012) has been proposed, which calls for familiarity with local language and culture with deep consideration of power dynamics and reflexivity for researchers' identity and cultural humility. While researchers may not necessarily identify as "cross-cultural", the principles used in IPRF align with those that would be employed by cross-cultural scholarship, although they may not go as far as a CBPR approach in empowering the local community.

9.3.4. *Cultural competency/critical theory/DEI*

Finally, cross-cultural work in neuroethics may take the form of cultural competency, the goal of which is to educate practitioners to be able to work with members of different kinds of cultural groups (Betancourt et al. 2003). Early developers of the skill set argued that cultural competency is especially important in multicultural, pluralistic contexts such as North America or Europe, although in the United States, structural competency is arguably the more relevant skill (Metzl and Hansen 2014). In recent years, cultural competency has been criticized in the American hospital and social work setting as:

> [...] a myth of [...] American "know-how". It is consistent with the belief that knowledge brings control and effectiveness, and that this is an ideal to be achieved above all else. I question the notion that one could become "competent" at the culture of another (Dean 2001, p. 624).

Diversity, equity and inclusion efforts can also be "cross-cultural" if they study how best to create welcoming research environments for people with different perspectives, including African and low-income and middle-income countries in neuroethics research (Matshabane 2021).

9.4. Challenges in cross-cultural neuroethics

There are clear benefits to doing cross-cultural work, as explained in section 9.2, and ways to do this work well, as explored in section 9.3. This is not to say that cross-cultural neuroethics does not face challenges beyond questions of how to get started. In this section, we describe some of the challenges of doing genuine cross-cultural neuroethics work. "Genuine" is a term we have used a number of times throughout this chapter, and we understand it to mean cross-cultural work that aims, at the very least, at a level of intercultural understanding that leads to questioning assumptions made from our own cultural perspective. Cross-cultural work may be ungenuine or superficial if it only advocates for diverse perspectives as a symbolic commitment, but not as a source of self-critique and institutional change (Ahmed 2012).

9.4.1. *Intracultural diversity*

The first challenge to cross-cultural neuroethics is not unique to neurothics, but is a challenge for all projects that aim to be cross-cultural: how do we define culture? If culture is taken as a proxy for nationality (Lombera and Illes 2009b), then there is profound intracultural diversity, especially in pluralistic cultures such as the United States. In section 9.1, we proposed that "cross-cultural" work can operate without a definition of culture as long as what is highlighted is the comparison itself, and not the cultural status of either group on each side of the comparison. Yet that solution only takes us so far: indeed, if the two cultures being compared are the United States and Japan, then researchers must make assumptions about what "counts" as American culture and what counts as Japanese. In other words, who speaks for the culture? We will raise this issue in more detail in section 9.4.2.

For now, the concern is whether cross-cultural work can be pursued without simplifying or essentializing cultures into stereotypical elements. We suggest that if the emphasis is on perspective-taking and reflexivity, then essentialization is less likely to occur. No one set of beliefs or ideas represent an entire culture, no one person can speak for a culture, and cultures are not defined by how they differ from a culture seen as "standard" for a discipline. Rather, cross-cultural work asks how participation in social groups and group identity shapes beliefs, values and practices, and further how these beliefs, values and practices influence neuroethical concerns. Cultures are not uniform, nor are they static, and so the challenge is to continually refresh and reconsider what is described as cultural understanding.

9.4.2. Domestic politics and power

As we begin to explain above, a major concern in cross-cultural projects is the question of cultural representation, or who speaks for the culture. For instance, while much work in Japanese studies begins from the premise that "Japanese" refers to an ethnically homogeneous group within the country of Japan, this overlooks ethnically Korean and Filipino immigrants in Japan, not to mention the indigenous Ainu population in Hokkaido or the residents of Okinawa, many of whom identify as culturally distinct from people who are ethnically Japanese. In cross-cultural work that identifies "Japan" as a point of comparison, who speaks for Japan?

One option is to say that cross-cultural projects begin by identifying stakeholders in the enterprise who identify as part of different cultural groups, and taking care to be precise about what role that person plays in their group. Individuals within cultures have different vantage points on the culture; an older person may have a very different cultural understanding than a younger one. If the approach described in 9.4.1 is taken, such that cultures are understood as dynamic and diverse, then it will not be possible for just one person to represent a culture's interests.

9.4.3. International politics and power

Yet another challenge for cross-cultural work is identifying the stakes of the endeavor. For projects that involve countries with disparate economic statuses, there may be more involved in cross-cultural work than merely the free exchange of ideas. Some countries may view cross-cultural work as an opportunity to "flex their political muscle" by urging a consensus in their favor. Others may see these projects as chances to redefine their presence on the world stage. From this perspective, small decisions, such as who leads the project and what its aims are can have immense ramifications.

For instance, the UN has routinely been criticized for its Western bias. The international challenge of using this "rights" framing becomes interpreting and enforcing laws in diverse contexts with this monolithic view of "rights". While a non-hierarchical and democratic approach to cross-cultural work is best, in practice it can be difficult to achieve as no one researcher can control their culture's broader status. Nevertheless, if participants routinely consider the question of who benefits from any cross-cultural arrangement, this may be one means of acknowledging potential power imbalances and, ideally, addressing them.

Along similar lines, the framework of "cultural competence" has been criticized for failing to address critical dimensions of power dynamics, differentials and tacit acceptance of colonialist structures, where mastery of competence benefits the already powerful (Gopalkrishnan 2019). In addition, in a competence framework, cultures are often defined as fairly static, against a norm of "Whiteness" (Pon 2009). Gopalkrishnan (2019) instead advocates for a culturally dynamic partnership that is characterized by equitable partnerships of mutual learning and in doing so "rejects notions of developing competencies to do something to the 'other' and focuses on people working with each other".

Cultural competency is also characterized as a prominent framework in the Global North making its way to the Global South yet fraught with an orientation of "othering" clients and collaborators with little attention for self-reflection of the practitioner (Gopalkrishnan 2019; Racine and Senghor 2022). This orientation also fails to acknowledge power dynamics and differentials that occur when cultures do interact.

9.4.4. *The dilemma of cross-cultural neuroethics*

Finally, the last challenge is an issue we term "the dilemma of cross-cultural neuroethics". As addressed in the introduction, one of the premises of mainstream neuroethics is that the brain sciences are unique and deserve special ethical consideration. One of the major reasons for this is thought to be that the brain is the locus of cognition, and thus the seat of identity and agency. If we modify the brain, we modify the person; if people are due respect due to their rational capacities (an idea perhaps best captured by Immanuel Kant's dictum that people should be treated as ends, and not as means), then we should be circumspect about any change to how we understand, use or develop these capacities.

Implicit (Vidal 2009) and explicit (Illes et al. 2010) in this account of the peculiarity of work into the brain is the idea that people are primarily located therein. However, this reduction of brain and identity is not universally held, as is perhaps most apparent by many cultures' resistance to brain death as a definition of death. The word *kokoro* in Japanese, for instance, describes an inextricable relationship between mind–heart–spirit (Swanson 2011). A number of religious traditions do not accept a brain death definition of death and rely on cardiac death, instead (Fins 1995). Many areas of research begin to have similar features to those of the brain death discussion the more they are studied: what at first seems to be agreement of the special status of the brain, gradually begins to look like just one cultural view among many.

This implies that implications of brain findings may not be so unique as assumed by many neuroethicists. Furthermore, this could suggest that neuroethics as a distinct field is overvalued and potentially superfluous to research at the intersection of ethics and the brain sciences, when extant disciplines such as sociology, anthropology, public health and philosophy can ably address neuroethical issues (Adam et al. 2015). Cross-cultural accounts of neuroethics, by acknowledging the roles of social groups and practices, seemingly elevate the role of these groups in determining the moral status of the individual (Sakura 2012). Thus, cross-cultural projects could potentially undermine what may be seen as a kind of reductionism in neuroethics, given its current pairing with contemporary neuroscience. This raises the question of whether neuroethics would disappear with the continued pursuit of genuinely cross-cultural work.

While we acknowledge this concern, cross-cultural work need not signal the end of contemporary neuroethics. Rather, cross-cultural work encourages neuroethics to grow to consider the cultural significance of the brain sciences, whether the brain is elevated to the status of cultural linchpin or not. As we have repeatedly emphasized, a benefit of cross-cultural work is the reflexive understanding of our own perspective and that of others that it provides. While this may call into question some neuroethical assumptions, this need not signal the end of neuroethics, as long as the field is willing to grow into a more inclusive and diverse field that acknowledges differences, even though it cannot resolve them.

9.5. Conclusion

In recognition of the global landscape of neuroscience research and the reverberating effects and implications of neuroscience findings, we advocate for a more robust cross-cultural approach to neuroethics scholarship and research. Global governments have collectively invested over $7 billion[52] (Rommelfanger et al. 2018) to specific neuroscience research efforts and industry investments for just 2020 were projected to be $11–14 billion (Cavuoto 2021). Cultural and cross-cultural considerations must be taken into account, and comparative research methods pursued, to maximize the potential of these significant investments and critically explore the ethical implications of this work. As we have argued, cross-cultural work is essential for identifying unacknowledged assumptions that shape research programs and determine their success.

While the field of neuroethics has seen rapid development alongside neuroscience research, cross-cultural studies have been lagging. We share some examples from neuroethics and adjacent fields to illustrate opportunities for cross-cultural neuroethics research. We also share pitfalls of uncritical appraisals of cross-cultural methodologies. We close by identifying a dilemma for neuroethics, such that embracing a cross-cultural approach may highlight the irrelevance of neuroethics for rich work in ethics and the brain sciences. In conclusion, we propose that a cross-cultural pursuit of neuroethics instead offers an opportunity to create a more inclusive, expansive and relevant field.

9.6. References

Adam, H., Obodaru, O., Galinsky, A.D. (2015). Who you are is where you are: Antecedents and consequences of locating the self in the brain or the heart. *Organizational Behavior and Human Decision Processes*, 128, 74–83.

Ahlgrim, N.S., Garza, K., Hoffman, C., Rommelfanger, K.S. (2019). Prodromes and preclinical detection of brain diseases: Surveying the ethical landscape of predicting brain health. *eNeuro*, 6(4).

Ahmed, S. (2012). *On Being Included Racism and Diversity in Institutional Life*. Duke University Press, Durham, NC.

Akabayashi, A., Fetters, M.D., Elwyn, T.S. (1999). Family consent, communication, and advance directives for cancer disclosure: A Japanese case and discussion. *Journal of Medical Ethics*, 25(4), 296–301.

Betancourt, J.R., Green, A.R., Carrillo, J.E., Ananeh-Firempong, O. (2003). Defining cultural competence: A practical framework for addressing racial/ethnic disparities in health and health care. *Public Health Rep.*, 118(4), 293–302.

Campbell, A.V. (1999). Presidential address: Global bioethics – dream or nightmare? *Bioethics*, 13(3/4), 1.

Cavuoto, J. (2021). The market for neurotechnology: 2020–2024. *Neurotech Reports* [Online]. Available at: https://www.neurotechreports.com/pages/execsum.html.

Chattopadhyay, S., and De Vries, R. (2008). Bioethical concerns are global, bioethics is Western. *Eubios J. Asian Int. Bioeth.*, 18(4), 106–109.

Dean, R.G. (2001). The myth of cross-cultural competence. *Fam. Soc.*, 82(6), 623–630.

Engelhardt, T.H.J. (2006). Global bioethics: An introduction to the collaspe of consensus. In *Global Bioethics: An Introduction to the Collaspe of Consensus*, Engelhardt, T.H.J. (ed.). M&M Scrivener Press, Salem, MA.

Etienne, A., Laroia, T., Weigle, H., Afelin, A., Kelly, S.K., Krishnan, A., Grover, P. (2020). Novel electrodes for reliable EEG recordings on coarse and curly hair. *Annual International Conference of the IEEE Engineering in Medicine and Biology Society. IEEE Engineering in Medicine and Biology Society. Annual International Conference*, 6151–6154. doi: 10.1109/EMBC44109.2020.9176067.

Fan, R. (1997). Self-determination vs. family-determination: Two incommensurable principles of autonomy: A report from East Asia. *Bioethics*, 11(3–4), 309–322.

Fins, J.J. (1995). Across the divide: Religious objections to brain death. *Journal of Religion and Health*, 34(1), 33–9. doi: 10.1007/BF02248636.

Fox, R.C. and Swazey, J.P. (1984). Medical morality is not bioethics–medical ethics in China and the United States. *Perspect. Biol. Med.*, 27(3), 24.

Gopalkrishnan, N. (2019). Cultural competence and beyond: Working across cultures in culturally dynamic partnerships. *Int. J. Commun. Soc. Dev.*, 1(1), 28–41.

Gottlieb, A. (2006). Think again. *The New Yorker* [Online]. Available at: https://www.newyorker.com/magazine/2006/11/20/think-again-2.

Hershock, P.D. (2021). *Buddhism and Intelligent Technology: Toward a More Humane Future*. Bloomsbury Academic, London.

Holm, S. and Williams-Jones, B. (2006). Global bioethics – myth or reality? *BMC Med. Ethics*, 7(1), 10.

Illes, J. and Lou, H. (2019). A cross-cultural neuroethics view on the language of disability. *AJOB Neurosci.*, 10(2), 75–84.

Illes, J., Moser, M.A., McCormick, J.B., Racine, E., Blakeslee, S., Caplan, A., Check Hayden, E., Ingram, J., Lohwater, T., McKnight, P. et al. (2010). Neurotalk: Improving the communication of neuroscience research. *Nat. Rev. Neurosci.*, 11(1), 61–69.

Jull, J., Giles, A., Graham, I.D. (2017). Community-based participatory research and integrated knowledge translation: Advancing the co-creation of knowledge. *Implementation Sci.*, 12(1), 150.

Lock, M.M. (2001). *Twice Dead: Organ Transplants and the Reinvention of Death*. University of California Press, Berkeley, CA.

Lombera, S. and Illes, J. (2009). The international dimensions of neuroethics. *Dev. World Bioeth.*, 9(2), 57–64.

Matshabane, O.P. (2021). Promoting diversity and inclusion in neuroscience and neuroethics. *EBioMedicine*, 67, 103359.

McMahan, J. (2006). Alternative to brain death. *J. Law Med. Ethics*, 34(1), 44–48.

McWeeny, J. and Butnor, A. (2014). *Asian and Feminist Philosophies in Dialogue Liberating Traditions*. Columbia University Press, New York.

Metzl, J.M. and Hansen, H. (2014). Structural competency: Theorizing a new medical engagement with stigma and inequality. *Social Sci. Med.*, 103, 126–133.

Mhlambi, S. (2020). From rationality to relationality: Ubuntu as an ethical and human rights framework for artificial intelligence governance. Paper, Carr Center Discussion Paper Series.

Moosa, E. (2012). Translating neuroethics: Reflections from Muslim ethics. *Science and Engineering Ethics*, 18(3), 519–528.

Morse, S.J. (2013). A good enough reason: Addiction, agency and criminal responsibility. *Inquiry*, 56(5), 490–518.

Musesengwa, R. (n.d.). Rosemary Musesengwa [Online]. Available at: https://www.psych.ox.ac.uk/team/rosemary-musesengwa.

Pinto, R.M., da Silva, S.B., Penido, C., Spector, A.Y. (2012). International participatory research framework: Triangulating procedures to build health research capacity in Brazil. *Health Promot. Int.*, 27(4), 435–444.

Pon, G. (2009). Cultural competency as new racism: An ontology of forgetting. *J. Progressive Hum. Services*, 20(1), 59–71.

Racine, E. and Sample, M. (2018). Two problematic foundations of neuroethics and pragmatist reconstructions. *Camb. Q. Healthc. Ethics*, 27(4), 566–577.

Rommelfanger, K.S., Jeong, S-J., Ema, A., Fukushi, T., Kasai, K., Ramos, K.M., Salles, A., Singh, I., Global Neuroethics Summit Delegates (2018). Neuroethics questions to guide ethical research in the International Brain Initiatives. *Neuron*, 100(1), 19-36. doi: 10.1016/j.neuron.2018.09.021.

Sakamoto, H. (1999). Towards a new "Global Bioethics". *Bioethics*, 13(3–4), 191–197. doi: 10.1111/1467-8519.00146.

Sakura, O. (2012). A view from the Far East: Neuroethics in Japan, Taiwan, and South Korea. *East Asian Science, Technology Society: Int. J.*, 6(3), 297–301.

SharpBrains (2020). Market report on pervasive neurotechnology: A groundbreaking analysis of 10,000+ patent filings transforming medicine, health, entertainment and business [Online]. Available at: https://sharpbrains.com/pervasive-neurotechnology/.

Shook, J.R. and Giordano, J. (2014). A principled and cosmopolitan neuroethics: Considerations for international relevance. *Philosophy, Ethics, and Humanities in Medicine*, 9(1), 1.

Sofaer, N. (2014). Reciprocity-based reasons for benefiting research participants: Most fail, the most plausible is problematic. *Bioethics*, 28(9), 456–471.

Sparrow, R. (2005). Defending deaf culture: The case of cochlear implants. *J. Political Philos.*, 13(2), 135–152.

Sullivan, LS. (2016). Uncovering metaethical assumptions in bioethical discourse across cultures. *Kennedy Inst. Ethics J.*, 26(1), 47–78.

Swanson, P.L. (ed.) (2011). *Brain Science and Kokoro: Asian Perspectives on Science and Religion*. Nanzan Institute for Religion & Culture, Nagoya.

Taylor, L. and Rommelfanger, K.S. (2022). Mitigating white Western individualistic bias and creating more inclusive neuroscience. *Nature Reviews. Neuroscience*, 23(7), 389–390. doi: 10.1038/s41583-022-00602-8.

Tsomo, K.L. (2012). Compassion, ethics, and neuroscience: Neuroethics through Buddhist eyes. *Science and Engineering Ethics*, 18(3), 529–537.

Türközer, H.B. and Öngür, D. (2020). A projection for psychiatry in the post-COVID-19 era: Potential trends, challenges, and directions. *Mol. Psychiatry*, 25(10), 2214–2219.

UNESCO (2001). UNESCO universal declaration on cultural diversity. UNESCO General Conference, Paris.

Van Norden, B.W. (2019). *Taking Back Philosophy*. Columbia University Press, New York.

Vidal, F. (2009). Brainhood, anthropological figure of modernity. *History of the Human Sciences*, 22(1), 5–36.

Volkow, N.D. and McLellan, A.T. (2016). Opioid abuse in chronic pain – Misconceptions and mitigation strategies. *N. Engl. J. Med.* 374(13), 1253–1263.

Wenner, D.M. (2017). The social value of knowledge and the responsiveness requirement for international research. *Bioethics*, 31(2), 97–104.

Wong, D.B. (2006). *Natural Moralities: A Defense of Pluralistic Relativism*. Oxford University Press, New York.

Yang, Q. and Miller, G. (2015). East–west differences in perception of brain death. *J. Bioethic. Inquiry*, 12(2), 211–225.

10

Neuroethics in Religion and Science: Hume's Law and Bodily Value

Denis LARRIVEE[1,2]

[1] *Mind and Brain Institute, University of Navarra Medical School, Pamplona, Spain*
[2] *Arts and Sciences Department, Loyola University Chicago, USA*

10.1. Introduction

Ethics is a pragmatic science. It asks questions such as "what should I do?" and "how do I know what I should do?" or even "why ought I do what I should do?". It is apparent from such questions that they are dependent on the context in which they are posed and that addressing them is therefore shaped by information content which affects this contextual relationship. Because these are normative questions, information reflecting on value appraisal can be expected to mold the ethical praxis these questions are intended to address.

In bioethics, and in the more recently emerging field of neuroethics, this information content falls outside the domain of external behaviors and is often understood as entailing questions about the physical reality of or the physical effects mediated on the body. Accordingly, how we appraise value for addressing neuroethical questions is closely related to the body's physical and material aspects; that is, the information content about what ought to be done originates in a physical dimension that shapes value content itself. Hence, the type of information needed for value appraisal originates from an appeal to *a physical dimension that is constitutive of the specific locus of value that comprises the human being*.

For this appeal, it is manifestly evident that the body is a contingent reality; this is, existentially, that its form and operation depend on physical aspects of reality independent of itself – often termed "meta" features – but which nonetheless underlie its emergence. Understanding these "metaphysical" aspects of the physical world, therefore, is fundamental to ascertaining bodily value. As a philosophical domain, the development of this understanding has involved a lengthy intellectual investment by science and engendered an even longer philosophical heritage in religion. The respective conceptions of contingency on these features that have developed in both have shaped the value content that accrues to the body, affecting the ethical theory and praxis of each domain.

Neuroethics, for example, has been defined by the International Neuroethics Society as "... a field that studies the implications of neuroscience for human self-understanding, ethics, and policy". By this definition, neuroscience lays claim to an epistemological content that guides neuroethical theory and praxis. Neuroethics considers, for example, the brain-related dimensions of normative work at the intersection of neuroscience or the medical ethics of clinical therapy. In laying claim to the epistemological content of neuroscience, neuroethics thus conceives of value shaping through neuroscience's specific capacity to address physical aspects of the neural complexity of the brain and nervous system. Because the interpretation of these findings is influenced by the manner in which metaphysical reality is conceived, what is also implicit in recourse to such findings is the effect on the value of the *manner of understanding* this more general aspect of the physical world.

In neuroscience's epistemological claim, there is, for instance, a latent presupposition that neural operation is contingent on a mechanistic understanding (Machamer et al. 2003; Craver and Taberry 2015; Kalkman 2015), which identifies efficient causal relations as primary aspects of nature, which has shaped the conception of what is valued in human cognition. In adopting this presupposition – its general application typically traced to Descartes – neuroscience premises its understanding on an accumulation of facts assembled according to the empirical, hypothetico-deductive method, which is designed to identify causal relationships that "explain" features of natural reality from an extrinsic causal perspective (Bunge 1959), and which, therefore, serve to validate a mechanist contingency. Scientific facts are thus taken to be truth "qualifiers", information that makes a proposition true. Truth statements in science have thereby been delimited by what is "constructed" rather than by that "which is", a revised understanding of the Aristotelian correspondence notion of truth[1], a philosophical revision that can be

---

1. "Veritas et adequatio rei et intellectus" loosely translated as "Truth resides in the "adequation" of the mind to the thing known" (Aquinas).

traced to Giambattista Vico (Costelloe 2021). Because "facta" in the Vicoian interpretation are conceived as elements of a whole that can be assembled by the mind, Hume regarded them as independent of a specific value content; hence, for cognition and brain operation neuroscientific "facts" corresponding to elements or processes in the brain lack value and it is therefore necessary to confer it in normative determinations, with accounting made for how the neuroscientific facts are used to interpret cognitive operation. Thus, neuroscientific facts isolate value from the body, brain and processes of cognition and must be conferred from outside the body by an extrinsic construal. The adoption of the hypothetico-deductive method for interrogating the physical contingency of the body and cognition has thus served as an epistemological verification of the independence of value from cognitive neuroscience, which underwrites Hume's fallacy principle, also known as the naturalistic fallacy or Hume's Law.

However, the a-normative posture of Hume's Law, which posits that values and facts are irreconcilable, poses a specific quandary for how value and physical reality are related, a relationship essential for bioethics. Because value is conferred extrinsically, its conferral is necessarily conditional, being either relative – here modified by "truth qualifiers" – or arbitrary. Indeed, the need to confer value is itself indicative of an absence of intrinsic worth. Hence, there is a question of whether value can, in fact, be imputed to the material world, specifically to the human body and brain.

The tension introduced by this question – between the sense of an inherent value attributable to the physical reality of the body and the denial of this presence that is exemplified by Hume's Law – has been the subject of considerable conjecture by those concerned with the implications of its absence in nature, the latter characterized by science's understanding of how the body is contingent on meta-features of physical reality. Hans Jonas, for example, identifies in the lack of value an estrangement between man and the natural world[2] (Jonas 1966).

---

2. "There is no overlooking one cardinal difference between the gnostic and the existentialist dualism: Gnostic man is thrown into an antagonistic, anti-divine, and therefore anti-human nature, modern man into an indifferent one. Only the latter case represents the absolute vacuum, the really bottomless pit. In the gnostic conception the hostile, the demonic, is still anthropomorphic, familiar even in its foreignness, and the contrast itself gives direction to existence … Not even this antagonistic quality is granted to the indifferent nature of modern science, and from that nature no direction at all can be elicited. This makes modern nihilism infinitely more radical and more desperate than gnostic nihilism could ever be for all its panic terror of the world."

In Jonas' view, therefore, what is precisely introduced by science's value neutrality is a division between nature, conceived as that which comprises the essence of physical reality, and the individual person, *who* is composed from this reality – which at its bottom reflects a carefully redacted meta-physical conception of contingency, a circumscription that has given rise to existentialist thinking and its nihilistic implications.

The notion of a contingency that separates the individual, who is constituted by his body, from his worth has been of specific interest to Christian religious thinking. Karol Wojtyla (Pope John Paul II), for example, has expressed sentiments similar to those of Jonas and developed a strikingly similar line of thought[3] (Waldstein 2006).

By grounding value in the physical properties of the human body, the Christian religion in the words of the Pope thus introduces a challenge to the widely accepted Hume fallacy principle, which separates normative praxis from an epistemological approach described by facts and their associated mechanist interpretation. Hence, the Christian religion presupposes an understanding of physical contingency that is distinguished from that of neuroscience.

Nonetheless, this distinctive understanding is not predicated on a dismissal of the epistemological content claimed by neuroethics. For example, both share a common recognition of the validity of the methodological approach pursued by neuroscience in their shared understanding of the causal associations endorsed by science that underpin numerous observable phenomena. However, it is also apparent that the explanatory utility of these meta-features alone – the causal modes constitutive of (neuro)scientific accounts that emerge from hypothetico-deductive, experimental reasoning – is insufficient to account for all physical phenomena.

This insufficiency has led, for example, to modern attempts in philosophy of science to fill lacunae in causality accounts with so called deductive nomological (DN) explanations, which have sought to "cover" non-efficient, physical occurrences by addressing the question of why phenomena occur (Hempel and Oppenheim 1948). The DN model holds to a view of scientific explanation whose

---

3. The human family is facing the challenge of a new Manichaeism, in which body and spirit are put in radical opposition; the body does not receive life from the spirit, and the spirit does not give life to the body. Man thus ceases to live as a person and a subject. Regardless of all intentions and declarations to the contrary, he becomes merely an object. Religion, particularly the Christian religion, opposes this radical separation of the material world and the value-laden world of the spirit.

conditions of adequacy are derivability, law-likeness and empirical content, and has together with Hempel's inductive statistical model constituted scientific explanation's covering law model. Braillard (2010), for example, cites the Design Principle as an a priori, formal state required for the extrinsic causal interactions occurring within the bacterial flagellar motor.

As described by Hempel and Oppenheim (1948), these explanatory accounts reintroduce analogous Greek and Scholastic metaphysical causal accounts of form and teleology but comprise a lower level of abstraction due to their attempt to fulfill the adequacy conditions of the covering law. Broadly, these seek to account for physical phenomena precisely through the physical contingency of natural phenomena on meta-features of physical reality often regarded as intrinsic or relational and that address the 'why' as to the contingent relation, as, for example, formal and teleological causality.

The need to provide explanatory sufficiency for the occurrence of physical events is therefore indicative of a contingency relation between the events and an explanatory premise that is resident in natural reality but not accounted for by asymmetric and temporally successive, extrinsic interactions. That is, the insufficiency of the mechanistic account springs from an absence of grounding metaphysical features or regularities that undergird the physical world, as for example, a principle of individuation and/or relations between entities, which are incorporated within neo-Aristotelian metaphysical descriptions. *Hence, the need to posit additional explanatory accounts implicates additional meta-features on which the physical world is contingent.*

The understanding that has evolved in Western religion situates physical contingency within this broader framework. In contrast to the Vicoian conception of truth, which coincides with "facta" accumulated by the scientific method, physical contingency is here understood to be existentially grounded[4], where truth is coincident with that "which is" – here approximated in teleologically oriented, operational holisms – an encapsulation that extends the Aristotelian and Scholastic correspondence notion of truth to reality's essence. In the physical order, this framework specifically embraces an asymmetric causal dependency that originates *within* the personal subject, who occupies the apex of an ontological hierarchy. Contingent to meta-features of reality, the personal subject comprises an explanatory locus for teleological orientation and value origin.

Conceived as a holistic, self-directed, causal origin, the personal subject recapitulates the radical contingency of physical reality on a personal causal origin

---

4. "Now everything in so far as it has being, ... so is the true." (Aquinas).

constituted by order, rationality and value, which is the ground of existence itself; that is, a self-contained subject, "whose relations penetrate the world of physical reality" (Pieper 1989), identified as God. By contrast, the mechanist understanding that has evolved in neuroscience excludes a physical contingency grounded on a metaphysics of agency that is associated with a personal causal origin. Hence, by extension, its endorsement of the Vicoian factual truth notion similarly excludes a resident value in the personal subject; that is, it excludes a value intrinsicity that is constitutive of the subject who is identified with the reality of his body.

The value neutrality characteristic of scientific understanding, indeed, the preclusion of bodily value, has notably propelled some scientific efforts to identify supplementary metaphysical features capable of anchoring the body's value. This pursuit has chiefly emphasized a retrieval of the Kantian notion of autonomy, with its focus on a metaphysical causal origin sited within the subject. The intrinsicity of autonomy, for instance, has served in bioethics to anchor the rights of the subject; that is, the capacity for autonomy is currently regarded as a value content capable of grounding rights entitlement (Rothaar 2010). The emergence of the autonomous subject from contingent metaphysical reality thus marks a common conceptual pathway for religion and science in a theoretical evolution of bodily value.

In the Christian understanding, nonetheless, the metaphysical basis of bodily value entails not only autonomy, a position characterizing the Kantian notion, but also a dyadic relational exchange meted at the level of the personal subject. That is, it views the subject as possessing an existential relational ground as a sustaining subject in a field of reference (Pieper 1989). The value of the personal subject in the Christian view is thus here understood to entail more than the capacity for autonomy, which grounds rights entitlement, and to also emerge from a personal relational exchange with God that includes the body. Hence, this latter Christian notion also structures its normative understanding, governing ethical praxis concerned with the body. Its praxis, accordingly, is anchored in the recently termed, personalistic norm (Wojtyla 1991).

This chapter will discuss aspects of a Christian understanding of bodily value originating in the physical contingency of the *autonomy* of the personal subject so as to present common points from metaphysical thinking that have emerged more recently within science. These aspects are taken primarily from Wojtyla's conception of the personal subject's physical embodiment. Its presentation, accordingly, will not address the contingency of bodily value in a relational context, which emerges from a metaphysics of relations.

10.2. Contingency, autonomy and bodily value

"I give thanks that I am fearfully, wonderfully made" (Psalm 139)

The exultant sentiment found in the 139th psalm is a direct acknowledgement not only of a radical personal contingency upon God, but also of a latent sense of personal worth associated with the physical reality of the individual. The admiration for the individual who is identified with his body[5] is, with rare exceptions, a sentiment common with all people.

10.2.1. *Identifying in human freedom bodily value: Newtonian regularity and quantum chaos*

Accordingly, it is perhaps not surprising that in a human reaction to science's value neutrality, individual scientists have attempted to construct a conceptual route for the emergence of bodily value that traces an alternative metaphysical trajectory through physical contingency. In his Arthur Holly Compton lecture given at Washington University in 1965, Karl Popper, for instance – like Compton before him – sought to establish a basis for bodily value in the physical capacity for responsibility and the freedom to make moral choices (Popper 1978). Entitled *Of Clouds and Clocks: An Approach to the Problem of Rationality and the Freedom of Man*, the talk was intended in part to reflect on how Compton's physical discoveries had bridged wave particle duality and quantum mechanics with Newtonian determinism. Compton, as Popper pointed out, had notably refuted the short-lived quantum theory of Bohr, Kramers and Slater, paving the way for the "new quantum theory" of Born, Heisenberg, Schrodinger and Dirac that had ushered in new thinking on the role of indeterminism in structuring the physical world. The talk was thus also meant to illustrate Compton's thinking on the metaphysical concepts of determinism and chance that arose out of his discoveries and how this might underpin the materiality of cognition and the immateriality of rationality and freedom. Delivered some 30 years after Compton's similar investigation of the topic, the talk's thesis thereby *raised the question of how a physically contingent being could acquire a capacity for free choice given* these *dual metaphysical* aspects, determinism and chaos, that then comprised science's understanding of the metaphysics of nature.

---

5. The identification of the person with his body has been a source of reflection in science, in philosophy and in theology, with notable theories having been advanced in each domain: embodied cognition in science (Thelen et al. 2001), in the thought of Merleau Ponti in philosophy (Ponti-Merleau 1964) and in the thought of Karol Wojtyla (Pope John Paul II) in theology (Waldstein 2006).

In a deferral to the 1935 publication of Compton's Terry Lectures, *The Human Meaning of Science* (Compton 1935), Popper opened with Compton's wording of the well-known causal closure argument[6]. The point of presenting this argument was its manifestation of the striking discrepancy between the universal human sense of freedom and that of a physical contingency circumscribed by extrinsic causal regularities, an allusion Popper characterized as "the nightmare of the physical determinist" in a Newtonian world. This inference was premised on the notion of a "clockwork" mechanism that, above all, was completely self-contained; that is, in the perfectly deterministic physical world, there was no room for outside intervention since everything that happened therein was physically predetermined. Thoughts, feelings and efforts could therefore have no practical influence upon external happenings. If not illusory, these were at best superfluous "epiphenomena". Naturally, Compton welcomed the new theory of quantum mechanics that rescued him from all of this[7].

Popper's investigation, like Compton's, thus began with the metaphysical premise of a cloudlike, probabilistic chaos, which had been revealed by the recently discovered quantum mechanical aspects of nature (given the meaningless of a purely deterministic order, this apparently left only a consideration of a contingency on a chance like disorder). Should chaos be the predominant feature of physical reality, then sheer chance would be expected to play a major role in the physical world. However, this would also require the explanation of physical regularities to be shifted from the deterministic mechanics of Newton to an indeterminate chaos; that is, the removal of "physical necessity" must result in the same outcome with chance as with determinism. Yet, by statistical standards, this was highly improbable; indeed, Popper concluded that such a conclusion would be absurd.

As noted by Popper, the implausibility of this solution therefore imposed on Compton another conclusion, which he drew from personal experience based on professional commitments and the propositional beliefs shared by others who were aware of those commitments[8]. Based on this experiential reasoning, Compton

---

6. The fundamental question of morality, a vital problem in religion, and a subject of active investigation in science: Is man a free agent? If [...] the atoms of our bodies follow physical laws...? What difference can it make how great the effort if our actions are already predetermined by mechanical laws...? (Compton 1935).

7. In my own thinking on this vital subject, I am thus in a much more satisfied state of mind (Compton 1935).

8. It was some time ago when I wrote to the secretary of Yale University agreeing to give a lecture on November 10 at 5 p.m. He had such faith in me that it was announced publicly that I should be there, and the audience had such confidence in his word that they came to the hall at the specified time. But consider the great physical improbability that their confidence was

concluded not only that mere physical indeterminism was not enough but also that motivations supplied by aims, purposes, rules or agreements enlisted a capacity capable of ensuring adherence to these aims. Hence, his conclusion led to the following novel proposal that (a) the metaphysical features of clock-like determinism and cloud-like indeterminism only partially reflected meta-properties imbued within the physical world, which must also include that of a "self-directed" causal origin and (b) the metaphysical character of the act of freely choosing was inherently meaningful for the human being.

10.2.2. *Biological autonomy and physical contingency – the organism*

10.2.2.1. *Autonomy and contingency in living systems*

Compton's conclusion is notably consonant with recent thinking on a more widely dispersed metapresence constitutive to living systems. This meta-feature is taken up as a causal source which, among other effects, contributes to evolutionary thrust (Odling-Smee et al. 2003; Moreno and Mossio 2015). Its biological role is that of an innate organismal ability for "structuring an organism's circumstances of survival", which, unlike purely physical systems, sustains the operational integrity of the organism (Ruiz-Mirazo and Moreno 2012).

Although confined to the subdomain of living systems within the broader physical world, its physical contingency is nonetheless manifested in its recapitulation at multiple performance scales; that is, as a thematically consistent application regardless of the circumstances of its expression. This consistency is independent of whether an organism is single celled or multicellular. In both cases, the organism is pitted between the need to sustain its internal operation and the circumstances under which it can autonomously conduct its existential program. Indeed, such pitting has the incipient effect of amplifying molding pressures that operate on organisms and that effect within them greater autonomy over interior and exterior domains (Hooker 2008) in existential pursuits.

In ascribing autonomy only to living systems, not only is there the claim for an intrinsic capacity for action, but also for the presence of a source that functions to

---

justified. In the meanwhile, my work called me to the Rocky Mountains and across the ocean to sunny Italy. A phototropic organism [such as I happen to be, would not easily]… tear himself away from there to go to chilly New Haven… Considered as a physical event, the probability of meeting my engagement would have been fantastically small. Why then was the audience's belief justified? They knew my purpose, and it was my purpose [which] determined that I should be there. "The act of choice [adds] a factor not supplied by the physical conditions […] determining what will occur".

engage this capacity. Conceptualized in terms like self-agency or self-governance, this source is identified with the whole organism and so is a predicable property of the whole (Ulanowicz 1986; Fong 1996; Smithers 1997; Ruiz-Mirazo and Moreno 2012). In humans, the capacity for autonomy is thus linked to the individual through the identification of his physical body as the source of action initiation.

Supporting the identification of this source with the whole individual, impairments in global brain activity have become evident in cognitive diseases like schizophrenia (Jeannerod 2009), which affect self-recognition. The inability of first rank schizophrenia patients to monitor their own actions, for instance, led Frith to identify schizophrenia as a disease of agency and to propose that its etiological basis lay in a defective central monitoring system that normally functioned to attribute self-generated events, motor actions or even thoughts, to the individual (Frith, Blakemore and Wolpert 2000). Dysfunctional monitoring, according to the Frith model, resulted in an inability to correctly attribute the origin of self-made actions to oneself, which was revealed in clinical symptoms of psychosis. Among these symptoms included, for example, acoustic-verbal hallucinations, thought insertion or withdrawal, or delusions of alien control. These are the so-called "first rank symptoms" according to Schneider (1955), which refer to feelings or experiences of losing control of oneself and/or being controlled or influenced by other agents.

The clarification of the causal aspects of this model was later resolved in several elegant experiments by Jeannerod (2009) and colleagues. In the experiment of Fourneret et al. (2002), subjects with schizophrenia were able to automatically negociate a discrepancy between visual and kinesthetic cues when the conflict between the two sources of afferent input was modest and the subjects were unaware of the discrepancy. This indicated that the comparator function was intact in these subjects and, therefore, the impairment was not due to a sensorimotor failure. However, when the conflict became sufficiently large and experimental subjects became consciously aware of the disparity, those with schizophrenia were significantly impaired in adjusting to the discrepancy. In separate tests, the schizophrenia subjects were unable to recognize self-initiated actions as their own in a variety of paradigms (Jeannerod 2009). Taken together, these results suggested that the conscious affirmation of the self as the source of executed actions was an essential step in the execution of such actions, or else they were not done. The apparent causal requirement for affirmation in humans is thus revelatory of a metaphysical causal order grounding the autonomy of humans (and to limited extents other living organisms).

10.3. Autonomy as a constituent ground of nature: a metaphysical composition

Consonant with these biological observations, religion likewise ascribes to the autonomous subject a distinct physical substratum; that is, as source and capacity free will is constitutive to the human being who is a body. Indeed, this identification comprises an important element of the philosophical heritage used by Western religion to explicate that which exists and, correlatively, that which has value. Among the key concepts identified, incorporated and/or developed in religious philosophical thinking which expand on this understanding are those of individuation, unity and substance, and, on its characterization, the notion of an ontological structure that is teleologically, hence, purposefully, oriented. These concepts and their varied understandings help in conceiving how contingent metaphysical features may be taken up within a physical locus; that is, their conceptual content explicates how autonomy is physically grounded in these fundamental features of reality.

10.3.1. *Substances and bodily unification*

Substance and its modern concepts, entity or holism, for example, connote that the natural world is composed of "things" that are distinct from all other realities; that is, the physical world is populated by entities that are constitutively individuated. Such entities are independently characterized by two aspects. They, first, "exist in themselves", that is, they are not predicated of any other thing and so, second, their attributes qualify wholes that are independent from all other things. The contemporary Thomist philosopher Pieper adds that because entities are not predicated by any other thing they are therefore able to relate as wholes to every other thing (Pieper 1989). This means that in the real world, entities relate to other things as wholes that are qualified by their holistic properties. Modern physics concurs in these grounding features. Esfeld, for example, points out that in modern mainstream, metaphysical thought, the physical world consists of independent things embedded in space-time (Esfeld 2004). As in the religious conception, these things are independent because they are spatio-temporally unique and are entities because they are (a) each the subject of the predication of properties and (b) qualitatively distinct from all other entities.

The substance notion, however, also entails that "things" are not merely physically distinct, as in a physical separation based on space-time points alone, but that their independence is due to an internal process of active unification, which is coextensive with the whole. That is, in the substance notion, entities are understood to be individuated because they are actively unified. Hence, individuation reveals, especially, that unity is constitutive, not solely for property predication, but insofar as what things "are" and the basis for their persistence; accordingly, only because things are unified can we speak of a source of distinguishing properties.

While Christian philosophy does not specify a mediating process for unification – a domain left to science – it identifies the presence of a principle that is the basis for the structuring of unity; that is, it posits that for unity to exist there must be an actuating principle that is both the source of its mediation and the form of its adoption. Hence, Christian philosophy also imputes to this principle a dynamic promotion of unity. In line with this notion, meta understandings of the physical world notably feature an a priori operational dynamic (Gilson 1952; Phelan 1967) that is a principle of unification, recapitulating Aquinas' notion of unity in operation through action[9].

This dynamic dimension that undergirds the unity of entities, in fact, pervades natural reality, imbuing it with an effusive quality (Gilson 1952). Understood in this dynamic sense, "things" are inherently active; hence, their unity presupposes an operation that is ordered within themselves and with respect to themselves. Indeed, evidenced as a normative standard, what is "good" or "bad" may be evaluated according to its contribution towards the persistence of a "thing's" operation (Nicholson 2014) through its adherence to this principle, a point originally made by Aquinas[10].

Much neuroscientific evidence, in fact, now demonstrates the active promotion of an operational unity in living systems through the selection of actions from among a range of possibilities available to an organism; that is, by their autonomous selection of executable actions, living systems (Fourneret et al. 2002; Farrar et al. 2004; Jeannerod 2009) reveal a self-organizing principle that functions as a dynamic locus of action origin. This principle is widely evident, for example, in a panorama of processes that neuroscience has characterized, which generate the unified subject of action. As for example: in the referencing of discrete motions to the whole body (Mimica 2018); in a global activity state regulating motor signal delivery (Kato et al. 2015); in mechanisms of total body integration involving action planning and execution (Mimica et al. 2018); in the somatic and multisensory integration of the individual as the subject of experience (Noel et al. 2018) and in the designation of the body as an agent of goal directed actions (Orban and Caruana 2008; Rizzolatti and Febri-Destro 2008; Jeannerod 2009; Brinkers 2015).

---

9. "Every thing exists for the sake of its operation" (Aquinas).
10. "Operation is the ultimate perfection of each thing" (Aquinas).

10.3.2. *Ontological subsistence: the personal subject as a causal origin*

Accordingly, this principle is determinative for ontological distinction; that is, because it mediates the unity of organismal operation, it anchors the predication of distinctive properties. On this basis, Wojtyla claims that the presence of a unique ontological principle is operative in each personal subject[11]. Hence, in Wojtyla's reading, the human ontological structure is resident in a physically circumscribed entity ("existent"), from which human properties predicate. This ontological structure is uniquely qualified not only by its capacity for self-awareness, but also by the recognition of its distinctiveness from all other entities who are similarly "subsistent".

The personal subject is therefore able to affirm the uniqueness of his "substance" from that of all other human beings. In agreement with Aquinas' notion that "every 'thing' exists for the sake of its operation" and the neuroscience of motor activity, Wojtyla concludes that the unity of the subject is the active outcome of his intentions; that is, his ontological structure is both structured and manifested through autonomous activity[11]. Aquinas identifies in the autonomy of the personal subject the perfecting of a distinctive metaphysical feature that is resident in reality and on which the subject is contingent (Clark 1993); that is, agency.

10.3.3. *Newtonian determinism and organismal autonomy: from an extrinsic to an intrinsic causal order*

Accordingly, Aquinas identifies in agency a fundamental distinction in the causal order between deterministic actions and autonomous activity, as Compton and Popper subsequently noted. This distinction is evident, for example, in the difference between an action done on oneself, an extrinsic cause and an action done by oneself, an intrinsic one. The former stipulates only that a causal relationship between a cause and its effect be asymmetric, interactive and extrinsic (Bunge 1959). In the former, the entity causing the effect must not be predicated of any part of another entity on which the effect impinges; that is, each must be wholly independent of the other. The implication of these constraints, therefore, is that the entity causing the

---

11. It is impossible to question the unity and identity of man at the foundation of his acting, nor is it possible to question the unity and identity at the roots of the efficacy and subjectiveness structurally contained in the acting that occurs in man. For the human being is a dynamic unity […] so much so that we called him outright the dynamic subject. Man's acting and the human efficacy which constitutes it experientially […] combine together as if issued from a common root. For it is the human being as a dynamic subject, who is their origin (Wojtyla 1979).

extrinsic effect need not be delimited by a particular material or formal qualification. Any matter or form will suffice. Hence, there are no formal requirements for extrinsic causal interactions, only that they be independent. This class of contingency is characteristic of Newtonian determinism, which is predicated on the independence of the respective entities.

Autonomous activity, by contrast, requires a unique physical substratum. This is to say that its effects require the properties predicated of the agent for its causal effect to occur. Understood in the context of Aquinas's dynamic, operationally defined "substances"[12], *this latter causal order is thus characterized by a contingency on a formal organization*, which is uniquely specified by its ontological nature. Therefore, the effects induced by this order of causality are restricted to the operational range delimited by the ontology of the agent. Due to its formal operational requirement, autonomous activity is characterized by its intrinsicity.

Due to this formal requirement, moreover, autonomous activity is further distinguished by the breaching of causal contiguity, which is constitutive to the efficient and mechanistic causal order. Accordingly, autonomous activity is necessarily multi-potent (Asaro 2009), requiring the determination of a range of "affordances" – i.e., attainable modes of behavior, (Gibson 1979) – prior to the selection of executable action. The selection of an affordance from within a range of possibilities thus implicates the need for a source not only spatially and temporally coincident with the autonomous substratum but also capable of modulating its internal operations, which Aquinas and others identify with an intrinsic dynamic principle.

Gibson's account of unified perception rests on two notions, however, which are distinct from Aquinas' concept. The first is that of self-awareness, which consists of proprioception and what he calls egoreception, or an awareness of the internal states of the body. The second notion is that of orientation – how these relate to the external states of the environment. Gibson's model thus stages a physical awareness of the body as a referential object in space and time, which is needed for assessing behavioral options. Accordingly, it lacks the active process of unification identified with Aquinas' dynamic principle hence, it also lacks the latter's self-determined and self-initiated causal role. As mentioned, various motor paradigms, for example, action identification, identify the autonomy capacity with an active top-down control and the source with the whole individual. In cases of operant conditioning, for

---

12. "Considered as a fundamental condition of the actual existence of every existing being it may be said that this structural/ontological basis […] is the subject of existing and acting […] It is not a passive substratum but, on the contrary, the first and fundamental level for the dynamization of being that is the personal subject…" (Wojtyla 1979).

example, voluntary control notably modifies the collective motor movements of the whole individual (Wolpaw 2018).

The identification of the agent with the personal subject is widely echoed, expressed in Compton's conclusion ("My purpose, they knew"), and extending across time, beginning with Descartes "ego", Kant's autonomous subject, and, recently, with Wojtyla's personal subject. Nonetheless, the association between the subject and his physical substratum has been problematic, due to the varying metaphysical understandings of its causal features. For example, despite Descartes' acknowledgement of the personal subject as an "existent" (Onishi 2011), his failure to ascertain a distinction in the causal order between the self-propelled agent and external, efficient causal modes led him to postulate a distinction in the substance of the personal subject from that of his body. He thereby introduced a fundamental division between the agent as a causal origin and its physical substrate; that is, the source and capacity were now disjointed due to his dualist premise. By contrast, the characterization of the agent as possessing a distinct causal order has more recently enabled Wojtyla to preserve the "substantial" unity between the subject as source and the physical operation that is expressive of agency through his identification of the subject with the dynamic form of the body.

10.4. The personal subject and intrinsic corporal value

10.4.1. *Wojtyla's metaphysical subject*

Wojtyla's emphasis on the metaphysical subject, particularly, has acquired support in numerous contemporary observations on living systems in its identification of the personal subject with the dynamic principle of unification. These studies identify a parallel principle(s) replicated across the evolutionary spectrum and arriving at an ontological crescendo in humans. At its human apex, this principle is manifested in autonomous activity, which predicates from a global and distinct, operative order (Rizzolatti and Febri-Destro 2008; Orban and Caruana 2014), revealing a formal contingency shared and deployed across the full range of living organisms. Along evolution's expanding arc of neural complexity and range of action, this principle functions to unify individual organisms operationally by means of action selection. The significance of Wojtyla's metaphysical humanism for neuroethics, therefore, is that by linking the subject with an autonomous and intrinsic causal order, *value content emerges from within the physical domain of the individual rather than from an external and contextually defined exterior.*

10.4.2. *Kant, intrinsic value and the categorical imperative*

The notion of a value content grounded solely in the possession of an innate autonomy traces its lineage chiefly to Immanuel Kant (Rohlfe 2020). In bioethics, it has been argued, for example, that human rights are grounded in a human dignity that flows from the possession of the capacity for autonomous choice, a position directly extrapolated from Kantian moral philosophy. Indeed, the fundamental idea anchoring Kant's moral philosophy, seen in Kant's "critical philosophy" – especially in his three "Critiques": the *Critique of Pure Reason* (1781, 1787), the *Critique of Practical Reason* (1788) and the *Critique of the Power of Judgment* (1790) – is that of human autonomy, and correspondingly, its value implications for the human being.

In the Kantian view, this innate capacity is widely present in nature. Organisms are regarded as purposive, with humans comprising a self-reflective end point; that is, Kant imputes to nature a teleological worldview in which autonomy is progressively elaborated through its replication in multiple biological forms, attaining an apex in humans, who are characterized by their capacity for "reflecting judgment". This is the final end of nature according to Kant. Such concepts have been taken up in the contemporary, biological autonomy models of Maturana and Varela (1979) and of Mossio and Moreno (2015) and by evolutionists like Laublicher (2000) who see in the progressive elaboration of behavioral range a metaphysical contingency resulting in the acquisition of a capacity for self-determined and purposive causal modification of nature. As in Compton and Popper's later observation, Kant thus understands autonomy to be a contingent *meta*-feature of nature, made manifest in its evolutionary display, progressively shaping the forms that nature adopts.

The presence of the human will as the culmination of this elaboration establishes the norm of Kant's Categorical Imperative and designates the human being as an end with a dignity that cannot be manipulated in any way for any reason[13]. Hence, Kant's view is that of a value content that resides within the human being, which normatively supersedes any other objective reality. The inheritance of his metaphysical causal understanding from Descartes, however (empirically reinforced by Newton), effected a dissociation of the personal subject as a causal origin from a purposive end, a dissociation that led to Kant's constructivist insights (Costelloe 2021), now challenged by current understandings of sensory motor coupling (Mimica et al. 2018).

---

13. "Only man himself, and with him every rational creature is end in itself, for in virtue of the autonomy of his freedom, he is the subject of the moral law, which is holy" (Kant).

The immanency of the human will as a ground for value content is nonetheless pervasive in modern bio- and neuroethics, where value is understood to flow from the will's experiential reality. Recognition of the will's immanency is shared with Christian religion, which, in contrast to Kant, metaphysically grounds the will in its distinctive causal order. Religion further anchors human value content in an alternative pillar identified in Wojtyla's teleological (and metaphysical) perspective of intersubjective encounter, the basis for his personalistic norm[14].

10.5. Conclusion

A key issue in neuroethical theory is that of ascribing value to the human body, with the inclusion of its constituent neural systems. Because the body comprises a physical locus situated in space and time, its physical nature constitutes a fundamental element in value appraisal, a domain of philosophical interest to science and religion alike. Remarkably, religion and science have developed philosophical legacies underpinning the understanding of physical reality that share common conceptual approaches, despite a colloquial perception of their adversarial relationship. On the contrary, they also diverge in notable respects that affect their respective normative conclusions. These differences are especially acute in regard to Hume's fallacy principle, which may succinctly be described as science's imputing of value from without and religion's appraisal of value from within. While the normative tension introduced by Hume's Law will likely persist into the future, the clarification of the tension's theoretical origin will undoubtedly help in establishing points of ethical approach common to both religion and science in the many forthcoming challenges posed by neurotechnology and other medical interventions in the brain.

10.6. References

Asaro, P.M. (2009). Information and regulation in robots, perception and consciousness: Ashby's embodied minds. *Int. J. Gen. Sys.*, 38(2), 111–128.

Braillard, P.A. (2010). Systems biology and the mechanistic framework. *Hist. Phil. Life Sci.*, 32(1), 43–62.

Brinkers, M. (2015). Beyond sensorimotor segregation: On mirror neurons and social affordance space tracking. *Cog. Syst. Res.*, 34, 18–34.

---

14. "[…] For above all […] what is proper (to ethics) is the formulation of all of reality *in light of ultimate causes* […] the personalistic norm is the primary principle of human acts…" (Wojtyla 1991).

Bunge, M. (1979). *Causality and Modern Science*. Transaction Publishers, New Brunswick, NJ.

Clark, N. (1993). *Person and Being*. Marquette University Press, Marquette, WI.

Compton, A. (1935). *The Freedom of Man*. Yale University Press, New Haven, CT.

Conant, R. and Ashby, W.R. (1970). Every good regulator of a system must be a model of that system. *Int. J. Sys. Sci.*, 1(2), 89–97.

Costelloe, T. (2021). Giambattista Vico. In *The Stanford Encyclopedia of Philosophy*, Zalta, E.N. (ed.) [Online]. Available at: https://plato.stanford.edu/archives/spr2021/entries/vico.

Craver, C. and Taberry, J. (2015). Mechanisms in science. In *The Stanford Encyclopedia of Philosophy*, Zalta, E.N. (ed.) [Online]. Available at: https://plato.stanford.edu/archives/spr2017/entries/science-mechanisms/.

Esfeld, M. (2004). Quantum entanglement and a metaphysics of relations. *Stud. Hist. Phil. Mod Phys.*, 35, 601–617.

Farrer, C., Franck, N., Georgie, V.N., Frith, C.D., Decety, J., Jeannerod, M. (2003). Modulating the experience of agency: A PET study. *Neuroimage*, 18, 324–333.

Fong, P. (1996). *The Unification Science and Humanity*. New Forums Press, Stillwater, OK.

Fourneret, P., Vignemont, F.A., Franck, N., Slachevsky, A., Dubois, B., Jeannerod, M. (2002). Perception of self-generated action in schizophrenia. *Cog. Neuropsych.*, 7(2), 139–156.

Frith, C., Blakemore, S.J., Wolpert, D.N. (2000). Explaining the symptoms of schizophrenia: Abnormalities in the awareness of action. *Brain Res. Rev.*, 31, 357–363.

Fuchs, T. (2018). *Ecology of the Brain*. Oxford University Press, Oxford.

Gibson, J.J. (1979). *The Ecological Approach to Visual Perception*. Houghton Mifflin, Boston, MA.

Gilson, E. (1952). *Being and Some Philosophers*. Pontifical Institute of Medieval Studies, Toronto.

Hempel, C.G. and Oppenheim, P. (1948). Studies in the logic of explanation. *Phil. Sci.*, 55, 135–175.

Hooker, C.A. (2008). Interaction and bio-cognitive order. *Synthese*, 166(3), 513–546.

Jeannerod, M. (2009). The sense of agency and its disturbances in schizophrenia: A reappraisal. *Exp. Brain. Res.*, 192, 527–532.

Jonas, H. (1966). *The Phenomenon of Life: Toward a Philosophical Biology*. Harper and Row Publishing, New York.

Kalkman, D. (2015). Unifying biology under the search for mechanisms. *Biol. Phil.*, 30, 447–458.

Kant, I. (2000). *Critique of the Power of Judgment*, Guyer, P. and Matthews, E. (trans.). Cambridge University Press, Cambridge (first published 1790).

Kato, S., Kaplan, H.S., Schrödel, T., Skora, S., Lindsay, T.H., Yemini, E., Lockery, S., Zimmer, M. (2015). Global brain dynamics embed the motor command sequence of *Caenorhabditis elegans*. *Cell*, 163, 656–669. DOI: 10.1016/j.cell.2015.09.034.

Laubichler, M.D. (2000). The organism in philosophical focus [special issue]. *Phil. Sci.*, 67(3), S256–S321.

Machamer, P.L., Darden, L., Craver, K. (2000). Thinking about mechanisms. *Phil. Sci.*, 67(1), 1–25.

Maturana, H.R. and Varela, F.J. (1979). *Autopoiesis and Cognition*. Reidel Publishing, Boston, MA.

Merleau-Ponty, M. (1964). *Signs*, McCleary, R.C. (trans.). Northwestern University Press, Evanston, IL.

Mimica, B., Dunn, D.A., Tombaz, T., Bojja, C., Whitlock, J.R. (2018). Efficient cortical coding of 3D posture in freely behaving rats. *Science*, 362, 584–589.

Moreno, A. and Mossio, M. (2015). *Biological Autonomy: A Philosophical and Theoretical Inquiry*. Springer Publishing, Dordrecht.

Nicholson, J. (2014). The return of the organism as a fundamental explanatory concept in biology. *Phil. Compass*, 9(5), 347–359. DOI: 10.1111/phc3.12128.

Noel, J.P., Blanke, O., Serino, A. (2018). From multisensory integration in peripersonal space to bodily self-consciousness: From statistical regularities to statistical inference. *Ann. New York Acad. Sci. Special Issue: The Year in Cognitive Neuroscience*, 1426, 146–165.

Odling-Smee, J., Laland, K., Feldman, M. (2003). *Niche Construction: The Neglected Process in Evolution*. Princeton University Press, Princeton, NJ.

Onishi, B. (2011). Tracing visions of the posthuman. *Sophia*, 50, 101–111.

Orban, G.A. and Caruana, F. (2014). The neural basis of human tool use. *Front. Psychol.*, 5(310), 1.

Phelan, G. (1967). *The Existentialism of St Thomas, Selected Papers*. Pontifical Institute of Medieval Studies, Toronto.

Pieper, J. (1989). *The Truth of all Things*. Ignatius Press, San Francisco, CA.

Popper, K. (ed.) (1978). Of clouds and clocks. In *Objective Knowledge: An Evolutionary Approach*. Oxford University Press, Oxford.

Rizzolatti, G. and Fabbri-Destro, M. (2008). The mirror system and its role in social cognition. *Curr. Opin. Neurobiol.*, 18, 179–184.

Rohlf, M. (2020). Immanuel Kant. In *The Stanford Encyclopedia of Philosophy*, Zalta, E.N. (ed.). Available at: https://plato.stanford.edu/archives/fall2020/entries/kant.

Rothaar, M. (2010). Human dignity and human rights in bioethics. The Kantian approach. *Med. Health Care Philos.*, 13(3), 251–257.

Ruiz-Mirazo, K. and Moreno, A. (2012). Autonomy in evolution: From minimal to complex life. *Synthese*, 185, 21–52.

Schneider, K. (1955). *Klinische Psychopathologie*. Thieme, Stuttgart.

Sgreccia, E. (2012). *Personalist Bioethics: Foundations and Applications*. Vita y Pensiero, Milan.

Smithers, T. (1997). Autonomy in robots and other agents. *Brain Cog.*, 34(1), 88–106.

Thelen, E., Schöner, G., Scheier, C., Smith, L.B. (2001). The dynamics of embodiment: A field theory of infant perseverative reaching. *Behav. Brain Sci.*, 24, 1–86.

Ulanowicz, R.E. (1986). *Growth and Development: Ecosystems Phenomenology*. Springer-Verlag, Berlin.

Wojtyla K. (Pope John Paul II). (1979). *The Acting Person*. D. Reidel Publishing Company, Dordrecht.

Wojtyla K. (Pope John Paul II). (1991). *Man in the Field of Responsibility*. Libreria Editrice Vaticana, Vatican City.

Wojtyla, K. (Pope John Paul II). (2006). In *Man and Woman He Created Them: A Theology of the Body*, Waldstein, M. (ed.). Pauline Press, Boston, MA.

Wolpaw, J.R. (2018). The negociated equilibrium model of spinal cord function. *J. Physiol.*, 596, 3469–3491.

11

How Would Neo-Confucians Value Moral Neuroenhancement?

Jie YIN
School of Philosophy, Fudan University, Shanghai, China

Current moral enhancement through neuropharmacology, as well as cutting-edge neurotechnology, either enhances pro-social attitudes or aims for a weaker goal, namely, facilitating moral decision-making. Therefore, no available methods can guarantee a change in moral character or even moral tendency. From a Confucian perspective, moral (neuro)enhancement may be doubtful, not only because the feasibility of moral enhancement as such is problematic, but also because the goal of moral enhancement seems to be set in the wrong direction. By comparison, what has been proposed by Confucians as *gongfu*, if practicable, may be closer to the goal of "learning to be human". Although there seems to be a sharp contrast between the ambitious yet unrealistic goal of current moral enhancement through technological innovation and the traditional Eastern way of cultivating virtue through daily experience, reflection and practice, neither the Confucian perspective nor the views held by neo-Confucians have to necessarily exclude the utility of moral neuroenhancement. Neo-Confucians could embrace neurotechnology and share the same goal of moral enhancement with incremental steps.

11.1. Moral neuroenhancement: the scenario and the conceptual challenge

Despite appearances, moral enhancement is not a new concept, but is simply dressed differently. Traditional moral education also aims at moral enhancement, albeit without the aid of biochemical technology. Children are often told to "be good", and in certain cases, they are even told exactly what and what not to do. However, such moral indoctrination can backfire. When children become adults and face complications or difficulties, they may not be able to easily differentiate "right" from "wrong". Moral indoctrination is often accompanied by stories depicting role models, and these stories, as well as the multiple forms of expression, such as movies and TV shows, all contribute to the expected result.

Moral enhancement, regardless of its form, faces serious challenges. The first question is, why do we need moral enhancement? Persson and Savulescu (2008) posit that moral enhancement is an imperative derived from human nature. However, this only further complicates the question. First, there is no consensus on what constitutes human nature, and even if there was a unified view of human nature, why should we have to accept that it urges us to enhance ourselves through neurotechnology?

There could be a long debate centered on whether moral enhancement is morally desirable. However, the focus of this paper does not completely overlap with such a debate. Moral enhancement, understood with an agential definition in mind (Pacholczyk 2011), promotes change in moral subjects by applying neurotechnology in an "important" way,[1] and this will lead to (or be expected to lead to) the result that this very moral subject will be improved. Therefore, the important question to be asked is not whether human nature requires us to enhance ourselves, since this would presuppose that we should aim at a certain goal regarding morality.

As "morally better" is quite a vague expression, different interpretations apply. From a Kantian deontological perspective, "morally better" may imply better compliance with our motive towards duty, which obviously frames it as an imperative request. Meanwhile, from a virtue ethics viewpoint, it could mean more robust moral character traits, and from a Humean point of view, it could mean someone is more sympathetic or more capable of empathizing with others. Lara and Deckers (2020) proposed a weak moral enhancement model through the innovation of AI technology, and their basic assumption is that better moral reasoning ability is a reasonable goal for moral enhancement.

---

1. But which way? And why is it important?

However, if all these accounts sound reasonable, perhaps we do not have to view "morally better" as a unified goal but rather a cluster of attainable goals, including those in both the short and long terms.[2] Some of them may soon be realized, but some, for instance, a character trait enhancement, may take time, and may not even be feasible, at least in the "medium-term future" (Beck 2015). The benefit of clarifying the goal in this way is that we no longer have to clash on the premises of moral enhancement just because agreement about what constitutes "morally better" cannot be reached. A hasty conclusion does not follow, simply because there are too many variations in terms of the goals of moral enhancement. As I will show later, this understanding will lay the foundation for a Confucian view to embrace moral neuroenhancement through technological innovation.

In contrast, for cognitive enhancement, the question does not seem to be complicated. Cognitive enhancement merely aims at a functional enhancement, so there is no need to be concerned about the ambiguous goals, although it remains controversial whether augmentation of a single cognitive function would be beneficial for the agents. For example, Glannon (2008) notes that augmentation does not always lead to an expected outcome. Memory enhancement alone, for example, can interfere with normal informational processes. Memory is useful only when it relates to the future in a meaningful way; otherwise, too much memory would be

---

2. I do not aim to show that moral act is composed of multiple components, but there are some distinctions that should be made clear. From a philosophical point of view, it seems that we have not had a clear idea on the relation among moral acts, moral motivation and moral judgment. These are all meta-ethical questions, but they do affect how we construe the feasibility of moral enhancement. Maybe we do not need to reconcile different moral perspectives regarding what is "morally good" (in an ultimate sense), but we can reach an agreement on how incremental change counts as heading at purported direction. For example, when we use serotonin to induce pro-social attitudes, it seems to be an efficient way of morally enhancing human beings for a single purpose but not for others (see Crockett (2010)). No matter how debates on theoretical level proceed, the practice of moral enhancement would nevertheless happen. Most of the research on moral enhancement by using neurotechnology do not aim for very complicated moral traits, but only aim for goals such as "pro-social attitudes" or "harm aversion (to others)". By saying that "morally better" is not a unified goal but a cluster of attainable goals, I mean to say that although we have different versions of what counts as "morally better", for example, some endorse stronger motivation, some suggest better function of moral agent (which likely relies on an agential definition of moral enhancement) and some claim we should aim for better judgment. All of these may facilitate moral acts, but none of these are clearly identified. If so, any single enhancement can be viewed as incremental moral progress, even though we do not have a clear idea on what counts as (ultimate) moral enhancement with which long-term goals may be achievable, such as dealing with the problem of nuclear weapon mass destruction or environmental crisis.

burdensome rather than convenient. Another example is listening. Augmenting listening on its own would likely create more disturbances than benefits.

Cognitive enhancement aims for a specific function, for example, the enhancement of memory or attention. Earp et al. (2018) write "interventions are considered enhancement [...] insofar as they [augment] some capacity of function (such as cognition, vision, hearing, alertness) by increasing the ability of the function to do what it normally does". Unfortunately, this does not apply to moral enhancement because it is not easy to name a single function that would result in moral enhancement. Better moral reasoning and more sympathetic feelings or emotions do not guarantee better moral action, if not leading to a contrary result. In addition, we have not reached a consensus on what kind of "benefit" the enhancement realizes. One specific factor regarding moral enhancement is that there is no single criteria for evaluating what is good. For cognitive or mood enhancement, what generates enhanced cognitive ability or a cheerful mood can be seen as good. On the contrary, moral enhancement does not necessarily improve the moral agent. Being nice and willing to help will benefit others rather than the agent, rendering the motivation to conduct moral enhancement suspicious. Indeed, Dees (2011) and Beck (2019) raised doubts about the effect of moral enhancement. It appears to be erroneous to simply assume that drugs such as oxytocin would be useful for moral enhancement, if we do not first ask the question whether a pro-social attitude would definitely contribute to the ideal of "morally better". And since "morally better" is vague or at least difficult to define at the moment, we are facing the same obstacle as when discussing the conceptual problem of what exactly represents moral enhancement. A further concern is that a large group of people participating in moral enhancement would raise more serious ethical concerns regarding coercion and exploitation engendered by possible brainwashing.[3]

---

3. The concern here is simply that a collective moral enhancement would not usually come directly from individual's consent but be imposed by power. I assume that autonomy is essential for any moral agent. But I agree with one reviewer that this requires argument or at least clarification. The thing is, the notion "morally good" is not clearly defined. What is good for the society does not necessarily imply the same goodness for the individual. The purpose of moral enhancement, as least as envisaged by Persson and Savulescu at the time when they proposed the idea, is it achieve pro-social attitudes, especially among people outside the group. As they see it, nepotism can be overcome, and mass destruction caused by nuclear weapons or environmental crisis can be avoided. On the contrary, moral enhancement, unlike cognitive enhancement, may not be beneficial for individuals. We need to consider the threshold for the very practice that may cause harm or sacrifice for individuals. The threshold could be, for example, retaining autonomy of the subject, which is deemed essential for any

Consequently, we are left with uncertainty regarding what constitutes moral enhancement. Perhaps we could set the question aside and ask again whether moral enhancement is desirable. Some argue that the controversy centers on the distinction between therapy and enhancement. For conservatives, enhancement is not morally desirable, and we should stop at the boundary where a distinction between therapy and enhancement can be drawn. As such, the debate goes much further than what can be accommodated here, but the point is, in short, most discussion shows that moral desirability of enhancement hinges on how we conceive and value the notion of human nature, since this helps to distinguish therapy from enhancement. Conservatives argue against biomedical enhancement because they believe that human nature cannot be altered. Their opponents do not see why we should stumble on the notion of human nature as such, since it is unclear.[4]

So, what is human nature? Conservatives like Fukuyama (2003) and Sandel (2007) do not offer a definition, although they both rigorously criticize biomedical enhancement by relying on what they insist on for the sake of human nature (Sandel specifically uses "giftedness" instead). Buchanan (2011, p. 118), on the contrary, offers a definition of human nature that may work for the purpose of discussion:

> [...] human nature is a set of characteristics (1) that (at least) most individuals who are uncontroversially regarded as mature human beings have; (2) that are recalcitrant to being expunged or significantly altered by education, training, and indoctrination; and (3) that play a significant role in explanations of widespread behavior and in explanations of differences between humans and other animals.

In a sense, human nature is like an acorn. If an acorn does not grow into an oak tree, then it certainly is not a good acorn. It seems that the evaluation of what is a good acorn is embedded in the descriptive account of it. Biological limitations apply to both. However, if a genetically modified acorn finally grows into a pine, then we cannot say what has been done is still "enhancement". It is so radical a change that

---

moral agent. It is another long story if we discuss why autonomy matters for moral agents at all, but this simply cannot be accommodated in this article.

4. One anonymous reviewer raises the point that this argument does not implicate an endorsement of intervention. Yes, it is true that by arguing that the notion of human nature is unclear and therefore we do not have to stumble on it, the opponents have not yet positively endorsed intervention. However, those who endorse moral enhancement or any kind of enhancement (e.g. cognition and mood) usually simply appeal to the reason that resembles those proposed by Persson and Savulescu (2008). They simply think humans have inherent moral deficits that would stop them from solving global crisis nowadays and in the future, and that there is no way out except adopting biomedical enhancement.

the nature of the oak tree has been altered; therefore, it is not an enhanced oak tree that is simply taller and more robust in a hostile environment. In summary, the nature of the acorn should be something like that depicted in Buchanan's definition, namely, "recalcitrant to being expunged or significantly altered by education, training, and indoctrination".

Take it a step further and consider that if we cannot conceive a concept of human nature without bringing value into the picture, then the "is-ought" problem does not remain. In other words, there is no such thing as "naturalistic fallacy", or as Sandel puts it, it is just "fallacious natural fallacy". We cannot separate the question of what human nature is from what human nature ought to be. There is always some perspective from which the view of human nature is shaped, be it functional or teleological.

Contemporary Confucian scholars echo this point. For example, Ni (2021) argues that the fact that human nature in Chinese philosophy cannot be understood as purely descriptive metaphysical accounts does not imply that Confucian doctrines fail. The reason why there is no distinct description regarding naturalistic accounts and normative recommendations is that, from the very beginning of proposing the "four shoots of moral conducts" (*siduan*, 四端), Mencius as well as his Song-Ming followers, never aimed to deduce "ought" from "is" (see also (Xu 2019, p. 51)). The ontological account must be integrated with the normative account, but from a Western philosophical point of view, the normative recommendation seems to stand on an unwarranted premise. How would Chinese philosophy prove this? The criticism does not only apply to the account of human nature alone, but also to Chinese philosophy in general. As we may be able to see from the discussion below, unlike Western philosophy, which attempts to establish a normative claim based on metaphysical claim, Eastern philosophy reverses the order, that is, it starts from the moral ideal and then seeks the metaphysical underpinnings for the whole endeavor.

11.2. How would neo-Confucians value moral neuroenhancement?

It may be an understatement to say that Chinese philosophy has a tradition of moral enhancement, as long as the term is broadly construed. Ni (2014) mentions that, for Confucians, especially neo-Confucians, *gongfu* (功夫 or 工夫) was widely used as a term to generally describe the method for Confucian "learning" (*xue*, 学). However, the specific aim of Confucian recommendation is not that we have to conform to a set of moral norms, but "more broadly a matter of mastering the art of life" (Ni 2014, p. 341). In addition, for neo-Confucians, the *gongfu* method is not for everyone; rather, it has a demanding requirement, namely, achieving excellence in ability. In contrast, moral norms can be used as standards to regulate people's behavior at an acceptable level. Hence, although every rational person can obey moral norms for the purpose of conforming to moral norms, that does not mean everybody can be or needs to be a sage that "strives for the perfection" (*zhiyuzhishan*, 止于至善, from *daxue*, 大学), which is set as the moral ideal for sages. This is not to say that moral ideal has two different sets of rules for normative application. Achieving moral ideal requires not only effort, but also endowment and luck. By advocating for "perfection", *daxue* names the direction for not only strictly moral progress, but also for daily practice that does not always require a fixed answer to right and wrong.

Zhu Xi (also called Zhu Zi, 朱熹/朱子), for example, believed that we must reflect both inwardly and outwardly to achieve moral excellence, as is usually articulated in his famous slogan: "*gewu zhizhi*" (格物致知). If both inward reflection and outward learning are indispensable, then for neo-Confucians, the feasibility of moral enhancement through neurotechnology seems to hinge on whether an inward reflection could be accomplished just as traditional moral education or self-cultivation. In other words, the way of striving for perfection presupposes the internalization of moral motivation. Mere conformity or compliance without an internal moral drive, for instance, would be insufficient. If we, at least for the purpose of discussion, treat Confucian moral self-cultivation as a traditional moral enhancement, then we could say that these two forms of moral enhancement have quite different goals.[5]

The goal of learning to be human follows directly from our understanding ourselves as being in the cosmos, and moral destiny is an integrated part of our cosmological understanding of ourselves. The background for this is that Song-Ming Confucians try to link their claim on cosmology and metaphysics with their moral ideals, thereby forming a more systematic and unified theory to rebut the overwhelming influence of Buddhism and Daoism. Pre-Qin Confucianism, on the contrary, places too much emphasis on ethical doctrines but is short of metaphysical underpinnings for its specific moral claims. Unlike Western philosophy, Chinese philosophy has an innate tendency to view the cosmos and humans as both being

---

5. I thank one anonymous reviewer for pointing out that "moral motivation" can be understood as an intellectual apprehension of a moral "sense of duty". However, such "sense of duty" does not render a moral act necessarily. It merely means that we work on cultivating moral tendencies that are prone to generate appropriate deeds or behaviors. It does not purport to a specific moral act that would be deemed right or good, as is usually confirmed by moral reasoning in the Western philosophical sense.

dominated by the highest governing entity, heaven (*tian*, 天道,天[6]). As we can see in the book Zhongyong (中庸):

What is inspired by heaven is human nature,

To manifest this nature is called the Way,

And to cultivate the way is called education. ("天命之谓性, 率性之谓道, 修道之谓教")

In short, in Chinese philosophy, human nature is believed to be inherited from heaven and so must obey the same principle governing both the cosmos and humans. To obey the rule and cultivate the character through education is to respond to the call of heaven. Humans are, like everything else in nature, part of the cosmos. Chinese people are not used to conceptualized thinking modes or habits, so "heaven" is not considered as an entity, and it does not have specific property. It is considered as an ideal that cannot be governed by properties. Human nature is contingent on heaven, and not in a casual way. The nature of the human being is not identical to that of heaven, but derives from heaven. Heaven is seen as a moral ideal that manifests itself in the operation of nature orders, and humans have to obey the rules so that we can achieve the morality that can be called "the sage". Again, we work towards being sage, but not necessarily being able to achieve it. Thus, we can see that for Confucians, human nature is not understood from a naturalistic perspective, as is usually stated in Western intellectual tradition. For Confucians, human nature is more like a metaphysical concept that cannot be detached from the moral ideal of "learning to be human" and is never a comparatively independent concept that can be used as a tool for debating the moral desirability of neuroenhancement. Following this distinction, we may be able to see that the current debate in Western literature now centering around the moral desirability of enhancement, and especially those hinged on the notion of human nature, does not apply in the context of Chinese culture, especially if we adopt a fundamentally Confucian point of view on what it means to be human at all.

Following Mencius, who holds that all humans have good intentions (*siduan*, 四端), Zhu Xi firmly believed that if someone sincerely devoted themselves to moral self-cultivation, then they would make moral progress, even though sagacity is not possible. Among Confucians, Zhu Xi has a comparatively systematic account of human nature. Although he agrees with Mencius that all humans are born with good intentions and are sensitive to the suffering of others, he does insist that due to

---

6. It is worth noting that the English word "heaven" does not fully capture the metaphysical meaning of "*tian*" in ancient Chinese.

"*qi*" (气) endowment, individuals may vary in their moral achievement due to the fact that they are endowed with different levels of talents and gifts. Thus, it makes sense that even though all humans have good intentions, some can become evil or bad. The possibility of becoming bad makes Confucian learning indispensable as a constant practice.

However, if the genuine difference between a moral practice of learning to be human and moral enhancement through neurotechnology is that the Confucian moral self-cultivation process does not merely aim for moral compliance, then moral enhancement, in the long run, does not have to exclude the goal of improving moral character. Ideally, moral enhancement should align more with Confucian practices rather than being distinct. However, given the current stage where moral enhancement is technically feasible, we should bear in mind that a temporary "enhancement" may backfire instead of being an incremental change towards moral character. From a *gongfu* practice point of view, pro-social attitudes, for example, do not necessarily improve together with mood changes. It could be hypocrisy, or a cover used by people who merely desire recognition or other benefits, thereby going against the very aim of moral excellence.

Regarding how to achieve moral excellence, what Song-Ming Confucians have in mind reflects both inwardly and outwardly. Outward reflection is to know things, including how the world is, what it is currently and the state of human affairs. For a rationalist culture like Western logos tradition, it may be difficult to imagine why knowing things is connected to knowing the condition of human affairs. However, since for Confucians, human nature is inherited from heaven, which is the same governing source of the world, it makes sense to posit that knowing the world, humans and society are one and the same thing, or in other words, the knowledge regarding these seemingly different things is mutually connected and influenced. The worldview embedded in Confucianism is traditional rather than a modern version. With a modern and secular worldview, we would not be able to see why knowledge must be unified and systematic at all.

Zhu Xi proposed twofold self-cultivation, which is believed to be heavily influenced by Zhou Dunyi. Their predecessors saw the human mind as capable of switching between two radically different states, namely, stillness and activity. Zhu Xi found this untenable yet further developed this theory. By emphasizing our feeling of reverence (*jing*, 敬), we can prepare for the subsequent steps of acquiring knowledge. The spirit here echoes those in pre-Qin Confucianism, since the feeling of reverence (*jing*, 敬) is of vital importance in *quasi*-religious ceremonies. For example, in Analects, the section called "八佾" (*BA YI*) states:

[3:1] Confucius said of the Ji family, "They have eight rows of dancers performing in their courtyard. If they can condone this, what are they not capable of" ([3-1] 孔子谓季氏八佾舞于庭："是可忍也，孰不可忍也！").

By asking the question of "what are they not capable of", Confucius expresses his anger about an inappropriate ritual. The Ji family should show reverence for the ancestors and his superiors, and if they do, then they should have known that it is unacceptable to have eight rows of dancers in his court. Zhu Xi confirms the importance of this feeling of reverence by incorporating the practice of investigating things to determine their "defining features" (*li*, 理, different from another *li*, rituals, 礼). The things here refer not only to natural objects, but also to the world, people and society. In short, Zhu Xi's famous slogan "*gewu zhizhi*" (格物致知) refers to the investigation and exploration practice that aims to determine the mechanisms and relationships that regulate the cosmos, including the natural as well as the artificial world: society and its constituents and people.[7] To reach this goal, we need to allow the good human nature to emerge; therefore, we need to focus on their feeling of reverence. A common charge against Zhu Xi's approach of self-cultivation is that "*gewu*" (格物) does not always lead to "*zhizhi*" (致知), that is, an investigation into the mechanism and relationship of things or objects obviously does not enable us to master knowledge about human affairs and society.

The epistemological doubt makes sense because we must first warrant the belief that: (1) knowledge regarding things would cohere with that regarding human affairs

---

7. One anonymous reviewer asked whether I am suggesting an equation of morality with natural causality. No, I am not. At a first glance, the Confucian way of thinking, if viewed through the lenses of western philosophy, does not have a very clear view on what counts as good or bad (or, right or wrong), since it does not hinge on a standard or criteria of moral evaluation based on principles. In *Analects*, Confucius himself gives different suggestions as to what to do regarding similar cases. We can be quite confused at how equivocal Confucius could be. However, given our experience of being brought up and living in this culture, we may be able to detect the difference between the way in which people perceive morality or ethics. There is no definitely morally right act unless we can act appropriately by fully considering the situation, and in order to do so, we have to learn how to deal with constantly changing environment, people with different character traits and personalities, and social or cultural backgrounds. Perceiving the non-hierarchical normative order is more like mastering art rather than being acquainted with propositional knowledge. It is one thing for people to believe in "heavenly hierarchy" in an ontological and evaluative sense, and it is another to master the art of a fluid and non-hierarchical normative order. The metaphysical framework does not have to correlate with normative order, since the former can be viewed as postulation (e.g. in a *quasi*-Kantian sense) and the latter as practical knowledge (*know-how*).

and (2) our knowledge of things would lead to, or at least facilitate, our knowledge of human affairs and society. However, as previously discussed, Chinese philosophy does not tear apart the metaphysical account from the epistemological. Thus, if we take a weak and charitable interpretation, the doctrine of Confucianism that cosmos, people and society obey the same guiding principles at least implies the possibility of a coherent system of knowledge, even though it appears that this does not entail the second half of the belief stated above. We may wonder why it is that through knowing external things/objects, we can then know human affairs and society and thereby accomplish cultivating moral character?

Yet, if Confucianism is right for moral excellence, then the second half of the belief stated above is of vital importance. It shows a fundamental view of the feasibility of moral enhancement. For Song-Ming Confucians such as Zhu Xi, moral excellence can never be achieved merely through a one-dimensional improvement about morality alone. In other words, aiming at moral improvement through the acquisition of moral ideas, beliefs or knowledge does not work. We must learn to be human in the very process of what Confucians call "*saisao yingdui jintui*" ("洒扫应对进退"), namely, the daily experience and practice of behaving ourselves, such as when and how to advance and retreat and how to answer and respond.

As discussed previously, moral neuroenhancement faces the dilemma of whether it ever has a clear aim of what and how to enhance. In contrast, it seems that Zhu Xi had a more robust account of how to conduct moral self-cultivation than the chaotic description we see in the current literature regarding moral neuroenhancement.[8] Zhu Xi's account may be promising in that a metaphysical underpinning of unified cosmos, humans and society, even though not necessarily entailing, is still providing an explanation of how moral agents know the mechanisms and relationships among everything, including human affairs and society, at least among the people who hold the absolute metaphysical view dear.

---

8. I am not declaring a clear-cut east–west distinction, since other traditions have been trying to work on broad systematic application of diverse moral perspectives. Mosa (2012) says, "Muslim ethics is a polyhedron: a hybrid of different genres of literature", in which they have duty-based rules, virtue ethics and utilitarianism. The broad systematic application of all these moral perspectives seems to me to be the endeavor of coping with modernity and science with a pragmatic point of view in mind, as well as in practice. By contrast, Confucianism, at least as is understood in tradition, does not show the integration of different moral perspectives via systematic application. That being said, I myself do not think contemporary Confucianism does not have such potential to explore how that can be achieved. Confucianism nowadays needs to develop itself and manifest its contemporary significance, only if it constantly dialogues with other philosophical, cultural and religious tradition, as well as engaging scientific inquiry and fact-value entanglement.

However, neo-Confucianism does not have to exclude moral enhancement once and for all, merely based on its central claim on the comparatively ideal aim of *gongfu* practice. My point would be, on the contrary, that neo-Confucians can certainly embrace moral enhancement through groundbreaking neurotechnology as well as AI technology, as long as moral neuroenhancement as such can augment those cognitive functions that would contribute to moral performance and, what is more, as long as those ethical concerns regarding safety, privacy and autonomy can be well cared for.

11.3. Concluding remarks: the complementary role of Chinese philosophy in applied ethics

For neo-Confucians and Chinese philosophy scholars, the meaning of reviving Chinese philosophy goes beyond the purpose of justifying Chinese philosophy, and thereby establishing its legality. As Ni (2021) states, as a practice, *gongfu* should be scrutinized by applying the Western philosophical method. The ideal of harmony between nature and humans has a profound influence on human beings, especially when environmental protection and sustainability have become of vital concern around the world.

My question is, specifically, what can Chinese philosophy offer applied ethics? I am not suggesting a grant proposal like that of Ni. Ni's hope is that Chinese philosophy can be world philosophy; while I admire such inspiration I would rather think that the beginning of mutual understanding would be the first step, after which we should know whether further dialog is possible, the answer to which I would very much like to give a "yes", but it depends on how this dialog goes. Contemporary Chinese philosophy scholars who have been trained mainly by Western approaches, especially those who have to use concepts and terms in Western philosophy, may struggle with both intellectual and cultural identity. The struggle is twofold because, on the one hand, they are trained to reason and argue like Western intellectuals, while, on the other hand, they live according to how their cultural identity nourishes them. This could easily lead to a state of *quasi*-schizophrenia where Chinese philosophy scholars, especially ethicists, find it difficult to use their intellectual resources to make sense of their own living practices. Eastern intellectual resources are not always available, not only because there is an international cultural merge or some historical reason that renders the loss of its own intellectual as well as cultural tradition, but also because the comparatively silent culture does not own the power of discourse on the world stage, and it is their job to initiate the dialog without first claiming dogmas. In this sense, I believe that, due to its practicality and the very nature of involving stakeholders from every corner, applied ethics provides a good opportunity to up open the channel of dialog to try and reach mutual understanding.

11.4. References

Ames, R.T. and Hall, D. (2001). *Focusing the Familiar: A Translation and Philosophical Interpretation of the Zhongyong*. University of Hawaii Press, Honolulu.

Beck, B. (2015). Conceptual and practical problems of moral enhancement. *Bioethics*, 29(4), 233–240.

Crockett, M.J., Clark, L., Hauser, M.D., Robbins, T.W. (2010). Serotonin selectively influences moral judgment and behavior through effects on harm aversion. *PNAS*, 107(40), 17433–17438.

Dees, R.H. (2011). Moral philosophy and moral enhancements. *AJOB Neuroscience*, 2(4), 12–13.

Earp, B.D., Douglas, T., Savulescu, J. (2018). Moral neuroenhancement. In *The Routledge Handbook of Neuroethics*, Johnson, L.S.M. and Rommelfanger, K.S. (ed.). Routledge, New York.

Fukuyama, F. (2003). *Our Posthuman Future: Consequences of the Biotechnology Revolution*. Picador, New York.

Glannon, W. (2008). *Brain, Body and Mind: Neuroethics with a Human Face*. Oxford University Press, New York.

Lara, F. and Deckers, J. (2020). Artificial intelligence as a Socratic assistant for moral enhancement. *Neuroethics*, 13, 275–287.

Lau, D.C. (1970). *Mencius*. Penguin Books, New York.

Moosa, E. (2012). Translating neuroethics: Reflections from Muslim ethics. *Science and Engineering Ethics*, 18, 519–528.

Ni, P.M. (2014). Rectify the heart-mind for the art of living: A *gongfu* perspective on the Confucian approach to desire. *Philosophy East and West*, 64(2), 340–359.

Ni, P.M. (2021). Theories of the heart-mind and human nature in the context of globalization of Confucianism today. *Dao*, 20, 25–47.

Pacholczyk, A. (2011). Moral enhancement: What is it and do we want it? *Law, Innovation and Technology*, 3(2), 251–277.

Persson, I. and Savulescu, J. (2008). The perils of cognitive enhancement and the urgent imperative to enhance the moral character of humanity. *Journal of Applied Philosophy*, 25(3), 162–177.

Qian, M. 钱穆. (2012). *A New Interpretation of Analects 论语新解*. Sanlian Shudian 三联书店, Beijing 北京.

Sandel, M. (2007). *Against Perfection*. Belknap Press of Harvard University Press, Cambridge, MA.

Slingerland, E. (2003). *Confucius Analects*. Hackett Publishing Company Inc., Indianapolis, IL.

Xu, B. 徐波. (2019). *From the Metaphor of Turbulent Water to the Dark Consciousness 由湍水之喻到幽暗意识*. Shanghai Sanlian Shudian 上海三联书店, Shanghai 上海.

Zhu, X. 朱熹. (2012). *Collected Commentaries of the Four Books 四書章句集注*. Zhonghua Shuju 中华书局, Beijing 北京.

PART 3

Illustrative Cases

12

How Do Arabic Cultural and Ethical Perspectives Engage with New Neuro-technologies? A Scoping Review

Amal MATAR
Center for Research Ethics and Bioethics, Uppsala University, Sweden

12.1. Background

One of the major criticisms in bioethics in general and neuroethics in particular is its domination by Western ethical tradition (Barugahare 2018; Herrera-Ferrá et al. 2018). Among the under-represented voices in neuroethics is the Arab region. The Arab world is defined, according to UNESCO, as all Arab-speaking countries in the Middle East and North Africa, therefore, countries such as Turkey and Iran are excluded from our analysis (UNESCO 2012). The Arab region is far from being homogeneous in terms of ethnicity, religious practices, history and economic status, but the included countries share similar obstacles to academic enquiry and research such as a lack of academic freedom, corruption and low research funding and pay (Amer 2019). In addition, most, if not all, countries subscribe to conservative social and religious values (Abu-Rabi 1990; World Values Survey Association 2020).

Two important definitions should be elucidated for this chapter: culture and neuroethics. Regarding the first: there is a myriad of definitions of culture that range from abstract, overarching descriptions to more specific ones that describe what culture is made up of (Prinz 2020). For the purpose of this chapter, I define culture as encompassing a set of features that are shared and professed by a group of people. Such features include but are not limited to customs, language, religious practices, cuisine and music.

As for neuroethics, it was first elucidated by Roskies (2002b) who categorized it into ethics of neuroscience and neuroscience of ethics. She further clarified that the first category tackles the ethical and social implications during and as a result of neuroscientific practice. The latter deals with the neuroscientific foundations for ethical reasoning and morality, so it would shed light on the neural basis of ethical notions such as autonomy, for example.

Neuroethics was further characterized by Farisco et al. (2018b) into three main branches, which ideally complement each other. They are neurobioethics, empirical neuroethics and conceptual neuroethics. The first is viewed as an extension of mainstream bioethics, but focuses on bioethical issues that ensue from neuroscience, while empirical neuroethics primarily use neuroscience research to inform and explain philosophical concepts. Finally, conceptual neuroethics is a theoretical analytic field that defines ethical and philosophical notions such as consciousness and the mind (Farisco et al. 2018b). Thus, the first two categories proposed by Farisco and colleagues are comparable to Roskies' (2002b) definitions for ethics of neuroscience and neuroscience of ethics.

In this chapter, I aim to explore and describe ethical and cultural perspectives of new neuro-technologies as portrayed in and by the Arab world. This chapter is an exploratory study that endeavors to answer the following research questions:

1) How are novel neuro-technologies harnessed to investigate cultural elements in the Arab region?

2) What are the discussions, if there are any, pertaining to neuroethics in the Arab region?

To answer these questions, a scoping review was determined as the most suitable method. This type of review is used when researchers seek to map out and assess knowledge gaps as well as investigate how research is performed within a certain subject area (Munn et al. 2018). These two purposes are direct aims of the research questions proposed.

12.2. Methods

In order to ensure that all relevant literature is captured, I consulted the assistance of the library at Uppsala University to carry out the scoping review. The search was performed in five databases: PubMed, CINAHL, SCOPUS, APA PsycInfo and Cochrane and included a time period from January 1, 2000 to December 8, 2021 (refer to the appendix to this chapter (section 12.7) for search strategies as provided by Uppsala University Library). There were no limitations in terms of language (e.g. finding academic papers in Arabic) or full text availability. The SPIDER scheme (Methley et al. 2014) was adopted to identify the search terms for the study as follows:

S: sample, PI: phenomenon of interest, D: design, E: evaluation, R: research type.

The scoping review aimed to identify:

– **Sample**: peer-reviewed academic literature as published by Arab scholars or contributors, or conducted on Arab research participants or signifying Arab ethical/conceptual analyses.

– **Phenomenon of interest**:

1) novel neuro-technologies, including but not limited to brain computer interfaces, neuroimaging and transcranial magnetic stimulation;

2) neuroethics.

– **Design**: any study design such as surveys, interviews, focus groups, vignette studies and conceptual ethical analysis.

– **Evaluation:** empirical and non-empirical.

– **Research type**: quantitative or qualitative studies or neuroethical analysis involving these technologies.

I received a list of 421 papers of which 85 were duplicates (n = 336). The analysis process was as follows. First, all the titles were read through two times and relevant articles were marked. During the second round of review, I read through the abstracts and the method sections in order to ensure that only pertinent target groups were selected. I was left with a total of 33 articles. Second, the 33 papers were categorized and tabulated into ethics and culture abstracts. Ethics abstracts are literature that specifically addresses ethical and neuroethics topics or analyses, including conceptual discussions, whereas the culture category includes studies that examine cultural aspects of or by new neuro-technologies, for example, language, religious practices, customs and food.

12.3. Results

In my analysis, I did not evaluate the quality of evidence generated or the methods employed. I primarily focused on finding and describing papers that pertain to the aforementioned research questions. As such, the scoping review is descriptive in nature.

Since the majority of the articles (n = 303) can be broadly designated as medical literature that examined the various aspects of neurological diseases including therapies, testing or epidemiology, etc., they were excluded from the analysis. Thirty-three articles were deemed suitable under either of the two categories – *ethics* and *culture*.

Twenty-nine studies were categorized under *culture*, while only four papers fit the *ethics* subcategory. As mentioned above, the culture category encompasses all identified articles that examined different cultural phenomena by harnessing novel neuro-technologies. These cultural dimensions can be broadly grouped into *spiritual/religious* – where religious practices such as praying and religious fasting were studied, *language* – where aspects of the Arabic language were examined, and finally *behavior* – which included research on attitudes towards outgroups in terms of tolerance (Yogeeswaran et al. 2021) and violence (Domínguez D et al. 2018), as well as the extent people are convinced by health communication video campaigns (Burns et al. 2019), research on extremism (Pretus et al. 2018), and the study of neural underpinnings of creativity across cultures (Ivancovsky et al. 2018). For the purpose of this chapter, I will report on two important elements of culture (spiritual/religious and language subcategories) and the ethics literature (see Table 12.1).

Category	Ethics	Culture		
Number	4	29		
		Spiritual	Language	Behavior
		7	7	15
Total		33		

Table 12.1. *The categorization and number of the studies for each category and subcategory*

12.3.1. Culture

12.3.1.1. *Spiritual/religious studies*

There was a total of seven papers under the spiritual/religious subcategory (n = 7), of which three examined Islamic prayers, three addressed Ramadan fasting and a the final one was a case study on "spiritual possession".

Newberg et al. (2015) employed Single Photon Emission Computed Tomography (SPECT) to scan three Muslims while they prayed. The test subjects, who were strict observers of Islamic prayers for more than 15 years, performed two different types of Muslim prayers. Two (a man and a woman) were scanned while performing *Dhikr*, a Sufi form of supplication, and one prayed the obligatory Islamic daily prayer. The SPECT scans were captured, after injection of contrast material, first as a baseline before the event and 30 minutes after their deep prayers. The results showed diminished cerebral blood flow in frontal, parietal and temporal brain lobes but an increased firing of certain areas such as the "anterior cingulate, dorsal medial cortex, caudate, insula, thalamus, and globus pallidus". The findings were associated with strong sensations of attachment with God during both types of prayers. One of the main purposes of the research was to find the biological foundation for intense spiritual feelings and to identify the areas of the brain that are engaged (either activated or inhibited) while devotees experience such sensations.

Another group of researchers has similarly assessed brain activity of praying Muslims using EEG (Doufesh et al. 2014, 2016). A total of 50 men had their brain waves recorded before, during and after prayers. In Doufesh et al. (2014), electrocardiogram ECG data of participants were also simultaneously gathered. Neither study had control groups, however. Doufesh et al. (2016) compared recordings of mimic praying positions of participants to results of actual prayers. The results showed higher gamma power during actual prayers, which have been interpreted as an indication of increased concentration (Doufesh et al. 2016). The analysis of EEG and ECG data revealed "high levels of alpha activity" in occipital and parietal lobes, and diminished heart rate variability, which correspond to boosting relaxation, minimizing stress and better focus (Doufesh et al. 2014).

According to the teachings of Islam, Muslims are required to fast during the lunar month of Ramadan by abstaining from food and drinks from sunrise to sunset. Roky et al. (2003) examined eight Muslim men's sleeping patterns employing a polysomnography (PSG) and rectal body temperatures in a Moroccan study. PSG encompassed the recording of brain waves by an EEG, eye movements by electro-oculogram and muscle movements by an electromyogram. The measurements were taken at different time points: 12 days before Ramadan, on the 11th and 25th day of Ramadan and finally two weeks after the end of the fasting month. Furthermore, participants were asked to complete surveys to assess their daytime alertness and mood. The results indicated diminished core body temperatures and a rise in daytime sleepiness, particularly at the end of the month.

The impact of Ramadan fasting on sleep was also the subject of a review article by Trabelsi et al. (2019). The authors collected 13 relevant studies that assessed the total sleep time, its quality and quantity among physically active Muslim men, including professional athletes (from Qatar, Tunisia, Algeria) such as soccer players, martial arts athletes and cyclists. Some of the studies employed EEG and actigraphy to monitor sleep patterns while most used validated sleep assessment questionnaires that were translated into Arabic. Their analysis concluded that no change occurred to total sleep time, but the duration of naps has notably increased during Ramadan.

Not only were the effects of Ramadan fasting evaluated in healthy individuals, but also on patients with a neurological disease. A small study in Egypt has investigated whether fasting impacted multiple sclerosis (MS) patients positively or negatively. They compared 15 non-fasting MS controls with 15 fasting MS patients and used clinical outcomes and MRI to measure effects on the disease. The Expanded Disability Status Score (EDSS) and number of relapses were used to assess clinical outcomes, while contrast uptake by MS lesions was assessed by MRI. There was little evidence of improvement between the two groups per EDSS, relapse rate and MRI findings (Abd El-Dayem and Zyton 2012).

The final study in this group investigated a case of altered state of mind in the cultural context of "spirit possession of a human". A young 22-year-old Omani man was presented to the psychiatry clinic by his family members, who described the illness as a mix of "spirit possession, or envy and sorcery". They recounted the drastic deterioration of his cognitive functions, persistent auditory hallucinations and anti-social behavior. The patient's family first resorted to traditional non-medical treatment alternatives and a pilgrimage (*Umra*) to Mecca, to no avail. In the psychiatry clinic, the researchers SPECT-scanned the brain and performed medical blood tests to identify a pathological basis for the patient's psychiatric symptoms. There was strong evidence of decreased blood perfusion of the left temporal lobe, but the blood workout was all normal. He was prescribed antipsychotic drugs that have significantly improved his condition. In the discussion section, the authors attempt to analyze concepts of culture-bound diseases such as "spirit possession" and their biological correlation (Guenedi et al. 2009).

12.3.1.2. *Language*

This section describes the literature where neuro-technologies have been used to investigate another important aspect of culture, namely, language. The studies are pretty varying in purpose and goals.

Arabic letters have a distinct more complex orthography than Latin writing. In addition to its being written from right to left, a single letter can exist in isolation or connected mid word or at the beginning of a word, with differences in form for each of these categories. The letters in Arabic writing are customarily connected (Taha et al. 2013a). Researchers investigated the time it took 18 (15 male and 3 female) native Arabic speakers to visually recognize Arabic words. To record the visual recognition, the authors used EEG and monitored eye movements. Each participant was allowed 150 milliseconds to read a word, and there were a total of 180 words, with 60 connected, 60 partially connected and 60 unconnected words. By digitally analyzing the amplitudes of the visual-specific EEG waves (or the so-called Event Related Potential) and the reaction times for each participant, the researchers uncovered that non-connected words were more frequently incorrectly read or participants took a longer amount of time to recognize them (Taha et al. 2013a).

A more elaborate set of experiments was conducted to capture the neurophysiology that underlines the effects of cultural versus text hints on the readiness of a bilingual research subject to retrieve the correct word in the correct language (Arabic/English). An example of a cultural cue is an Emirati traditional dress whereas an example of a script cue is an Arabic number. The researchers utilized Magnetic Encephalography (MEG) on 20 (19 male, 1 female) right-handed, native Arabic speakers with medium English proficiency. Multiple MEG Images of the Left Inferior Prefrontal Cortex (LIPC) and Anterior Cingulate Cortex (ACC), the two regions involved in choice and conversion of language in the brain of bilinguals, as well as the time lapse to correct responses by participants, were recorded. A parallel control experiment was performed among a unilingual group (Blanco-Elorrieta and Pylkkänen 2015). The results support previous literature of the engagement of LIPC and ACC in language conversion and selection. Written cues were more effective in stimulating language selection compared to cultural ones in ACC (Blanco-Elorrieta and Pylkkänen 2015).

In another article where language was an important factor, authors reported on the collaboration between a technology research group and the local disability service that catered to patients with Specific Learning Disabilities (SpLD). The authors describe human and methodological elements that play a part in building and designing assistive interactive technologies for SpLD users. Users and experts are customarily engaged, via focus group format, in "user centered design cycles" to enhance relevance and applicability of assistive technologies. They are sometimes referred to as "design partners or informants". With regard to methodological considerations, the authors named tight budgets, cultural deliberations and planning for testing activities as examples (Alkhashrami et al. 2014).

In another study, Sakr (2011) describes how functional MRI was tailored to the local dialect for listening tests, Arabic text instead of English and suitable cultural cues (such as using a picture of a donkey rather than a bear) that resonate with

Egyptian patients undergoing pre-surgery fMRI to test language and memory functions. The research subjects included different ages (6–68 years old), education level (including non-alphabetical), genders (14 males, 2 females) and pathology. These measures rendered fMRI applicable for use by Egyptian and Arab-speaking patients that were both literate and non-alphabetical, ensuring better patients' cooperation and compliance (Sakr 2011).

Based on evidence from fMRI, MEG and EEG, Perlovsky (2011) was able to generate a mathematical model that links cognition and language. In his computation, called the dual model, he envisioned cognition and language as two paths that are distinct but nevertheless intimately connected. One aspect that influences this interaction is emotions. Emotions can be discerned via language sounds, *which are in turn impacted by culture.* Emotionality is integral for language evolvement since it reflects motivation to use the language, according to the author. He described the status of modern English language, which has been a vessel to transfer political ideas, science and technology. As such, there is little emotionality during use, which lead to "ambiguity of meanings and values" over time. The classical Arabic language, in contrast, is highly emotional and "the meaning of words changes with changing sounds" and thus entertains "beauty and affectivity", yet it exhibits rigidity when it comes to the evolution of new expressions and meanings (Perlovsky 2011).

12.3.2. *Ethics literature*

Articles that primarily discussed or addressed ethical, social or legal aspects of the neuro-technologies were singularly lacking in my search. There was very limited input from Arab countries, with a total of four articles under this category: two of which focused on Muslim/Islamic perspectives of neuroethics/new neuro-technologies, while the remaining two papers addressed attitudes and knowledge of a new neurotherapy among specialists and economic constraints encountered by low-income countries during neurosurgery procedures, respectively.

In Egypt, researchers from Alexandria University Medical Hospital conducted a retrospective study to assess the impact of limited resources and cutting cost corners in the treatment of 193 patients in their neurosurgery ward. In order to save money and thus allegedly benefit more patients, certain care strategies were not optimal. For example, employing lower quality but more cost-effective MRI imaging techniques, reusing disposable items after sterilization, repurposing cheaper tools than what they were designed for, for instance using an "insulin needle as an arachnoid knife", diminishing image guidance intraoperatively to instead rely on anatomical indicators and even resorting to making their own cotton strings to avoid additional costs. Furthermore, there was no specialized neuro-ICU care for any of the patients post operation. The researchers explained the need to sometimes lease neurosurgical and imaging equipment in certain operations as it is costly for the hospital to buy and maintain such hardware. The more advanced fMRI and diffusion tensor image are too cost-prohibitive to use. All these compromises affected pre-, during and post-operative care (Helal et al. 2018).

Repetitive Transcranial Magnetic Stimulation (rTMS) was approved as a new therapy for the treatment of drug-resistant major depression by the Saudi FDA (Food Drug Administration). AlHadi et al. (2017) surveyed a convenience sample of 96 psychiatrists in the Kingdom of Saudi Arabia to assess their knowledge and attitudes towards this new form of therapy. There were five times more male respondents than female ones (male = 80, female = 16). The results showed a good level of knowledge among the senior professionals and among those who frequently consulted recent literature in the field. Almost half of the respondents would prescribe the therapy to their patients. This positive attitude would be reflected, according to researchers, in the uptake and acceptance of the new therapy by patients (AlHadi et al. 2017).

There was very little literature that addressed neuroethics as a field. The articles were primarily written by presumably authors of Arab origin in the US. One such article outlined an Islamic method for neuroethics to address novel imaging technology such as fMRI. The author was interested in presenting a distinct cultural and religious perspective that would be relevant to observant Muslims. Referencing religious sources such as the Quran and Hadiths (prophet's quotes), he argued for the respect and sanctity of one's private thoughts regardless of their "wrongness", since only actions are judged by God. He founded his reasoning on metaphysical Islamic concepts such as the "soul, spiritual consciousness, inner consciousness and mind" where only God can have insight into. Thus, the risk of an access to such deep ponderings by an outsider via new neuro-technologies is ethically and religiously questionable (Al-Delaimy 2012).

To create an Islamic framework for neuroethics, the author proposed the Islamic Jurisprudence rules of *"necessities overrule prohibitions"*, *"the sanctity of life"* prevails, *"there shall be no harm inflicted or reciprocated in Islam"*, free will, advancing community interests over individual benefits and finally consulting a body of Muslim scholars (e.g. International Islamic Juristic Council). During such consultation, medical and scientific professionals usually play a role in guiding and explaining the science behind the new technologies. The author compared these rules to principles of bioethics as proposed by Beauchamp and Childress (2001). Then, they were employed to assess the use of fMRI as an example. If the usage of fMRI is for intruding on one's private thoughts for no medical benefit or to single out Muslims as a group and compare them to other groups, the rulings would be against such use or research (Al-Delaimy 2012).

In response, a Professor of Islamic Studies at Duke University countered the simplistic view of approaching neuroethics from an Islamic perspective as proposed by Al-Delaimy (2012). He questioned the extent of reconciliation of science and faith since each present their own interpretation of knowledge and truth. In fact, people of different faiths, including Muslims, have long interacted with science, scientific discoveries and methods, synthesized these concepts and created their own meaning, which they deemed congruent to their faith. This "meaning-making" pursuit is what the author called "culture" (Moosa 2012). He criticized the current approach of Muslim legal experts (*faqih*), who addressed novel neuroscientific questions mostly from a deontological rule-forming approach that is informed by old jurisprudence concepts. According to the author, they fail to acknowledge and incorporate the more dynamic and intellectual processes of "meaning-making" that occur within a community. To illustrate this, the author presented the Islamic view of death versus the medical definition that has been adopted. To Muslim theologians, a person dies when the soul departs the body. This has sparked a debate among Muslim ethicists when medicine redefined a person's demise in terms of brain death and the eligibility of the deceased person to subsequently donate their organs. Other contentious concepts that may pose neuroethical debate among Muslim ethicists/practitioners are a lack of an accepted definition of consciousness or a neuroscientific characterization of the "*soul*" (Moosa 2012).

12.4. Discussion

In this chapter, I attempt to present an overview of the current research where novel neuro-technologies have been utilized to examine cultural elements such as language and religious practices. Furthermore, I describe some of the published neuroethical discussions that have been articulated by Arab academics. This was performed through a scoping review that attempted to capture all the relevant articles.

Per the results, there seems to be little neuroethical debate, which is a relatively young field, within the Arab world. Most of the literature was primarily medical studies addressing neuroscientific, neurological and neurosurgical topics. This can be explained by pervasive poor financing of research in the region, where more than 80% of the researchers resort to self-financing their research. Even researchers in oil-rich countries such as the Kingdom of Saudi Arabia or the UAE, which are afflicted with less money problems, complain of a lack of academic freedom and attractive career opportunities. Other challenges are armed conflict, corruption and lack of adequate research infrastructure such as internet access, subscription to academic journals, etc. (Amer 2019). Therefore, this may justify the prioritization to fund medical research and the disproportionate publishing of neuromedical studies.

One interesting finding, in the analysis, was the relatively larger number of published studies, despite limited resources, that addressed cultural aspects such as religious practices, language and behavior compared to neuroethics research. Most of the Arab countries are not secular and religious authority conjoins with political power, sometimes at the expense of other religious and ethnic minorities (Buchanan 1997; Rørbæk 2019). In addition, religious values infiltrate much of the day-to-day aspects of life such as work and choice of medical care (Robertson et al. 2002; Serour and Serour 2021). But investigation into religious practices and experience is not unique to the region. In fact, spiritual neuroscience and neurotheology are on the rise as research fields in other regions. Researchers in these areas attempt to harness neuroscience to prove or disprove religious concepts (such as that of the soul) and experience (Beauregard 2011; Jastrzebski 2018; Rickabaugh and Evans 2018; Rim et al. 2019). For example, novel neuro-technologies were used to capture spiritual experience (Beauregard and Paquette 2006).

In another context, the adoption of a new neuro-technology such as fMRI is aimed to accommodate the local culture and conditions, such as using the local dialect for listening tests, and using pictures of a donkey instead of a bear, adjusting fMRI tests to assist analphabetic patients and replacing the default English text with an Arabic one. These measures would ensure early adoption of novel neuro-technologies and the smooth diffusion of innovation within the society (Orr 2003).

The Arabic language plays a pivotal aspect in both the culture and religion of the region (Ajami 2016). The interaction between language and novel neuro-technologies in this review varies. Some neuro-technologies were used to examine language and cultural cues' effects on the brain while others were used to create a mathematical model to best capture the link between cognition and language (Perlovsky 2011; Taha et al. 2013b; Blanco-Elorrieta and Pylkkänen 2015). Incorporating local Arabic dialects to customize novel neuro-technologies such as fMRI and speech assistive tools were the focus of some studies in the review (Sakr 2011; Alkhashrami et al. 2014). The articles show how technology has been used to understand and interpret the neurophysiology of language, on the one hand, and that language was vital to enhance the utility and functionality of neuro-technologies, on the other hand. Thus, these are empirical evidence of how integral culture is for the discourse of and by novel technologies.

Another notable finding is that female research subjects were exceptionally outnumbered by men in all studies involving research participants in the review. The finding, it can be argued, reflects the general underrepresentation of women in public life such as in employment and education. This is something that dominates, at varying degrees, all the Arab countries (Karam and Afiouni 2014; OECD 2014). Indeed, female representation in research has also been criticized elsewhere and in other medical fields (Vitale et al. 2017; Jin et al. 2020) and this underrepresentation has negatively impacted their health (Norris et al. 2020).

Other voices from the region are singularly lacking in the results. There was no published literature that reviewed or examined input or considered views from religious or ethnic minorities such as Christians in Egypt, Lebanon or Syria or Nubians in Egypt's South. The discussions are dominated by Islamic views and religious practices such as praying and fasting related to Ramadan.

Some of the studies revealed the difficult choices healthcare professionals sometimes undertake to minimize healthcare costs and accommodate more patients (Helal et al. 2018). The researchers attempted to highlight priority setting and a utilitarian view of adjusting resources to maximize the number of patients treated. It also called into question the prohibitively expensive costs associated with advanced neuro-technologies, which are only affordable to high-income countries, further augmenting the healthcare equity divide between the haves and have nots.

An interesting observation was how scholars debating Islamic perspectives of neuroethics (Al-Delaimy 2012) have tried to reconcile the Islamic Jurisprudence rules to the four bioethical principles of autonomy, non-maleficence, beneficence and justice of Beauchamp and Childress (2001). "*Free will*" in Sharia/Islamic Jurisprudence corresponds to autonomy, "*the sanctity of life*" and "*there shall be no harm inflicted or reciprocated in Islam*" are analogous to non-maleficence, and last but not least "*necessities overrule prohibitions*" can be viewed as a form of supporting beneficence. The latter can also be interpreted as a utilitarian view of championing the good even in the face of defined strict deontology rules.

12.5. Conclusion

This chapter is one of the earliest review studies that maps out the perspectives from the Arab region on neuroethics and novel neuro-technologies. The study revealed a curious interplay between culture, ethics and novel neuro-technologies. The main bulk of the studies addressed the culture elements of language, religious practices and behavior. In this review, two elements, namely language and religious practices, were addressed.

The analysis also identified a wide gap between the current neuroethics debates in the Arab region and the in-depth deliberations in other parts of the world (Sakura 2012; Illes et al. 2019; Evers and Salles 2021). Neuroethics exists in the region as a field that does not seem to be well-established and that has little debate surrounding it, as it is dominated/infiltrated by religious concepts and argumentations.

There are a few limitations to this scoping review. One major shortfall is the failure to capture Arabic academic sources and literature, even though no language restrictions were imposed when performing the literature search. This is relevant because neuroethics as a field lies at the intersection of neuroscience, bioengineering, medicine, philosophy and ethics. While the former three disciplines are English-dominated in the region, the latter are primarily Arabic language subjects, so we can presume some academic scrutiny is carried out in Arabic. Furthermore, considering the lack of proper research infrastructure in most Arab countries and limited funds for academic publications in international journals, many researchers may resort to other forms of dissemination for their research such as online platforms (Amer 2019). Unfortunately, these alternative sources may not undergo traditional peer review processes. Therefore, future research endeavors should expand to include primarily Arabic sources and databases if they exist. In addition, other non-traditional or less established sources should be considered as well.

12.6. Acknowledgments

I am thankful for the feedback and suggestions given by colleagues at the Centre for Research Ethics and Bioethics at Uppsala University, in particular, that given by Michele Farisco and Arleen Salles.

12.7. Appendix

UPPSALA
UNIVERSITET

Uppsala University Library

APA PsycInfo, December 8, 2021		
Search number	Search term	Results
1	(DE "Neuroimaging") OR (DE "Electrical Brain Stimulation") OR (DE "Deep Brain Stimulation") OR (DE "Transcranial Direct Current Stimulation") OR (DE "Neuroimaging") OR (DE "Encephalography") OR (DE "Magnetic Resonance Imaging") OR (DE "Electroencephalography") OR (DE "Magnetoencephalography") OR (DE "Optogenetics")	137,706
2	TI OR AB ("brain-computer interface*" OR BCIs OR "brain imaging*" OR "brain-machine interface*" OR "brain mapping" OR "deep brain stimulation" OR electroencephalogra* OR EEG OR "magnetic resonance imaging" OR MRI OR "neural interface" OR "neural technolog*" OR "neuro-imag*" OR neuroimag* OR neuroprosthetic* OR "neural prosthetic*" OR neuroradiograph* OR neurostimulator* OR "neuro-technolog*" OR neurotechnolog* OR "non-invasive brain stimulation" OR "noninvasive brain stimulation" OR optogenetic* OR "transcranial direct current stimulation" OR "transcranial magnetic stimulation*")	151,037
3	1 OR 2	195,663
4	(DE "Arabs") OR (DE "Muslims") OR (DE "Islam")	5,123
5	TI OR AB (arab OR arabic OR arabs OR islam* OR maghreb* OR "middle east*" OR muslim* OR "north africa*" OR "northern africa*" OR shia OR sunni)	22,867
6	4 OR 5	23,805
7	(DE "Ethics") OR (DE "Bioethics") OR (DE "Morality") OR (DE "Religious Beliefs") OR (DE "Values") OR (DE "Social Approval") OR (DE "Social Desirability") OR (DE "Social Norms") OR (DE "Social Values") OR (DE "Sociocultural Factors") OR (DE "Philosophies")	158,984
8	TI OR AB (belief* OR bioethic* OR "community values" OR cultur* OR dignity OR ethic* OR moral* OR neuroethic* OR philosoph* OR religio* OR "socio-cultural*" OR "social norm*" OR "social stigma*" OR "social value*" OR "social view*" OR "societal norm*" OR "societal stigma*" OR "societal value*" OR "societal view*" OR theolog*)	647,618
9	7 OR 8	690,592
10	3 AND 6 AND 9	28
11	*filter: publication year 2000–*	28

\multicolumn{3}{c}{CINAHL, December 8, 2021}		
Search number	Search term	Results
1	(MH "Neuroradiography+") OR (MH "Neuroradiography") OR (MH "Transcranial Magnetic Stimulation") OR (MH "Brain-Computer Interfaces") OR (MH "Deep Brain Stimulation") OR (MH "Electroencephalography") OR (MH "Magnetic Resonance Imaging") OR (MH "Optogenetics")	160,316
2	TI OR AB ("brain-computer interface*" OR BCIs OR "brain imaging*" OR "brain-machine interface*" OR "brain mapping" OR "deep brain stimulation" OR electroencephalogra* OR EEG OR "magnetic resonance imaging" OR MRI OR "neural interface" OR "neural technolog*" OR "neuro-imag*" OR neuroimag* OR neuroprosthetic* OR "neural prosthetic*" OR neuroradiograph* OR neurostimulator* OR "neuro-technolog*" OR neurotechnolog* OR "non-invasive brain stimulation" OR "noninvasive brain stimulation" OR optogenetic* OR "transcranial direct current stimulation" OR "transcranial magnetic stimulation*")	128,408
3	1 OR 2	206,377
4	(MH "Middle East") OR (MH "Arabs") OR (MH "Africa, Northern") OR (MH "Islam")	10,267
5	TI OR AB (arab OR arabic OR arabs OR islam* OR maghreb* OR "middle east*" OR muslim* OR "north africa*" OR "northern africa*" OR shia OR sunni)	15,865
6	4 OR 5	20,204
7	(MH "Ethics") OR (MH "Bioethics") OR (MH "Decision Making, Ethical") OR (MH "Culture+") OR (MH "Morals") OR (MH "Social Values") OR (MH "Cultural Values") OR (MH "Stigma") OR (MH "Social Norms") OR (MH "Social Attitudes") OR (MH "Public Opinion") OR (MH "Philosophy")	265,976
8	TI OR AB (belief* OR bioethic* OR "community values" OR cultur* OR dignity OR ethic* OR moral* OR neuroethic* OR philosoph* OR religio* OR "socio-cultural*" OR "social norm*" OR "social stigma*" OR "social value*" OR "social view*" OR "societal norm*" OR "societal stigma*" OR "societal value*" OR "societal view*" OR theolog*)	319,925
9	7 OR 8	508,607
10	3 AND 6 AND 9	21
11	*filter: publication year 2000–*	21

\multicolumn{3}{c}{Cochrane Library, December 8, 2021}		
Search number	Search term	Results
1	"Brain-Computer Interfaces"[Mesh] OR "Deep Brain Stimulation"[Mesh] OR "Electroencephalography"[Mesh] OR "Implantable Neurostimulators"[Mesh] OR "Magnetic Resonance Imaging"[Mesh] OR "Neuroimaging"[Mesh] OR "Optogenetics"[Mesh] OR "Transcranial Magnetic Stimulation"[Mesh]	16,610
2	("brain-computer interface*" OR BCIs OR "brain imaging*" OR "brain-machine interface*" OR "brain mapping" OR "deep brain stimulation" OR electroencephalogra* OR EEG OR "magnetic resonance imaging" OR MRI OR "neural interface" OR "neural technolog*" OR "neuro-imag*" OR neuroimag* OR neuroprosthetic* OR "neural prosthetic*" OR neuroradiograph* OR neurostimulator* OR "neuro-technolog*" OR neurotechnolog* OR "non-invasive brain stimulation" OR "noninvasive brain stimulation" OR optogenetic* OR "transcranial direct current stimulation" OR "transcranial magnetic stimulation*"):ti,ab,kw	58,237
3	1 OR 2	59,701
4	"Middle East"[Mesh] OR "Africa, Northern"[Mesh] OR "Arabs"[Mesh] OR "Islam"[Mesh]	4,338
5	(arab OR arabic OR arabs OR islam* OR maghreb* OR "middle east*" OR muslim* OR "north africa*" OR "northern africa*" OR shia OR sunni):ti,ab,kw	2,141
6	4 OR 5	6,191
7	"Ethics"[Mesh] OR "Culture"[Mesh] OR "Sociological Factors"[Mesh] OR "Philosophy"[Mesh]	47,256
8	(belief* OR bioethic* OR "community values" OR cultur* OR dignity OR ethic* OR moral* OR neuroethic* OR philosoph* OR religio* OR "socio-cultural*" OR "social norm*" OR "social stigma*" OR "social value*" OR "social view*" OR "societal norm*" OR "societal stigma*" OR "societal value*" OR "societal view*" OR theolog*):ti,ab,kw	59,136
9	7 OR 8	102,403
10	3 AND 6 AND 9	17
11	*filter: publication year 2000–*	2

PubMed, NCBI, December 8, 2021		
Search number	Search term	Results
1	"Brain-Computer Interfaces"[Mesh] OR "Deep Brain Stimulation"[Mesh] OR "Electroencephalography"[Mesh] OR "Implantable Neurostimulators"[Mesh] OR "Magnetic Resonance Imaging"[Mesh] OR "Neuroimaging"[Mesh] OR "Optogenetics"[Mesh] OR "Transcranial Magnetic Stimulation"[Mesh]	782,904
2	"brain-computer interface*" OR BCIs OR "brain imaging*" OR "brain-machine interface*" OR "brain mapping" OR "deep brain stimulation" OR electroencephalogra* OR EEG OR "magnetic resonance imaging" OR MRI OR "neural interface" OR "neural technolog*" OR "neuro-imag*" OR neuroimag* OR neuroprosthetic* OR "neural prosthetic*" OR neuroradiograph* OR neurostimulator* OR "neuro-technolog*" OR neurotechnolog* OR "non-invasive brain stimulation" OR "noninvasive brain stimulation" OR optogenetic* OR "transcranial direct current stimulation" OR "transcranial magnetic stimulation*"[Title/Abstract]	618,770
3	1 OR 2	1,003,499
4	"Middle East"[Mesh] OR "Africa, Northern"[Mesh] OR "Arabs"[Mesh] OR "Islam"[Mesh]	192,803
5	arab OR arabic OR arabs OR islam* OR maghreb* OR "middle east*" OR muslim* OR "north africa*" OR "northern africa*" OR shia OR sunni[Title/Abstract]	50,415
6	4 OR 5	220,404
7	"Ethics"[Mesh] OR "Culture"[Mesh] OR "Sociological Factors"[Mesh] OR "Philosophy"[Mesh]	1,179,835
8	belief* OR bioethic* OR "community values" OR cultur* OR dignity OR ethic* OR moral* OR neuroethic* OR philosoph* OR religio* OR "socio-cultural*" OR "social norm*" OR "social stigma*" OR "social value*" OR "social view*" OR "societal norm*" OR "societal stigma*" OR "societal value*" OR "societal view*" OR theolog*[Title/Abstract]	1,531,970
9	7 OR 8	2,511,658
10	3 AND 6 AND 9	255
11	*filter: publication year 2000–*	237

Scopus, December 8, 2021		
Search number	Search term	Results
1	TITLE-ABS-KEY ("brain-computer interface*" OR BCIs OR "brain imaging*" OR "brain-machine interface*" OR "brain mapping" OR "deep brain stimulation" OR electroencephalogra* OR EEG OR "magnetic resonance imaging" OR MRI OR "neural interface" OR "neural technolog*" OR "neuro-imag*" OR neuroimag* OR neuroprosthetic* OR "neural prosthetic*" OR neuroradiograph* OR neurostimulator* OR "neuro-technolog*" OR neurotechnolog* OR "non-invasive brain stimulation" OR "noninvasive brain stimulation" OR optogenetic* OR "transcranial direct current stimulation" OR "transcranial magnetic stimulation*")	1,380,439
2	TITLE-ABS-KEY (arab OR arabic OR arabs OR islam* OR maghreb* OR "middle east*" OR muslim* OR "north africa*" OR "northern africa*" OR shia OR sunni)	275,370
3	TITLE-ABS-KEY (belief* OR bioethic* OR "community values" OR cultur* OR dignity OR ethic* OR moral* OR neuroethic* OR philosoph* OR religio* OR "socio-cultural*" OR "social norm*" OR "social stigma*" OR "social value*" OR "social view*" OR "societal norm*" OR "societal stigma*" OR "societal value*" OR "societal view*" OR theolog*)	4,568,045
4	1 AND 2 AND 3	137
5	*filter: publication year 2000–*	133

12.8. References

Abd El-Dayem, S.M. and Zyton, H.A.H. (2012). The effect of Ramadan fasting on multiple sclerosis. *Egyptian Journal of Neurology, Psychiatry & Neurosurgery*, 49(4).

Abu-Rabi, I.M. (1990). Contemporary Arab culture and the necessity of religious dialogue. *Islamic Studies*, 29(3), 287–301.

Ajami, H. (2016). Arabic language, culture, and communication. *International Journal of Linguistics and Communication*, 4(1), 120–123.

Al-Delaimy, W.K. (2012). Ethical concepts and future challenges of neuroimaging: An Islamic perspective. *Science and Engineering Ethics*, 18(3), 509–518.

AlHadi, A.N., AlShiban, A.M., Alomar, M.A., Aljadoa, O.F., AlSayegh, A.M., Jameel, M.A. (2017). Knowledge of and attitude toward repetitive transcranial magnetic stimulation among psychiatrists in Saudi Arabia. *The Journal of ECT*, 33(1), 30.

Alkhashrami, S., Alghamdi, H., Al-Wabil, A. (2014) Human factors in the design of Arabic-language interfaces in assistive technologies for learning difficulties. In *Human-Computer Interaction: Advanced Interaction, Modalities, and Techniques*, Kurosu, M. (ed.). Springer, Cham.

Amer, P. (2019). Not just money: Arab-region researchers face a complex web of barriers [Online]. Available at: https://www.al-fanarmedia.org/2019/12/not-just-money-arab-region-researchers-face-a-complex-web-of-barriers/.

Barugahare, J. (2018). African bioethics: Methodological doubts and insights. *BMC Medical Ethics*, 19(1), 1–10.

Beauchamp, T.L. and Childress, J.F. (2001). *Principles of Biomedical Ethics*. Oxford University Press, Oxford.

Beauregard, M. (2011). Neuroscience and spirituality – Findings and consequences. In *Neuroscience, Consciousness and Spirituality*, Walach, H., Schmidt, S., Jonas, W.B. (eds). Springer, Dordrecht.

Beauregard, M. and Paquette, V. (2006). Neural correlates of a mystical experience in Carmelite nuns. *Neuroscience Letters*, 405(3), 186–190.

Blanco-Elorrieta, E. and Pylkkänen, L. (2015). Brain bases of language selection: MEG evidence from Arabic-English bilingual language production. *Frontiers in Human Neuroscience*, 9. doi:10.3389/fnhum.2015.00027.

Buchanan, E.A. (1997). Cultural heritage, social values, and information in the Arab world. *Journal of Education for Library and Information Science*, 38(3), 215–220.

Burns, S.M., Barnes, L.N., McCulloh, I.A., Dagher, M.M., Falk, E.B., Storey, J.D., Lieberman, M.D. (2019). Making social neuroscience less WEIRD: Using fNIRS to measure neural signatures of persuasive influence in a Middle East participant sample. *Journal of Personality and Social Psychology*, 116(3), e1–e11. doi:10.1037/pspa0000144 10.1037/pspa0000144.supp.

Domínguez D.J.F., van Nunspeet, F., Gupta, A., Eres, R., Louis, W.R., Decety, J., Molenberghs, P. (2018). Lateral orbitofrontal cortex activity is modulated by group membership in situations of justified and unjustified violence. *Social Neuroscience*, 13(6), 739–755. doi:10.1080/17470919.2017.1392342.

Doufesh, H., Ibrahim, F., Ismail, N.A., Wan Ahmad, W.A. (2014). Effect of Muslim prayer (Salat) on α electroencephalography and its relationship with autonomic nervous system activity. *The Journal of Alternative and Complementary Medicine*, 20(7), 558–562.

Doufesh, H., Ibrahim, F., Safari, M. (2016). Effects of Muslims praying (Salat) on EEG gamma activity. *Complementary Therapies in Clinical Practice*, 24, 6–10.

Evers, K. and Salles, A. (2021). Epistemic challenges of digital twins & virtual brains perspectives from fundamental neuroethics. *Scio*, (21), 27–53.

Farisco, M., Salles, A., Evers, K. (2018). Neuroethics: A conceptual approach. *Cambridge Quarterly of Healthcare Ethics*, 27(4), 717–727.

Guenedi, A.A., Obeid, Y.A., Hussain, S., Al-Azri, F., Al-Adawi, S. (2009). Investigation of the cerebral blood flow of an Omani man with supposed "spirit possession" associated with an altered mental state: A case report. *Journal of Medical Case Reports*, 3(1), 1–5.

Helal, A.E., Abouzahra, H., Fayed, A.A., Rayan, T., Abbassy, M. (2018). Socioeconomic restraints and brain tumor surgery in low-income countries. *Neurosurgical Focus*, 45(4), E11.

Herrera-Ferrá, K., Salles, A., Cabrera, L. (2018). Global neuroethics and cultural diversity: Some challenges to consider. *The Neuroethics Blog* [Online]. Available at: http://www.theneuroethicsblog.com/2018/04/global-neuroethics-and-cultural.html [Accessed 10 April 2018].

Illes, J., Weiss, S., Bains, J., Chandler, J.A., Conrod, P., De Koninck, Y., Fellows, L., Groetzinger, D., Racine, E., Robillard, J. et al. (2019). A neuroethics backbone for the evolving Canadian brain research strategy. *Neuron*, 101(3), 370–374.

Ivancovsky, T., Kleinmintz, O., Lee, J., Kurman, J., Shamay-Tsoory, S.G. (2018). The neural underpinnings of cross-cultural differences in creativity. *Human Brain Mapping*, 39(11), 4493–4508. doi:10.1002/hbm.10.1002/hbm.24288.

Jastrzebski, A.K. (2018). The neuroscience of spirituality. *Pastoral Psychology*, 67(5), 515–524.

Jin, X., Chandramouli, C., Allocco, B., Gong, E., Lam, C.S., Yan, L.L. (2020). Women's participation in cardiovascular clinical trials from 2010 to 2017. *Circulation*, 141(7), 540–548.

Karam, C.M. and Afiouni, F. (2014). Localizing women's experiences in academia: Multilevel factors at play in the Arab Middle East and North Africa. *The International Journal of Human Resource Management*, 25(4), 500–538.

Methley, A.M., Campbell, S., Chew-Graham, C., McNally, R., Cheraghi-Sohi, S. (2014). PICO, PICOS and SPIDER: A comparison study of specificity and sensitivity in three search tools for qualitative systematic reviews. *BMC Health Services Research*, 14(1), 1–10.

Moosa, E. (2012). Translating neuroethics: Reflections from Muslim ethics. *Science and Engineering Ethics*, 18(3), 519–528.

Munn, Z., Peters, M.D., Stern, C., Tufanaru, C., McArthur, A., Aromataris, E. (2018). Systematic review or scoping review? Guidance for authors when choosing between a systematic or scoping review approach. *BMC Medical Research Methodology*, 18(1), 1–7.

Newberg, A.B., Wintering, N.A., Yaden, D.B., Waldman, M.R., Reddin, J., Alavi, A. (2015). A case series study of the neurophysiological effects of altered states of mind during intense Islamic prayer. *Journal of Physiology – Paris*, 109(4–6), 214–220.

Norris, C.M., Yip, C.Y., Nerenberg, K.A., Clavel, M.A., Pacheco, C., Foulds, H.J., Hardy, M., Gonsalves, C.A., Jaffer, S., Parry, M. et al. (2020). State of the science in women's cardiovascular disease: A Canadian perspective on the influence of sex and gender. *Journal of the American Heart Association*, 9(4), e015634.

OECD (2014). *Women in Public Life: Gender, Law and Policy in the Middle East and North Africa*. OECD Publishing, Berlin.

Orr, G. (2003). *Diffusion of Innovations*, by Everett Rogers (1995) [Online]. Available at: https://web.stanford.edu/class/symbsys205/Diffusion%20of%20Innovations.htm [Accessed 21 January 2005].

Perlovsky, L. (2011). Language and cognition interaction neural mechanisms. *Computational Intelligence and Neuroscience*, doi:10.1155/2011/454587.

Pretus, C., Hamid, N., Sheikh, H., Ginges, J., Tobeña, A., Davis, R., Vilarroya, O., Atran, S. (2018). Neural and behavioral correlates of sacred values and vulnerability to violent extremism. *Frontiers in Psychology*, 9. doi:10.3389/fpsyg.2018.02462.

Prinz, J. (2020). Culture and cognitive science. In *Metaphysics Research Lab*, Zalta, E.N. (ed.). Stanford University, CA.

Rickabaugh, B. and Evans, C.S. (2018). Neuroscience, spiritual formation, and bodily souls: A critique of Christian physicalism. In *Christian Physicalism? Philosophical Theological Criticisms*, Loftin, R.K. and Farris, J. (eds). Lexington, Lanham, MD.

Rim, J.I., Ojeda, J.C., Svob, C., Kayser, J., Drews, E., Kim, Y., Tenke, C.E., Skipper, J., Weissman, M.M. (2019). Current understanding of religion, spirituality, and their neurobiological correlates. *Harvard Review of Psychiatry*, 27(5), 303–316. doi:10.1097/HRP.0000000000000232.

Robertson, C.J., Al-Khatib, J.A., Al-Habib, M. (2002). The relationship between Arab values and work beliefs: An exploratory examination. *Thunderbird International Business Review*, 44(5), 583–601.

Roky, R., Chapotot, F., Benchekroun, M.T., Benaji, B., Hakkou, F., Elkhalifi, H., Buguet, A. (2003). Daytime sleepiness during Ramadan intermittent fasting: Polysomnographic and quantitative waking EEG study. *Journal of Sleep Research*, 12(2), 95–101. doi:10.1046/j.1365-2869.2003.00341.x.

Rørbæk, L.L. (2019). Religion, political power, and the "sectarian surge": Middle Eastern identity politics in comparative perspective. *Studies in Ethnicity and Nationalism*, 19(1), 23–40.

Roskies, A. (2002). Neuroethics for the new millenium. *Neuron*, 35(1), 21–23.

Sakr, H.M. (2011). Language and memory lateralization and localization using different fMRI paradigms in Arabic speaking patients: Initial experience. *Egyptian Journal of Radiology and Nuclear Medicine*, 42(2), 223–231. doi:10.1016/j.ejrnm.2011.05.005.

Sakura, O. (2012). A view from the far east: Neuroethics in Japan, Taiwan, and South Korea. *East Asian Science, Technology and Society: An International Journal*, 6(3), 297–301.

Serour, G.I. and Serour, A.G. (2021). The impact of religion and culture on medically assisted reproduction in the Middle East and Europe. *Reproductive BioMedicine Online*, 43(3), 421–433.

Taha, H., Ibrahim, R., Khateb, A. (2013). How does Arabic orthographic connectivity modulate brain activity during visual word recognition: An ERP study. *Brain Topography*, 26(2), 292–302.

Trabelsi, K., Ammar, A., Zlitni, S., Boukhris, O., Khacharem, A., El-Abed, K., Khanfir, S., Shephard, R.J., Stannard, S.R., Bragazzi, N.L. et al. (2019). Practical recommendations to improve sleep during Ramadan observance in healthy practitioners of physical activity. *La Tunisie Médicale*, 97(10), 1077–1086.

UNESCO (2012). States parties [Online]. Available at: https://whc.unesco.org/en/statesparties/?searchStates=&id=&region=4&submit=Search.

Vitale, C., Fini, M., Spoletini, I., Lainscak, M., Seferovic, P., Rosano, G.M. (2017). Under-representation of elderly and women in clinical trials. *International Journal of Cardiology*, 232, 216–221.

World Values Survey Association. (2020). The Inglehart-Welzel world cultural map – World values survey 7 (Provisional version) [Online]. Available at: https://www.worldvaluessurvey.org/WVSNewsShow.jsp?ID=428.

Yogeeswaran, K., Nash, K., Jia, H., Adelman, L., Verkuyten, M. (2021). Intolerant of being tolerant? Examining the impact of intergroup toleration on relative left frontal activity and outgroup attitudes. *Current Psychology: A Journal for Diverse Perspectives on Diverse Psychological Issues*. doi:10.1007/s12144-020-01290-2.

13

The Binary Illusion

Karin GRASENICK
Convelop Cooperative Knowledge Design GmbH, Graz, Austria

13.1. A brain is still a brain

Neuroscience, similar to all fields of science, especially those whose research objects are human beings, is embedded in a specific societal context, which determines how individuals who are assigned or identify with a specific gender are perceived and which opportunities are offered to them. As such, neuroscience has intentionally and unintentionally contributed to the neglect of health issues, the reproduction of binary roles for women and men and the justification of societal power relations. The question arises how neuroethical reflection can contribute to careful contextualization of methodological designs and neuroscientific conclusions drawn for the benefit of society, as well as the benefit of each individual affected by its findings.

To introduce the complex interplay of science and medical practices with societal and personal factors, a fictional example is combined with an actual historical example from the time when science was used specifically for ideological purposes, the effects of which can still be observed to this day. Generally, historical and recent findings in brain research, as well as critical reviews on prevailing biases, have led to a heated debate regarding the extent to which the search for sex/gender differences supports a political agenda justifying and enhancing gender roles that are considered outdated. This critical reflection has led to two different lines of argument:

– Feminist neuroscience has focused on proving that dimorphic differences between men and women, especially in relation to cognitive abilities, do not exist. In addition, it has been demonstrated that studies have often reproduced stereotypes to biologically ground what can actually be considered as the result of societal factors.

– In contrast, gender medicine (GM) has focused on proving that biological differences between men and women are, in fact, relevant and that researchers should not generalize conclusions drawn solely from studying one group as a valid result for all individuals.

Both lines of argument started with a binary distinction of sex/gender and a focus on Western societal perspectives before moving towards models that acknowledge the complex interplay of biological, societal, cultural and socio-economic factors. In a century where individualism prevails in Western culture, they advocate an individualized medical approach. These developments illustrate that while it is possible to distinguish scientific progress from biased research or unethical medical practice with the benefit of hindsight, any critical reflection has also overlooked factors and omitted findings that could disrupt reasoning based on a particular research interest. Based on the analysis of the well-founded and at the same time controversial contributions to sex/gender in neuroscience, the author of this chapter derives seven questions for a neuroethical reflection which specifically encourage us to contextualize actual scientific practices and invite an interdisciplinarity dialog.

13.2. Imagine all the people living life in vain

Imagine that you have a highly talented friend who is experiencing professional and financial hardships and is financially insecure. This friend's lifestyle and choices are unorthodox and considered societally scandalous and damaging to their family's reputation, causing their family to reject them. Recently, this friend has not been responding to your messages. Therefore, you decide to pay them a visit, only to find them in a very shocking and disturbing state, not eating and constantly talking about being persecuted. The windows are shut, the furniture is shattered, and the entire flat is filthy. You try to reach out to their family, but they refuse to take note of the situation, let alone get involved. Therefore, you consult with mutual friends and a psychiatric clinic of excellent reputation, and the conclusion is that hospitalization seems to be the only advisable solution. Your friend is forcibly committed to a clinic; the doctors confirm the seriousness of the situation and diagnose them with schizophrenia. You leave the clinic relieved, convinced that, with the right therapy and enough time to rest, they will certainly recover. However, despite all hopes, and although they do not want to stay, they never leave the clinic again. Much later on, you read a note left by your friend saying "I live in a world that is so curious, so strange. Of the dream, which was my life, this is the nightmare" (Mathews 1999, p. 84).

Actually, these sentences were written by Camille Claudel, a 19th-century woman born in 1864 who wanted to be a sculptor and indeed became known for her exceptional art (Souter 2018). Her talent was recognized by her father, as well as other artists and mentors, at a time when striving for an artistic career was inappropriate for women, especially sculpturing. Yet, Claudel attended a private art school and became a scholar and later a colleague, co-creator and lover of Auguste Rodin, although he continued to pursue his partnership with the mother of his son. At the time, this life was considered extremely scandalous, and Claudel was abandoned by her family. She struggled to further develop her own independent work; a lack of public commission made her financial situation difficult. She became increasingly isolated and started to destroy her own work. In 1909, after visiting his sister, Paul Claudel made the following note: "In Paris, insane Camille. The paper on the walls torn to shreds, a single armchair broken and torn, horrible filth. She was enormous and her face soiled, talking incessantly in a monotonous, metallic voice" (Claudel 1995, p. 105).

Following the death of her father, who had always protected and sponsored her, Camille's family immediately requested that she be admitted to a mental asylum. Her certificate of admission described her as paranoid, suffering from "systematic persecution delirium based on false interpretations and imaginations". Over the remaining 30 years of her life, she asked her family several times to release her from the asylum. Her doctors even recommended her release at least twice. However, her family refused.

13.3. This land is my land, from the asylum to the last island

Claudel's story raises awareness of how life circumstances and societal and cultural factors can contribute to a mental condition and a diagnosis with far-reaching consequences. This story took place in an era when a hierarchical classification of binary sexes and races was strongly supported by scientific reasoning.

Claudel lived in a time of scientific discoveries and societal change. The first light bulb shone when she was 14 years old. The technological revolution of 1870–1914 (Wolfe and Poolos 2015, pp. 86–89) was characterized by increasing industrialization and standardization processes. Simultaneously, imperialistic nation states expanded their global territories (Berlin Conference 1884–1885, also known as the Scramble for Africa) (Pakenham 1991; Williams and Chrisman 1994). The

European culture also came to be considered as superior, and newly developed scientific methods promised to prove it.

During this new era of increasingly industrialized and imperialistic nation states, territorial claims over citizens and colonies were scientifically justified. At the same time, sex differences were explored to argue that the specific gender roles within society were biologically grounded. In her book "The Gendered Brain", Gina Rippon quotes Gustave Le Bon, who was interested in scientifically proving a hierarchy of races and sexes and has influenced other scientists and powerful societal members beyond the 19th century, including those supporting National Socialism. According to him, "without doubt there exist some distinguished women, very superior to the average man, but they are as exceptional as the birth of any monstrosity, as, for example, of a gorilla with two heads; consequently, we may neglect them entirely" (Rippon 2020, p. 6).

To prove the inferiority of women, several scientific measures were taken, with the focus shifting from the reproduction function to cognitive abilities. As a starting point, it was proven that the brains of women were on average smaller than those of men, and this difference in size and weight was directly related to intelligence: "In general, the brain is larger in mature adults than in the elderly, in men than in women, in eminent men than in men of mediocre talent, in superior races than in inferior races" (Gould 1981, p. 83). Such arguments, which were supposedly based on scientific facts, were used to ethically justify discrimination against those identified as most unfortunately disadvantaged by nature itself.

> Sex and gender have been differentiated as binary categories. It is thereby assumed that sex is determined by biological factors. If the combination of chromosomes, hormones and physiological factors cannot be determined as either male or female, the term "intersex" is used (Ainsworth 2015). In this context, the terminology of hermaphroditism was used until the 21st century, regardless of its questionable historical roots (Dreger et al. 2005; Vilain et al. 2007). The term "gender" refers to the social construction of women, men and non-binary individuals by associating competences, behaviors and attitudes with a person's biological sex. Moreover, "sex/gender" has been used to raise awareness regarding the notion that both categories are socially constructed and that both affect the brain in a complex interplay (Kaiser 2012).

Box 13.1. *Definitions of sex, gender and sex/gender*

For such scientific and ideological ambitions, a demarcation line between the two sexes would indeed have been most helpful, to clearly assign rights and duties precisely to one kind of person, or the other. However, even when just the inspection of genitalia was the method of choice, it was well known that sexes showed a high variability in terms of physical appearance, talents and aspirations, as well as perceived identity, which was additionally mixed up with a sexual orientation called "psychic hermaphroditism" (Chauncey 1982; Dreger 1998).

13.4. Little bits of history repeating

Throughout history, medical research has been used either directly by scientists or indirectly by politicians and influential members of society to rationalize and legitimize a cultural status quo. It has also been occasionally used to justify societal power relations, with severe consequences for those who do not (want to) fit into foreseen roles. Scientists, particularly neuroscientists, have been criticized for the methodological designs and conclusions driven by their conscious or unconscious desire to biologically rationalize an ideology or at least unreflective societal norms.

In this sense, awareness of historical roots explains the vigilant and critical reflection by feminist scholars, which resists the justification of societal power relations based on biological differences and scientific reasoning for gender roles. As such, over the centuries, the findings of brain research have served as arguments in societal discussions on sex and gender. For example, over the last decade, in cognitive and emotional behavior studies, functional magnetic resonance imaging (fMRI) has been used to describe anatomical and functional differences as evidence of sexual dimorphism. For instance, Satterthwaite et al. (2015) analyzed the patterns of functional connectivity in the resting state using fMRI to demonstrate that men outperform women in motor and spatial cognitive tasks and that women outperform men in tasks that require emotion identification and non-verbal reasoning.

These findings based on MRI techniques were, however, criticized. For instance, several feminist neuroscientists analyzed the research designs and outcomes of such studies and concluded that they were biased and discriminatory. There has also been some doubt regarding the extent to which resulting images can be clearly categorized as sex specific. Moreover, other scientists have demonstrated that some studies have overemphasized the differences between women and men as statistically significant without reporting the effect size or the commonalities of the participants (Joel et al. 2015). Several feminists have explored the unjustified emphasis on such differences and criticized the conclusions derived from the binary differentiation of mental processes such as cognition or empathy (Studholme et al. 2020). Even conflicting data have been interpreted to conform to actual gender stereotypes (Bluhm 2013).

However, the popular science literature has adopted such neuroscientific announcements to explain the behavioral differences between men and women as deterministic and biologically grounded. Books such as "Men Are from Mars, Women Are from Venus" (Gray 2012) and "Why Men Can Only Do One Thing at a Time and Women Never Stop Talking" (Pease and Pease 2017) have promised to improve personal relationships. Such stereotypes imply that there is a "naturally" good or bad behavior according to sex/gender. These highly popular guidebooks also ascribe specific talents to one sex, or the other in a stereotypical way, which counteracts the measures taken to enhance gender equality in education, professional life and health care.

Feminist research has relentlessly striven to elucidate how scientific questions and interpretation of results have served to biologically justify social inequalities. Neurosexism and neurogendering were introduced to describe the misuse of neuroscience to contribute to the reproduction of stereotypes and the justification of power relations (Fine 2010). Methodological approaches, the acquisition of data and the interpretation of results in search of dimorphic cognitive differences have been criticized (Cipolla and Gupta 2017; David et al. 2018; Eliot et al. 2021). Accordingly, it has been requested that neuroscientists should acknowledge the high variability of individual brains, rather than assume binary sex-specific brain differences (Jäncke 2018; Joel and Vikhanski 2019).

13.5. What's good for me is good enough for you

While feminist neuroscience mainly focuses on apparent biological differences in the cognitive abilities of men and women, GM has become a subject taught at medical universities that has drawn attention to the disturbing consequences of a lack of sex/gender differentiation in the diagnosis and treatment of diseases (Lippi et al. 2020). Medical research has ignored sex/gender differences and their influence on symptoms, disease progression and treatment effects. Healy (1991) was the first to pay attention to this "one-size-fits-all" approach of medical research and practice. As a cardiologist, Healy focused on the different symptoms of heart attacks in men and women and the consequences of a wrong diagnosis. In their meta-analysis on actual GM publications, Shai et al. showed that "gender disparities and their relation to health are still addressed in a minority of the GM publications" (Shai et al. 2021, p. 6). The results from studies based solely on men have been considered valid for *mankind* in general (Cahill 2014; Cleghorn 2021).

Neuroscientific findings have also been derived from studies with only male subjects, thereby neglecting potential sex/gender differences and further diversity traits. Twenty years after Healy's ground-breaking study, Beery and Zucker reviewed studies based on mammals across several disciplines and reported that in 2009 "male bias was evident in 8 disciplines and most prominent in neuroscience, with single-sex studies of male animals outnumbering those of females 5.5 to 1" (Beery and Zucker 2011, p. 565).

Thus, research design should pay more attention to sex differences that can be traced down to the level of stem cells (Shah et al. 2014). In mammals, biological sex differences are understood as based on sex chromosomes and steroid hormones. Many studies have shown that sex steroids are relevant for the course of various diseases, especially neurodegenerative diseases, such as Parkinson's disease (Bryant et al. 2019) and multiple sclerosis (Du et al. 2014). Sex differences in patients with Alzheimer's disease were demonstrated in brain regions critical for high cognitive function (Ferretti et al. 2018). Moreover, neurodegenerative diseases have been demonstrated to take different courses for women and men, who also react differently to drugs (Gobinath 2019; Mauvais-Jarvis et al. 2020). For example, for migraines, a pain condition that is more frequent in women than in men, increased pain sensitivity has been proven in women (Cahill 2014; Clayton and Collins 2014; Gobinath 2019; Mauvais-Jarvis et al. 2020).

These insights have contributed to increased awareness of and interest in exploring sex and gender differences in neuroscience (Clayton and Collins 2014). Given the findings obtained over the last decade, it has been argued that findings on sex differences in the brain are robust, an argument that has been followed by enthusiastic statements:

> Knowing that the entire brain is affected by sex hormones with subtle sex differences, we are entering a new era in our ability to understand and appreciate the diversity of gender-related behaviours and brain functions. This will enhance "personalized medicine" that recognizes sex differences in disorders and their treatment and will improve understanding of how men and women differ, not by ability, but by the "strategies" they use in their daily lives (Marrocco and McEwen 2016).

13.6. They keep saying they have something for you

Generally, GM has advocated for the relevance of sex differences, thereby including brain-related diseases and mental disorders. However, as feminist research has demonstrated, sex as a biological factor has been difficult to differentiate from gender and cultural and societal factors. For example, living conditions and life experiences such as violence and abuse influence several biological factors, such as hormone levels, and response to medical treatment. By neglecting the intersectionality between biological and cultural and societal factors, GM is at risk of drawing incorrect conclusions. In addition, the consequences of reductionism not considering the complex interplay of biological and socio-cultural factors, combined with a lack of scientific self-reflection on our own societal embeddedness, are illustrated by the historical development of mental health research and treatment.

In the 19th century, new asylums promised to cure "the insane" for the benefit of society and patients alike. As a result, in many nations, the number of asylums grew rapidly (Scull 2021). The so-called moral treatment was considered to enhance the behavioral and moral self-control (Wallace and Gach 2008, p. 536) of "abnormal" individuals. Using hypnosis and emergent psychotherapies, many physicians optimistically claimed that "insanity" could be cured. However, by the end of the century, all types of cruel physical experiments had been performed on the inmates of the by-then-overcrowded asylums. Those diagnosed with hysteria, which was considered a biologically grounded mental disorder of women, were at a risk of having their ovaries or uteruses removed (Tasca et al. 2012, p. 118) before, in line with scientific progress, their brains were acknowledged as being of relevance.

While often merely diagnosed through symptoms, mental illness was further argued as biologically rooted and treated with seizures, shocks, coma therapy and invasive interventions, irreversibly damaging the brains of those identified as needing such treatment, that is, people suffering from hallucinations, depression, anxiety, disorientation and so forth, of whom 60–70% were women. For example, lobotomy, a surgery that permanently damages parts of the prefrontal lobe, won its so-called pioneers a Nobel Prize and was widely used until 1970 (Wallace and Gach 2008), despite eventually being recognized for its "side effects" rather than its supposedly positive outcomes (Berrios and Porter 1999).

In the 1950s, psychiatric drugs promised cures and created new markets for pharmaceutical companies. However, once again, apparent scientific progress has been embraced rapidly and too enthusiastically, as exemplified by the Clinical Antipsychotic Trials of Intervention Effectiveness (CATIE) study. In this study, the anti-psychotic drugs available for schizophrenia since the 1990s were compared, concluding that the new drugs were more expensive and, despite all prior announcements, had severe known and new side effects and that there was no clear evidence for better outcomes (Lieberman et al. 2005; National Institute of Mental Health 2005; Scull 2021, pp. 174–175).

Such careful questioning, as in the CATIE study, should not be restricted to the actual treatment of a disease, but should rather also include how a disease is diagnosed. Diagnosis has proven to be a rather difficult endeavor, and retrospectively, it has been strongly influenced by cultural perceptions of deviant behavior.

A standard guide for defining and classifying mental disorders in international use is the American Psychiatric Association's Diagnosis and Statistical Manual (DSM), which has been regularly revised, thereby offering insights into the history of re-classification and interpretations of symptoms (Scull 2021, p. 272). It was not until 1952 that hysteria was removed from classification systems. The first introductory note that the DSM might not be applicable across different cultures appeared in 1987. Moreover, in 1995, a reference to "culture-bound syndromes" and a framework to consider the cultural aspects of a person's mental illness were included (Dinos et al. 2017).

Several studies have raised concerns that the diagnosis of mental disorders, such as schizophrenia, differs depending on the cultural context (Viswanath and Chaturvedi 2012; Yarris and Ponting 2019). In general, cultural norms influence how affected people and their socio-cultural environment perceive their condition. Examples are the expected interactions between doctors, family members and patients and their individual responsibility. How individuals and families participate in treatment is influenced by religious beliefs that intersect further with gender-related roles in society (Yarris and Ponting 2019). "Culture is a conceptually distinct and potentially powerful environmental factor capable of exerting a significant effect on the course of schizophrenia or any other mental disorder" (Edgerton and Cohen 1994, p. 230).

Life experiences clearly differ depending on sex/gender intersecting with further diversity traits, such as race or ethnicity, in their specific socio-cultural context. Experiences of discrimination, domestic violence or abuse, as expected, increase the risk of mental disorders, such as anxiety or depression, and should therefore be considered in diagnosis and treatment. Underestimating or even ignoring diversity or individual and societal perceptions of behavior contributes to harmful conclusions and practices (Nasser et al. 2002). Hence, research focusing solely on biological factors and treatments relying on psychopharmaceuticals have been opposed (Oram et al. 2013; Riecher-Rössler 2017, p. 8).

While diversity traits influence (mental) health as well as the course and response to treatment, mental disorders such as schizophrenia, anxiety and depression are characterized by similar symptoms across sexes/genders and cultures. Moreover, as sex-related factors such as genes and hormones intersect with not only cultural factors but also individual psychological or behavioral characteristics, biology alone cannot sufficiently explain mental disorders or neurodegenerative diseases (Joel et al. 2015). As a result, Joel et al. advocated for individualized medicine, which broadens the perspective beyond sex/gender to include other factors. In doing so, they have once again highlighted the importance "to ensure that the focus of research remains on the individual with the disease, rather than the disease alone" (Nasser et al. 2002, p. 358).

13.7. Sign of the times

As history reveals, until now, brain research has had an impact on those classified as patients and treated "with the best of intentions" in accordance with the latest research. Findings that had their beginning with the comparison of brain sizes have either deliberately or unintentionally contributed to justify what opportunities a society offers its (non-) conforming members for personal development and participation. Moreover, research on the differences between women and men has been misused to portray the inferiority of women, with the line of arguments becoming subtler over time.

Throughout history, feminist scientists have helped reveal scientific weaknesses, specifically in the search for biological reasons for cognitive differences. In the fight for equal rights and social change, many attempts to scientifically justify inequalities have been unmasked as ideologies. Such critiques have challenged the notion of simple dimorphic sexual differentiation, particularly with respect to the brain, and have led to a better understanding of the complex interplay between biological and cultural factors making each brain unique.

Simultaneously, neuroscience has been criticized for ignoring potential differences between groups of individuals with respect to health issues, such as mental health and neurodegenerative diseases. Calls to analyze differences in symptoms, progression and treatment, as well as the psychological and social impact of diseases, are rooted in GM, an approach that traditionally focuses on acknowledging biological differences as a binary differentiation of women and men (Baggio et al. 2013; Shai et al. 2021). GM has led to the development of regulations stipulating the inclusion of (still binary) sex/gender in study designs and publications, thereby, among other effects, preventing findings based on studying specific aspects of homogeneous groups from being applied as valid for all (groups of) people (Grasenick 2019).

The definition of sex as binary was re-considered on the basis of insights into the complex interplay of genetic, hormonal and environmental factors. In addition, the interplay among biological, socio-economic, cultural and psychological factors has been acknowledged as relevant for cognitive and behavioral human variability, including regarding mental disorders and diseases. Hence, given the difficulties in differentiating between sex and gender and the combined effects of both sex and gender, the term "sex/gender" was introduced (Marrocco and McEwen 2016).

If cognitive abilities and behaviors cannot be categorized as dimorphic and solely biologically grounded, does that mean that the relationships between biological sex attributes, health and wellbeing should not be considered? Several neuroscientists focusing on biological aspects have protested all their research being labeled as flawed, one-sided or even neurosexist. There have also been claims that feminist neuroscientists recommend abandoning studying sex differences (Cahill 2014). However, this is an erroneous simplification that negates the necessity of developing a model that better operationalizes the complexity by which:

> [...] genetic, hormonal, and environmental factors [...] act via multiple partly independent mechanisms that may vary according to internal and external factors. These observations led to the "mosaic" hypothesis – the expectation of high variability in the degree of "maleness"/"femaleness" of different features within a single brain. [... C]onclud[ing] that co-analysis of several (preferably, many) features and going back from the group level to that of the individual would advance our understanding of the relations between sex and the brain in health and disease (Joel et al. 2020).

Scientists such as Joel or Rippon do not claim that differences are a myth, but rather that clear dimorphic differences in cognition can be called a myth. They emphasize that actual neuroscientific findings can contribute to broadening the understanding of human variability and open societal gender norms. To counteract potential misuse of findings, neurofeminists have drawn attention to the need for higher-quality standards for the reporting of statistical findings (Roy 2012). This means that neuroscientists should cautiously consider how their findings are communicated in public media and carefully reflect on potentially unexplored co-variables that may have caused a specific effect. Differences with a small effect size should not be labeled as "sex differences", but rather as "sex effects" (Maney 2015). According to their research on academic flaws in behavioral neuroscience, Fine et al. (2019) offered a public guideline to help people critically examine scientific findings before making further use of such findings by questioning statistics and explanatory models (Fine et al. 2019).

13.8. Just microscopic cogs for a neuroethics plan (conclusion)

The previous sections demonstrated that neuroscientific assumptions and findings are tightly linked to their contemporary culture and historical roots. Accordingly, we may ask ourselves: what if Camille Claudel lived today and we had to evaluate her condition and decide what would be best for her? Would the results be different? What if families and friends who are struggling to care for their loved ones ask for professional support? What influences the resulting treatment nowadays? Can such a thought experiment be applied to an actual story? Britney Spears is an example of how societal factors, in her case mass media and apparent public gloating, may have contributed to a personal mental breakdown and, over decades incapacitation and treatments without consent (Gilbert 2021).

We may think that references to historical and contemporary events are less suitable for neuroethical reflection and more appropriate for physicians, lawyers, journalists and politicians. However, such examples help to clarify actual, concrete consequences of scientific developments that may not be presented in a discussion on an abstract level. They raise awareness to the complex interplay of biological and socio-cultural factors and may therefore support a neuroethical reflection that contextualizes neuroscientific findings.

For scientific reflections on how to consider sex/gender, several guidelines have been developed – for neuroscience, as well as for research and development in general (e.g. Tannenbaum et al. 2019; Grasenick et al. 2020). Such guidelines support the integration of sex/gender and intersectional factors in research designs, methodological approaches and conclusions. These questions have been difficult to address within one discipline, especially for neurosciences, as they call for an interdisciplinary approach to better consider the biological, psychosocial, socio-cultural and ethical factors relevant for a phenomenon. In addition, ethical, legal and societal implications cannot be considered independently of the respective context. Neuroethics may provide the framework and guiding questions for a context sensitive reflection, which benefits from interdisciplinarity. Therefore, there have been calls for neuroethics to collaborate more deeply with further disciplines, such as the history of medicine or feminist neuroscience, to offer a forum for an interdisciplinary discourse and "create opportunities for shared perplexity" (Roy 2012). In this sense, Roy has suggested three guiding questions referring to the purpose of research on differences, biological complexity and conclusions drawn on structural and functional differences, especially within brain research (Roy 2012).

However, neuroethics should contribute even more profoundly to strengthen the receptiveness of interdisciplinary collaboration and willingness to consider the potential harmful consequences of neuroscientific practices. For such intentions, it is important to question which assumptions, values and norms scientists, members of their society and of specific schools of thought base their research on. Therefore, this chapter concludes with seven questions for neuroethical reflections (which can be abbreviated to SEQUENCIAL-reflection) that extend existing guidelines through the preceding analysis of the different lines of argumentation on sex/gender in neuroscience. SEQUENCIAL-reflection includes assumptions underlying current practices, as well as inter- and trans-disciplinary collaboration to further explore new alternatives to binary illusions.

---

1) What are the cultural roots of the actual scientific approach, and to what extent are research questions and designs based on societal values and norms?

2) Has the scientific focus been clearly contextualized, explaining which biological, social and psychological factors have not been considered and why?

3) Have models and variables in use been questioned for potentially different ways of conceptualization and disaggregation?

4) Have differences and commonalities been studied?

5) Have the findings been critically reflected on in the context of potentially alternative interpretations?

6) Has an inter- and trans-disciplinary discourse taken place across natural and social sciences, including those affected by the findings?

7) Have the conclusions and public communication of findings been critically evaluated against potential misinterpretations?

Box 13.2. *SEven QUEstions for NeuroETHICAL reflection (SEQUENCIAL-reflection)*

13.9. References

Ainsworth, C. (2015). Sex redefined. *Nature*, 518(7539), 288–291. DOI: 10.1038/518288a.

Baggio, G., Corsini, A., Floreani, A., Giannini, S., Zagonel, V. (2013). Gender medicine: A task for the third millennium. *Clinical Chemistry and Laboratory Medicine*, 51(4). DOI: 10.1515/cclm-2012-0849.

Beery, A.K. and Zucker, I. (2011). Sex bias in neuroscience and biomedical research. *Neuroscience and Biobehavioral Reviews*, 35(3), 565–572. DOI: 10.1016/j.neubiorev.2010.07.002.

Berrios, G.E. and Porter, R. (1999). *A History of Clinical Psychiatry. The Origin and History of Psychiatric Disorders*. Athlone, London.

Bluhm, R. (2013). New research, old problems: Methodological and ethical issues in fMRI research examining sex/gender differences in emotion processing. *Neuroethics*, 6(2), 319–330. DOI: 10.1007/s12152-011-9143-3.

Bryant, K., Grossi, G., Kaiser, A. (2019). Feminist interventions on the sex/gender question in Neuroimaging research. *S&F Online* 15(2), 19 [Online]. Available at: https://sfonline.barnard.edu/neurogenderings/feminist-interventions-on-the-sex-gender-question-in-neuroimaging-research/# [Accessed 13 August 2021].

Cahill, L. (2014). Equal ≠ the same: Sex differences in the human brain. *Cerebrum: The Dana Forum on Brain Science*, 5.

Chauncey Jr., G. (1982). From sexual inversion to homosexuality: Medicine and the changing conceptualization of female deviance. *Salmagundi* 58/59, 114–146 [Online]. Available at: https://www.jstor.org/stable/40547567.

Cipolla, C. and Gupta, K. (2017). Neurogenderings and neuroethics. In *The Routledge Handbook of Neuroethics*, 1st edition, Syd, L., Johnson, K.M., Rommelfanger, S. (eds). Routledge, Taylor & Francis Group, New York.

Claudel, P. (1995). *Journal Tome I: 1904–1932*. Gallimard – Pléiade, Paris.

Clayton, J.A. and Collins, F.S. (2014). Policy: NIH to balance sex in cell and animal studies. *Nature*, 509(7500), 282–283. DOI: 10.1038/509282a.

Cleghorn, E. (2021). *Unwell Women. Misdiagnosis and Myth in a Man-made World*. Dutton Penguin Random House LLC, New York.

David, S.P., Naudet, F., Laude, J., Radua, J., Fusar-Poli, P., Chu, I., Stefanick, M.L., Ioannidis, J.P.A. (2018). Potential reporting bias in neuroimaging studies of sex differences. *Nature Publishing Group* [Online]. Available at: https://www.nature.com/articles/s41598-018-23976-1 [Accessed 7 May 2021].

Dinos, S., Ascoli, M., Owiti, J.A., Bhui, K. (2017). Assessing explanatory models and health beliefs: An essential but overlooked competency for clinicians. *BJPsych Advances*, 23(2), 106–114. DOI: 10.1192/apt.bp.114.013680.

Dreger, A.D. (1998). *Hermaphrodites and the Medical Invention of Sex*. Harvard University Press, Cambridge, MA.

Dreger, A.D., Chase, C., Sousa, A., Gruppuso, P.A., Frader, J. (2005). Changing the nomenclature/taxonomy for intersex: A scientific and clinical rationale. *Journal of Pediatric Endocrinology and Metabolism*, 18(8), 729–733. DOI: 10.1515/JPEM.2005.18.8.729.

Du, S., Itoh, N., Askarinam, S., Hill, H., Arnold, A.P., Voskuhl, R.R. (2014). XY sex chromosome complement, compared with XX, in the CNS confers greater neurodegeneration during experimental autoimmune encephalomyelitis. *Proceedings of the National Academy of Sciences of the United States of America*, 111(7), 2806–2811. DOI: 10.1073/pnas.1307091111.

Edgerton, R.B. and Cohen, A. (1994). Culture and schizophrenia: The DOSMD challenge. *The British Journal of Psychiatry: The Journal of Mental Science*, 164(2), 222–231. DOI: 10.1192/bjp.164.2.222.

Eliot, L., Ahmed, A., Khan, H., Patel, J. (2021). Dump the "dimorphism": Comprehensive synthesis of human brain studies reveals few male-female differences beyond size. *Neuroscience & Biobehavioral Reviews*, 125, 667–697. DOI: 10.1016/j.neubiorev.2021.02.026.

Ferretti, M.T., Iulita, M.F., Cavedo, E., Chiesa, P.A., Schumacher, D.A., Santuccione, C.A., Baracchi, F., Girouard H., Misoch S., Giacobini, E. et al. (2018). Sex differences in Alzheimer disease – The gateway to precision medicine. *Nat Rev Neurol*, 14(8), 457–469. DOI: 10.1038/s41582-018-0032-9.

Fine, C. (2010). *Delusions of Gender. How Our Minds, Society, and Neurosexism Create Difference*, 1st edition. W.W. Norton, New York.

Fine, C., Joel, D., Rippon, G. (2019). Eight things you need to know about sex, gender, brains, and behavior: A guide for academics, journalists, parents, gender diversity advocates, social justice warriors, Tweeters, Facebookers, and everyone else. *The Scholar & Feminist Online*, 2(15) [Online]. Available at: http://sfonline.barnard.edu/neurogenderings/eight-things-you-need-to-know-about-sex-gender-brains-and-behavior-a-guide-for-academics-journalists-parents-gender-diversity-advocates-social-justice-warriors-tweeters-facebookers-and-ever/ [Accessed 19 July 2021].

Gilbert, S. (2021). Why were we so cruel to Britney spears? *The Atlantic* [Online]. Available at https://www.theatlantic.com/culture/archive/2021/02/britney-spears-cruelty-media-treatment/618018/ [Accessed 19 July 2021].

Gobinath, A. (2019). Neuroscience should take sex differences in the brain more seriously [Online]. Available at: https://massivesci.com/articles/neuroscience-sex-differences-feminism-stem-brain-research/ [Accessed 3 July 2021].

Gould, S.J. (1981). *The Mismeasure of Man*, Revised and Expanded edition. Norton, New York and London.

Grasenick, K. (2019). Same, same – or different? Common challenges in neuroscience, AI, medical informatics, robotics and new insights with diversity & ethics. *The Neuroethics Blog* [Online]. Available at: http://www.theneuroethicsblog.com/2019/09/same-same-or-different-common_10.html [Accessed 5 September 2021].

Grasenick, K., Kleinberger, P.M., Pilinger, A. (2020). Taking diversity in research projects into account. How to make it work. *Graz* [Online] Available at: https://openlib.tugraz.at/download.php?id=5fbba66e73887&location=browse.

Gray, J. (2012). *Men are from Mars, Women are from Venus. The Classic Guide to Understanding the Opposite Sex*, 1st edition, 20th anniversary edition. Harper, New York.

Healy, B. (1991). The Yentl syndrome. *The New England Journal of Medicine*, 325(4), 274–276. DOI: 10.1056/NEJM199107253250408.

Jäncke, L. (2018). Sex/gender differences in cognition, neurophysiology, and neuroanatomy. *F1000Research*, 7. DOI: 10.12688/f1000research.13917.1.

Joel, D. and Vikhanski, L. (2019). *Gender Mosaic: Beyond the Myth of the Male and Female Brain*. Endeavour, London.

Joel, D., Berman, Z., Tavor, I., Wexler, N., Gaber, O., Stein, Y., Shefi, N., Pool, J., Urchs, S., Daniel S.M. et al. (2015). Sex beyond the genitalia: The human brain mosaic. *Proceedings of the National Academy of Sciences of the United States of America*, 112(50), 15468–15473. DOI: 10.1073/pnas.1509654112.

Joel, D., Garcia-Falgueras, A., Swaab, D. (2020). The complex relationships between sex and the brain. *Neuroscientist*, 26(2), 156–169. DOI: 10.1177/1073858419867298.

Kaiser, A. (2012). Re-conceptualizing "Sex" and "Gender" in the human brain. *Zeitschrift für Psychologie*, 220(2), 130–136. DOI: 10.1027/2151-2604/a000104.

Lieberman, J.A., Stroup, T.S., McEvoy, J.P., Swartz, M.S., Rosenheck, R.A., Perkins, D.O., Keefe, R.S., Davis S.M., Davis, Cl.E., Barry D. et al. (2005). Effectiveness of antipsychotic drugs in patients with chronic schizophrenia. *The New England Journal of Medicine*, 353(12), 1209–1223. DOI: 10.1056/NEJMoa051688.

Lippi, D., Bianucci, R., Donell, S. (2020). Gender medicine: Its historical roots. *Postgraduate Medical Journal*, 96(1138), 480–486. DOI: 10.1136/postgradmedj-2019-137452.

Maney, D.L. (2015). Just like a circus: The public consumption of sex differences. *Current Topics in Behavioral Neurosciences*, 19, 279–296. DOI: 10.1007/7854_2014_339.

Marrocco, J. and McEwen, B.S. (2016). Sex in the brain: Hormones and sex differences. *Dialogues in Clinical Neuroscience*, 18(4), 373–383.

Mathews, P.T. (1999). *Passionate Discontent. Creativity, Gender, and French Symbolist Art/Patricia Mathews*. University of Chicago Press, Chicago, IL.

Mauvais-Jarvis, F., Bairey Merz, N., Barnes, P.J., Brinton, R.D., Carrero, J.-J., DeMeo, D.L., De Vries, G., Epperson, C.N., Govindan, R., Klein S.L. et al. (2020). Sex and gender: Modifiers of health, disease, and medicine. *The Lancet*, 396(10250), 565–582. DOI: 10.1016/S0140-6736(20)31561-0.

Nasser, E.H., Walders, N., Jenkins, J.H. (2002). The experience of schizophrenia: What's gender got to do with it? A critical review of the current status of research on schizophrenia. *Schizophrenia Bulletin*, 28(2), 351–362. DOI: 10.1093/oxfordjournals.schbul.a006944.

National Institute of Mental Health (2005). Questions and answers about the NIMH clinical antipsychotic trials of intervention effectiveness study (CATIE). Phase 1 results. *National Institutes of Health, U.S. Department of Health and Human Services* [Online]. Available at: https://www.nimh.nih.gov/funding/clinical-research/practical/catie/phase1results [Accessed 19 July 2021].

Oram, S., Trevillion, K., Feder, G., Howard, L.M. (2013). Prevalence of experiences of domestic violence among psychiatric patients: Systematic review. *The British Journal of Psychiatry: The Journal of Mental Science*, 202, 94–99. DOI: 10.1192/bjp.bp.112.109934.

Pakenham, T. (1991). *The Scramble for Africa: White Man's Conquest of the Dark Continent from 1876 to 1912*. Avon Books, New York.

Pease, A. and Pease, B. (2017). *Why Men Can Only Do One Thing at a Time and Women Never Stop Talking*. Orion Books, London.

Riecher-Rössler, A. (2017). Sex and gender differences in mental disorders. *The Lancet Psychiatry*, 4(1), 8–9. DOI: 10.1016/S2215-0366(16)30348-0.

Rippon, G. (2020). *The Gendered Brain. The New Neuroscience that Shatters the Myth of the Female Brain*. Vintage, London.

Roy, D. (2012). Neuroethics, gender and the response to difference. *Neuroethics*, 5(3), 217–230. DOI: 10.1007/s12152-011-9130-8.

Satterthwaite, T.D., Wolf, D.H., Roalf, D.R., Ruparel, K., Erus, G., Vandekar, S., Gennatas, E.D., Elliott, M.A., Smith, A., Hakonarson, H. et al. (2015). Linked sex differences in cognition and functional connectivity in youth. *Cereb. Cortex*, 25(9), 2383–2394. DOI: 10.1093/cercor/bhu036.

Scull, A. (2021). *Psychiatry and Its Discontents*. University of California Press, Oakland, CA.

Shah, K., McCormack, C.E., Bradbury, N.A. (2014). Do you know the sex of your cells? *American Journal of Physiology. Cell Physiology*, 306(1), C3–18. DOI: 10.1152/ajpcell.00281.2013.

Shai, A., Koffler, S., Hashiloni-Dolev, Y. (2021). Feminism, gender medicine and beyond: A feminist analysis of "gender medicine". *International Journal for Equity in Health*, 20(1), 177. DOI: 10.1186/s12939-021-01511-5.

Souter, J. (2018). *Camille Claudel*. Parkstone International (Mega Square), New York.

Studholme, C., Kroenke, C.D., Dighe, M. (2020). Motion corrected MRI differentiates male and female human brain growth trajectories from mid-gestation. *Nature Communications*, 11 [Online]. Available at: https://www.nature.com/articles/s41467-020-16763-y#citeas [Accessed 19 July 2021].

Tannenbaum, C., Ellis, R.P., Eyssel, F., Zou, J., Schiebinger, L., (2019). Sex and gender analysis improves science and engineering. *Nature*, 575(7781), 137–146. DOI: 10.1038/s41586-019-1657-6.

Tasca, C., Rapetti, M., Carta, M.G., Fadda, B. (2012). Women and hysteria in the history of mental health. *Clinical Practice and Epidemiology in Mental Health: CP & EMH*, 8, 110–119. DOI: 10.2174/1745017901208010110.

Vilain, E., Achermann, J.C., Eugster, E.A., Harley, V.R., Morel, Y., Wilson, J.D., Hiort, O. (2007). We used to call them hermaphrodites. *Genetics in Medicine*, 9(2), 65–66. DOI: 10.1097/GIM.0b013e31802cffcf.

Viswanath, B. and Chaturvedi, S.K. (2012). Cultural aspects of major mental disorders: A critical review from an Indian perspective. *Indian Journal of Psychological Medicine*, 34(4), 306–312. DOI: 10.4103/0253-7176.108193.

Wallace, E.R. and Gach, J. (2008). *History of Psychiatry and Medical Psychology. With an Epilogue on Psychiatry and the Mind-body Relation*. Springer, New York.

Williams, P. and Chrisman, L. (1994). *Colonial Discourse and Post-Colonial Theory. A Reader*. Routledge, London.

Wolfe, J. and Poolos, C. (2015). *The Industrial Revolution. (The Age of Revolution)*. Britannica Educational Publishing, Chicago, IL.

Yarris, K.E. and Ponting, C. (2019). Moral matters: Schizophrenia and masculinity in Mexico. *Ethos*, 47(1), 35–53. DOI: 10.1111/etho.12226.

14

What's Next? The Chilean Neuroprotection Initiative, in Light of the Historical Dynamics of Human Rights

Manuel GUERRERO[1,2]

[1] Center for Research Ethics and Bioethics, Uppsala University, Sweden
[2] Department of Bioethics and Medical Humanities, Faculty of Medicine, University of Chile, Santiago, Chile

14.1. Introduction

Human rights are dynamic. They synthesize collective efforts to get states to respect and protect every person's dignity and take positive action to facilitate a life worth living. At different times in history, changes have been demanded to give adequate protection and recognition to these rights and limit the power that positions some groups in a situation of vulnerability or weakness compared to others. Thus, there are times when the guarantees that protect rights are extended to new social groups. The idea of the minimums that allow a dignified life and the current possibilities of well-being also varies. The recent inclusion of the notion of neuroprotection into the Chilean Constitution, accompanied by the formulation of a Neuroprotection Bill (Law No. 21.383, Senate of the Republic of Chile n.d.b), can be read as new steps within this historical dynamics of human rights, in this case as a pioneering local effort to regulate neurotechnology development and applications. This chapter describes such attempts focusing on the Chilean human rights debate around these new brain-related issues.

Neuroethics and Cultural Diversity,
coordinated by Michele FARISCO. © ISTE Ltd 2023.

14.2. Battling on the "last frontier": the Chilean neuroprotection legislation

When introducing the Universal Declaration of Human Rights to the United Nations General Assembly on December 9, 1948, one of its authors, Chafik Malik, from Lebanon, highlighted the "negative roots" of the document – the atrocities of World War II. At that historical event, he emphasized that the Declaration should serve "as a potent critic of existing practice" to "transform reality" (Glendon 2001, p. 165). Historically, this has been the case: human rights in practice commonly appear as a reaction to violating some fundamental rights or liberties. Human rights usually have been addressed a posteriori. And here lies part of the novelty and also one of the challenges of the "neurorights" legal formulation in Chile: the issues it tries to regulate have been motivated more by what is potentially possible in terms of risks and harms of the advances in neurotechnology than by what is occurring at the present stage of its development. In this sense, the nature of the Chilean neuroprotection bill debate and formulation is precautionary, declaring some a priori conditions for further development.

In December 2020, two projects related to the advancement of neurotechnologies entered the Senate of the Republic of Chile for discussion: a constitutional reform project that proposed to modify the Chilean Fundamental Charter, to "protect the integrity and mental indemnity concerning the advancement of neurotechnologies" (Senate of the Republic of Chile n.d.a) and a "bill on the protection of neurorights and mental integrity, and the development of research and neurotechnologies" (Senate of the Republic of Chile n.d.b). Both projects had reports that substantiated the need for regulation developed by the Commission on Challenges of the Future, Science, Technology and Innovation of the Chilean Senate.

The constitutional reform project claims that advances in science and technology entail risks that often unpredictably impact societies. Scientific and technological development has the possibility of quickly reaching global applied consequences and, in doing so, restructuring "the ethical-value limits of a given society" (Senate of the Republic of Chile n.d.a, p. 1). As an illustration, the project reflects on the development of computer technology, in which privacy is in the present threatened as data is voluntarily transferred to systems whose domain is beyond the control of the person who provides it. The concern about the potential dual use of science and technology development is then extended to brain research. In that respect, the project claims that an increased "knowledge of the brain leads us to ask ourselves which and what control we want of that object of study called 'brain'", which is not only considered in its physical dimension but also "in its dimension of mental potentiality that envelops the mysteries of human existence, which is why it must have the maximum fundamental legal protection" (Senate of the Republic of Chile n.d.a, pp. 2 and 8).

Neurotechnological advances – understood by the project as the set of methods and instruments that allow a direct connection of technical devices with the nervous system – have helped to achieve huge medical successes and benefits[1]. Nevertheless, they also potentially enable "access to people's mental information and the possibility of its external manipulation", which could affect an individual's identity, integrity and behavior (Senate of the Republic of Chile n.d.a, p. 2). This new technological reality should then set off an alarm not only from an ethical point of view but also about how society restructures its social rules of coexistence. The reason for this is that if technology can make human beings act without their will being involved, the bases of law must be reformulated as its statute rests on the idea that human beings act freely, with the autonomy of will, in such a way that it is possible to assign standards of responsibility in externalized behavior. Consequently, it is claimed that the neurotechnological advances challenge the legal system's essential substratum, which relies on human agency and identity, considered fundamental expressions of human dignity.

The project adds that, due to its impact on the human, neurotechnology has the ability to affect critical societal values by redefining what is conventionally acceptable in social terms. Thus, for example, neurotechnological induced "brain enhancement can jeopardize the dignity of human beings as equal subjects". In this respect, the constitutional reform project refers to the preamble of the Universal Declaration of Human Rights, where it is stated that the fundamental rights and freedoms of people are based on the inherent value of each human being, and the exercise of such fundamental rights and freedoms are consubstantial and cannot be revoked by any state. As human beings possess an inherent value (dignity), all are holders of human rights and deserve to be protected against acts contrary to such intrinsic value. Therefore, human dignity sets the framework for what is considered as legitimate to do or not do in a democratic society, which is valid for people, institutions and the state. This should also apply to the case of neurotechnological development.

How to guarantee the protection of human dignity in the time of the "neurotechnological revolution"? In search of axiological and legal referents, the reform project quotes the Universal Declaration of Human Rights, which in Article 27 declares everyone's right to enjoy and share in scientific advancement and its benefits. From the perspective of medical science, the project also quotes the

---

1. Bulletin No. 13.827-19 (Senate of the Republic of Chile n.d.a) quotes Marta Farah's definition (2012).

Nuremberg Code (1947), the World Medical Association's Declaration of Helsinki (1964, 2013) and the International Ethical Guidelines for Health-related Research Involving Humans prepared by the Council for International Organizations of Medical Sciences in collaboration with the World Health Organization (CIOMS 1982, 2016). Those documents contain relevant ideas about the purpose of medicine, the protection of scientific evidence on human beings and the protection of consent. Regarding the incentive to development, the equitable benefit of scientific progress and wellbeing, the project reviews the International Covenant on Economic, Social and Cultural Rights (1966), the UNESCO's Recommendation on Science and Scientific Researchers (1974), UNESCO's Declaration on Science and the Use of Scientific Knowledge (1999), the International Declaration on Human Genetic Data (2003) and the Universal Declaration on Bioethics and Human Rights (2005).

Still, the authors of the constitutional reform project claim that none of these declarations contains a review of the practical impact that science can have on the physical and mental integrity of the human being and how this could affect the right to life. The issue is considered relevant because personal integrity could be affected by the disrupting potential capacities of neurotechnologies as they open "possibilities of auscultation and public exposure of what previously seemed the only redoubt of human intimacy, such as thoughts, desires, emotions, the subconscious and all that information produced by neuronal activity". The brain, in this sense the seat of those features, is understood as "the last frontier" of the human that needs to be protected in the face of neurotechnological progress (Senate of the Republic of Chile n.d.a).

Consequently, the constitutional reform project estimates the possibilities and potentialities that advances in neurotechnology have as an urgent opportunity to specify the protection of what it considers as "this new dimension of dignity", consecrating in the constitution what, until today, has only been left to ethical parameters self-imposed by the scientific community. And since the current international human rights documents related to science and technology seem not to be sufficient to take care of neurotechnologies' potential risks derived from their capacity of accessing and intervening "in that hitherto insurmountable limit that is the human brain", the reform concludes that these risks will have to be faced preventively through the creation of a new human right: the right to neuroprotection that derives from the need to protect human dignity against the use of new technologies (Senate of the Republic of Chile n.d.a).

Regarding the content of the right to neuroprotection, that is, the fundamental *ius* of the right enshrined, the reform indicates that it is a right anchored in human dignity, which has a set of prerogatives that translate into powers invoked by people against arbitrary attacks or transgressions while allowing the demand for positive actions by the state to provide such protection. In the absence of explicit international guidelines on human rights and neurotechnologies, the Chilean reform project chose the content of the new neuroprotection right to be based on the four ethical priorities for neurotechnologies and AI proposed by Rafael Yuste and Sara Goering et al. in *Nature* in 2017 (Senate of the Republic of Chile n.d.a)[2]:

– *Privacy and consent*: the right to privacy of information produced by brain activity, which can be accessed through neurotechnology. It is about the protection of "neurodata", information considered useful and valuable that would open the door to the annulment of privacy or informative self-determination without the proper safeguards and appropriate security measures.

– *Agency and identity*: the right to personal identity and self-determination. Neurotechnology is considered to open the possibility of annulling or altering the identity of people since it could be a tool to inhibit the consciousness and determination of the person's self.

– *Augmentation*: the right to equality in the access to brain enhancement. It is a prerogative of regulation to avoid inequity, and when it is technologically possible to artificially increase people's brain capacity, legal questions arise regarding who will have access to these technologies and whether this will generate new social asymmetries or exacerbate existing ones.

– *Bias*: the right to protection against biased control of algorithms. It refers to the due legal support against the biases that automated decision-making implies.

After one year of reviews and discussions, on October 14, 2021, Law No. 21.383 was promulgated by the President of the Republic of Chile, Mr Sebastián Pinera. It modifies the Chilean fundamental charter to establish scientific and technological development at the service of the people, adding a new paragraph to number 1 of Article 19 of the Political Constitution of the Republic, as follows:

---

2. Yuste and Goering et al. (2017). Rafael Yuste is a Spanish–American neurobiologist, Professor of Biological Sciences at Columbia University, researcher of the US BRAIN Initiative and leader of the Morningside Group. He has spoken several times about the brain, neurotechnologies and mental identity, integrity and privacy at the Future Congress organized by the Chilean Senate, which is referred by the history of the law as a source of inspiration for the legal discussion process. He attended and intervened in all the parliamentarian debates as an expert and his participation and influence in the structure and contents of constitutional reform and the neuroprotection bill was significant.

Scientific and technological development will be at the service of people and will be carried out with respect for life and physical and mental integrity. The law will regulate the requirements, conditions and restrictions for its use by people, and must especially protect brain activity, as well as the information from it[3].

In parallel to this reform process, a second project was discussed by the Chilean Congress, known as the Neuroprotection Bill, which operationalizes and gives content to the above-commented constitutional reform (Senate of the Republic of Chile n.d.b)[4]. The bill is also declared to be anchored in the concept of human dignity as an underlying principle and goal to which neurotechnological development must conform (Senate of the Republic of Chile n.d.c). Its declared purpose is to protect people's life and physical and mental integrity in the development of neurosciences, neurotechnologies and clinical applications. The law proposal acknowledges the freedom to carry out neuroscience procedures and to use neurotechnologies, having "the essential rights that emanate from human nature" as a limit recognized by the Chilean Constitution and international treaties.

The law assigns the state the duty to watch over the development of neuroscience and neurotechnologies and promote access to their advances without arbitrary discrimination. In addition, it offers a conceptual framework for those who must interpret the law, defining neural data as the "information obtained from the activities of people's neurons, which contain a representation of brain activity", and neurotechnologies as "a set of devices or instruments that allow a connection with the central nervous system, for reading, registering or modifying brain activity and the information coming from it" (Senate of the Republic of Chile n.d.b). People are free to use any permitted neurotechnology. However, when using them on others, we must have their free, prior and written informed consent, which must be given expressly, explicitly or by any legal representative that must supply their will according to the law. The law refers to other existing legal bodies that regulate this type of consent for therapeutic or medical purposes and scientific research goals. Consent forms must contain the information according to the available evidence on the possible effects of the respective neurotechnology and regarding personal neural data privacy regulations, when applicable.

The law in the discussion also indicates that neurotechnologies must be registered with the Institute of Public Health for use on people and that their operation must be reversible. The health authority may restrict or prohibit the use of

---

3. Official Journal of the Republic of Chile, October 25, 2021. Translated by Michael Russo.
4. At the present time, the bill is still under discussion in Congress in its second constitutional process.

neurotechnologies due to undermining fundamental rights, which triggers specific sanctions. Finally, neural data is typified as reserved and sensitive.

14.3. The Chilean neuroprotection initiative, an unfinished project

From the human rights perspective, the Chilean neuroprotection legislation entails an expansion effort of the notion of human dignity, linking brain activity and "neurodata" to physical and mental integrity. As it is known, the concept of dignity came up in ancient times in different cultures (An-Na'im 1992; Howard 1992; Ishay 2004), and in the case of the Western tradition, it was already present in the thought of Cicero, St. Thomas Aquinas, Pico Della Mirandola and Francis Bacon (Rosen 2012). However, it would not be until the 18th century that the consecration of such a notion was incorporated as a founding principle for the limitation of the power of the rulers, and legal frameworks were generated to actively respect it through its recognition and enforcement in new Constitutions, international conventions and declarations (Griffin 2008).

Despite the critics of the "vagueness" and the "unnecessary obscurity" of the concept (Griffin 2002; Macklin 2003), dignity, that is, the idea according to which human beings possess an intrinsic, inherent and unconditional value and, as a result, are holders of certain fundamental rights and freedoms, is presented by international law as the ultimate foundation of the global system of human rights that emerged in response to the horrors of World War II (Habermas 2012; Mégret et.al. 2014). In modern political thought, the raison d'être of the state is to protect people's dignity, rights and freedoms. In this sense, we speak of "the Rule of Law": the state is prohibited from acting arbitrarily and is subject to respect for constitutional and legal norms, particularly concerning its obligation to respect, promote and guarantee the rights and freedoms of people (Andorno 2014).

In this understanding, which inspired the Universal Declaration of Human Rights, human dignity is not a generic term that encompasses the sum of rights and freedoms of people but is the source of such rights and freedoms. According to the Preamble of the Universal Declaration, human beings possess an inherent value (dignity), all are holders of human rights and deserve to be protected against acts contrary to such intrinsic value. Therefore, human dignity sets the framework for what is considered legitimate to do or not do in a democratic society, which is valid for people, institutions and the state. For this set of reasons, human rights are formulated, in positive terms, under the formula "everyone has the right to…", and in negative terms, "no one shall be subject to…".

When the Declaration speaks of this level of human rights, of what must be guaranteed to be able to do and what is prohibited, according to the inherent value that society recognizes in each human being, it refers to the objective component of dignity, that is, to the legal-normative dimension of human rights. This dimension highlights human dignity as a founding principle of the legal system and, therefore, of the state, and it refers to the enforceability and inalienability of human rights, aspects that are endorsed in the obligations of states, institutions and individuals, as well as in the mechanisms of protection and enforceability. In its legal-normative dimension, human rights are expressed in a set of norms contained in international pacts, treaties and conventions, and the Constitution and the laws of each country.

However, human rights are not only exhausted in the legal-normative, but also have an ethical dimension (so, for example, Rorty 1993). Dignity, in this sense, indicates a moral standard. It operates as a kind of super-principle that, along with providing the foundation for legal and social institutions, means the direction to which society, in general, and our individual attitudes should tend. And this in a double sense. On the one hand, as a minimum threshold that cannot be exceeded or transgressed and, on the other hand, as an indication of a desirable positive point to which we must tend. Indeed, human rights are absolute for the principle of dignity as the minimum inviolable threshold. Such is the case of the prohibition of certain practices that violate human dignity, which entail a sanction of legal and moral character, as occurs with the practice of torture and inhuman and degrading treatment. Regarding the principle of dignity as indicative of a positive direction towards which we have to move forward as a society and as individuals, the possibility of a certain degree of gradualness is admitted. Such is the case of respect for people's lifestyles.

Thus, with the legal-normative dimension of human rights, we can refer to the objective component of dignity, that is, to the intrinsic value that society recognizes in each human being; in the case of human rights as a moral standard, it refers to the subjective component of dignity, that is, to the inherent value that each one of us recognizes in themselves. This refers to the fact that we acknowledge ourselves as persons and not as things, subjects and not objects, so we expect certain attitudes and behaviors from others. In this subjective dimension of dignity, we are in the realm of attitudes and behaviors that can be perceived as respectful towards each person's unique and unrepeatable value or, on the contrary, contempt towards one's own or another's dignity. Consequently, human rights in their ethical dimension recognize guiding principles of action ensuring that the treatment between people is respectful and attentive so that no one feels their dignity affected, threatened, degraded or directly violated. The ethical dimension of human rights recognizes a set of values that permeate culture and guide social behavior.

As we have previously seen, the Chilean neurorights-inspired constitutional reform incorporated a new paragraph in Article 19 of the Chilean Constitution, which addresses the right to life and the physical and mental integrity of the person. The new section refers to the due protection of human rights in the face of the development of neurotechnologies as part of people's physical and mental integrity, considering them as constitutive elements of identity since they ensure the possibility of acting freely and self-determined, a value that is regarded as "intrinsic to human existence and the biological evolution that precedes it" (Senate of the Republic of Chile n.d.a). In this way, the right to neuroprotection became linked to the notion of human dignity, constituting a new human right. The regulation of the development and use of neurotechnologies is thus related to the duty to protect the identity and self-determination of human beings in an attempt to cover both the objective and the subjective components of dignity.

The proposed legal and philosophical reasoning that underlies the constitutional reform implies moving from considering access to "neurodata" as a matter of privacy to considering it as a matter of integrity. The argument runs as follows: if the development of neurotechnologies potentially allows reading and knowing what is in the mind of another without consent, mental privacy is being violated. However, neurotechnology also potentially makes it possible to determine, induce or cause another to think, feel, want or desire something. Consequently, it is not only a privacy matter but also of indemnity. Thus, privacy as well as the very integrity of the person are affected by non-consented neurotechnological intervention, since free will is compromised, which calls into question the ability of human beings to make decisions for themselves, to decide what to do, what not to do, what to want and what to wish for. In such a way, from a human rights perspective, an essential attribute that defines human beings is under threat, which is their free will, the ability and freedom to make one's own decisions. With this line of reasoning, the constitutional reform linked neuroprotection as a new dimension of human dignity related to the person's physical and mental integrity[5].

Besides the legal-normative and ethical dimensions of human rights, their political aspect must also be considered. Human rights are an ideal that has not been fully achieved but one towards which, according to the Universal Declaration, all peoples and nations must strive. The political dimension of human rights points to a horizon towards which to advance and indicates a gap between the ideal and reality. We can find this political dimension already stated in the Universal Declaration of Human Rights Preamble, with expressions such as aspiration, faith, promotion and

---

5. This line of thought was highlighted with particular clarity by de Chilean lawyer Mr Ciro Colombara (Senate of the Republic of Chile n.d.b).

commitment. The Preamble proclaims the Universal Declaration of Human Rights as "a common standard of achievement" for all peoples, nations, every individual and organ of society. In this dimension, what some authors have called "the human rights movement" is constituted in the sense that we all have an active role in its promotion and defense.

In the pioneering effort to regulate the development of neurotechnologies using the notion of "neuroprotection" and "neurorights", the Chilean case can be seen as a political attempt to deal with the dilemma of the anticipatory and participatory control of emerging technologies and the practice of the governance of innovation facing the co-evolution between technology and society (Collingridge 1980) by using the tradition of the human rights language in front of the new demands that the emergent technologies pose to the protection of certain novel rights. However, this political attempt has been restricted to the Chilean legislative power, with some active interactions with the academic and scientific communities but not with the general population. Still, there have not been many public discussions with other relevant social actors and societies in general. This is something that calls to attention the context of social revolt and a new constitutional process triggered by massive social disruptions, which ended in a referendum that decided to write a new constitution for the country (Rojas 2022). The Chilean neurorights initiative reformed the constitution of 1980, and it is possible that it will be replaced by a new one, which, in its present form, does not include the explicit brain-related paragraph. It was a political decision by the Chilean Senate not to link the neuroprotection discussion and bill to the ongoing new constitutional process[6].

Integrating human rights into the state's legal system is known as the dynamic of positivization. Through this process, rights acquire a legal objectivity, which allows them to be known, claimed and argued on the basis of the texts in which they are formalized. Positivization helps, in turn, to spread and educate on human rights. The Chilean neurorights experience can be seen as a first but not fully completed milestone in the positivization process of new neuro-specific human rights. Still, it lacks a broader and more active involvement of civil society, which puts the sustainability of such efforts in question. As Raffin (2006) has correctly pointed out, human rights comprise, in a dialectical relationship, both a theory (an explanation, an understanding, a theorization, and also a normative dimension and a hermeneutic process) and a practice (procedures, protection mechanisms, enforceability,

---

6. The new proposed Constitution, which will be a matter of referendum this year (Constitutional Convention of Chile 2022), says in Article 21: "Every person has the right to life and personal integrity. This includes physical, psychosocial, sexual and affective integrity". As we can see, there is no mention of protecting brain activity, as well as the information from it.

enforcement devices, at the same time as judicial practices). And taken in their multidimensional nature, contemporary human rights can even be understood as a "culture" that emerged after World War II, which implies individual and social processes of adhesion, promotion and defense of a set of core values (Rabossi 1991; Rorty 1993; Raffin 2006). This human rights culture challenges the prevalent national, regional or local cultures attempting to stop discriminatory practices (Arat 2006). In this sense, human rights go beyond the strictly legal sphere to become a symbolic device typical of democracy, understood as a regime and social construction of meaning (Lefort 1981; Ricoeur 1992; Schulz and Raman 2020). In its cultural dimension, human rights demand to be sustained not only by legislation but also by active societal actors who give them life. As Eleonor Roosevelt once said, human rights are the fruit of a long tradition of vigorous thinking and courageous action.

Human rights initial historical development assigned rights only to specific population segments. But it was followed in the 19th century by a series of movements and social revolutions that extended the notion of rights to new groups in society. The extension of rights to all individuals implied a dynamic of *generalization* of human rights. That is, the historical process through which human rights are recognised as universal, belonging to all people, by the mere fact of being human, without distinction of race, sex, color, gender, social or economic position, political ideas, philosophical, religious or of any kind. Such was, for example, the progress pressured by different social movements towards achieving universal suffrage and applying "man's" rights to society as a whole. The historical struggle carried out by women's movements for the right to vote, the recognition of children and adolescents as well as people with disabilities or belonging to indigenous peoples as subjects of rights are exemplary in this sense (Gould 2004; Ishay 2004).

The same has been the case with the process that implies the gradual and progressive appearance of new and more precise obligations regarding human rights and the historic *expansion* of human rights. Thus, for example, the Universal Declaration of Human Rights of 1948 was formulated, through debates that involved participants from different cultures, as "a common ideal" for all peoples and nations, which established for the first time in history fundamental civil, political, economic, social and cultural rights that all human beings, without exception, should be able to enjoy, based on the principles of equality and non-discrimination (United Nations 1948; Glendon 2001). Today, such a "common ideal" is no longer just a stated ideal but an entire body of international human rights law that continues to expand and enlarge, adding new international human rights standards to the Universal Declaration that address emerging human rights issues, such as the environment, water and science, among others. That will surely also be the case for neurorights, but for them to be sustainable, it is necessary they are elaborated together with a set of social actors which go beyond the legislative powers and the academia so that neurorights can eventually be pushed and sustained for society and with society. This participatory dimension is a path that still needs to be explored and traveled.

14.4. References

An-Naʿim, A.A. (ed.) (1992). *Human Rights in Cross-cultural Perspectives: A Quest for Consensus*. University Pennsylvania Press, Philadelphia.

Andorno, R. (2014). Human dignity and human rights. In *Handbook of Global Bioethics*, ten Have, H.A.M.J. and Gordijn, B. (eds). Springer, Dordrecht.

Arat, Z.F.K. (2006). Forging a global culture of human rights: Origins and prospects of the international bill of rights. *Human Rights Quarterly*, 28(2), 416–437.

Bell, D. and Coicaud, J.-M. (eds) (2007). *Ethics in Action: The Ethical Challenges of International Human Rights Nongovermental Organizations*. Cambridge University Press, New York.

Bobbio, N. (1996). *The Age of Rights*, Cameron, A. (trans.). Polity Press, Cambridge.

Castoriadis, C. (1997). *The Imaginary Institution of Society*. MIT Press, Cambridge.

Collingridge, D. (1980). *The Social Control of Technology*. Printer, London.

Constitutional Convention of Chile (2022). Propuesta Constitución Política de la República de Chile. Report, Constitutional Convention of Chile.

Farah, M. (2012). Neuroethics: The ethical, legal, and societal impact of neuroscience. *Annual Review of Psychology*, 63, 571–591. DOI: 10.1146/annurev.psych.093008.100438.

Glendon, M.A. (2001). *A World Made New: Eleanor Roosevelt and the Universal Declaration of Human Rights*. New York Random House, New York.

Goering, S. and Yuste, R. (2016). On the necessity of ethical guidelines for novel neurotechnologies. *Cell*, 167(4), 882–885.

Goering, S., Klein, E., Sullivan, L.S., Wexler, A., Arcas, B.A., Bi, G., Yuste, R. (2021). Recommendations for responsible development and application of neurotechnologies. *Neuroethics*. DOI: 10.1007/s12152-021-09468-6.

Gould, C.C. (2004). *Globalizing Democracy and Human Rights*. Cambridge University Press, Cambridge.

Griffin, J. (2002). A note on measuring well-being. In *Summary Measures of Population Health: Concepts, Ethics, Measurement and Applications*, Murray, J.L. (ed.). World Health Organisation, Geneva.

Griffin, J. (2008). *On Human Rights*. Oxford University Press, Oxford.

Habermas, J. (2012). *The Concept of Human Dignity and the Realistic Utopia of Human Rights. The Crisis of the European Union: A Response.* Polity, Cambridge.

Hunt, L. (2008). *Inventing Human Rights. A History.* Norton Paperback, New York.

Ienca, M. (2021). Common human rights challenges raised by different applications of neurotechnologies in the biomedical fields. Report commissioned by the Committee on Bioethics (DH-BIO) of the Council of Europe. Council of Europe, Geneva.

Ienca, M. and Andorno, R. (2017). Towards new human rights in the age of neuroscience and neurotechnology. *Life Sciences, Society and Policy*, 13(1), 1–27.

Ienca, M. and Haselager, P. (2016). Hacking the brain: Brain–computer interfacing technology and the ethics of neurosecurity. *Ethics and Information Technology*, 18(2), 117–129.

Ienca, M., Fins, J., Jox, R.J., Jotterand, F., Voeneky, S., Andorno, R., Ball, T., Castelluccia, C., Chavarriaga, R., Chneiweiss, H. et al. (2022). Towards a governance framework for brain data. *Neuroethics*, 15, 20. DOI: 10-1007/s12152-022-09498-8.

Ishay, M.R. (2004). *The History of Human Rights: From Ancient Times to the Globalization Era.* University California Press, Berkeley.

Lavazza, A. (2018). Freedom of thought and mental integrity: The moral requirements for any neural prosthesis. *Frontiers in Neuroscience*, 12, 82.

Lefort, C. (1981). *L'Invention démocratique. Les Limites de la domination totalitaire.* Fayard, Paris.

Macklin, R. (2003). Dignity is a useless concept. *British Medical Journal*, 1419–1420.

Rabossi, E. (1991). El fenómeno de los derechos humanos y la posibilidad de un nuevo paradigma teórico. In *El derecho, la política y la ética*, Sobrevilla, D. (ed.). Siglo XXI Editores, Mexico.

Raffin, M. (2006). *La experiencia del horror. Subjetividad y derechos humanos en las dictaduras y posdictaduras del Cono Sur.* Editorial del Puerto, Buenos Aires.

Rawls, J. (1999). *The Law of Peoples: With "The Idea of Public Reason Revisited".* Harvard University Press, Cambridge.

Ricoeur, P. (1992). *Oneself as Another.* University of Chicago Press, Chicago.

Rojas, H. (2022). Chile at the crossroads: From the 2019 social explosion to a new constitution. *Seattle Journal for Social Justice*, 20(4), 11.

Rorty, R. (1993). Human rights, rationality, and sentimentality. In *On Human Rights. The Oxford Amnesty Lectures*, Shute, S. and Hurley, S. (eds). Basic Books, New York.

Rosen, M. (2012). *Dignity: Its History and Meaning.* Harvard University Press, Cambridge.

Russo, M. (trans.) (2021). Official Journal of the Republic of Chile.

Schultz, W.F. and Raman, S. (2020). *The Coming Good Society. Why New Realities Demand New Rights.* Harvard University Press, Cambridge.

Senate of the Republic of Chile (n.d.a). Bulletin No. 13.827-19. Congreso de Chile, Santiago.

Senate of the Republic of Chile (n.d.b). Bulletin No. 13.828-19. Congreso de Chile, Santiago.

Senate of the Republic of Chile (n.d.c). Bulletin No. 578/SEC/21. Aprobación al Proyecto de ley sobre protección de los neuroderechos y la integridad mental, y el desarrollo de la investigación y las neurotecnologías, correspondiente al Boletín.

Yuste, R., Goering, S., Bi, G., Carmena, J.M., Carter, A., Fins, J.J., Kellmeyer, P. (2017). Four ethical priorities for neurotechnologies and AI. *Nature News*, 551(7679), 159.

Yuste, R., Genser, J., Herrmann, S. (2021). It's time for neuro-rights. *Horizons: Journal of International Relations and Sustainable Development*, 18, 154–165.

United Nations (1948). Universal Declaration of Human Rights. UN, Geneva.

United Nations (1993). Viena Declaration on Human Rights. UN, Viena.

15

Interrogating the Culture of Human Exceptionalism: Animal Research and the Neuroethics of Animal Minds and Brains

L. Syd M JOHNSON

Center for Bioethics and Humanities, SUNY Upstate Medical University, New York, USA

15.1. Introduction

Neuroscience has added to our understanding of the brains and minds of non-human animals (hereafter referred to as *animals*), and the many ways that they are similar to human brains and minds. At the same time, the study of animal behavior and minds in other sciences, including comparative psychology, animal cognition, ethology, and primatology have expanded our understanding of the sophistication and complexity, as well as the human-like qualities of animals, and the richness of their cognitive, social, emotional, cultural, and moral lives. Evolutionary biology and genetics have revealed the continuity and overlap of species, and that taxonomic boundaries are not based on biological differences, but rather on cultural classifications and convenience (Andrews et al. 2019). These scientific successes make more pressing the ethical dilemma of research with animals, and nowhere is that dilemma more acute than in the brain sciences. The more similar an animal's brain is structurally and functionally to a human brain, and the more like a human's that animal's behaviors and traits are, the more that animal appears to be a useful model organism for human-relevant research. Yet, the morally relevant similarities – many of which involve brain-based capacities like consciousness, autonomy, agency, sociality, and psychological complexity – make it harder to maintain that it is ethically permissible to use such animals in ways that are harmful to them (Ferdowsian and Gluck 2015; Johnson 2020b).

Neuroethics, having marked the ethical and societal implications of the brain sciences as its domain, has paid little attention to those implications for the many species of animals used in neuroscientific research. While the emergence and development of neuroethics as an interdisciplinary field coincided with the rapid growth of knowledge about animal brains, minds, intelligence, culture, behaviors, and capacities, that knowledge has not been integrated into neuroethics to a notable degree despite the centrality of the brain regarding all these features of animals and their lives. That rich body of research and thought has similarly had little impact on animal research practices and regulations. The philosophical discourse in animal ethics, meanwhile, has integrated and used empirical knowledge of animals from numerous and diverse scientific fields to bolster empirically informed ethical conclusions (Gruen 2014; Andrews and Beck 2019; Andrews et al. 2019). Neuroethics is in a position to lead in thoughtful and frank deliberations concerning the implications of our understanding of the minds of other animals for neuroscientific research, the moral status of these animals, and our moral obligations to them (Johnson 2020a). It ought to do so.

This chapter presents an argument for introducing more intellectual and ethical rigor into neuroethics, and calls for neuroethics to interrogate and confront the anthropocentric speciesism and human exceptionalism that have dominated its discussions on the use of animals in neuroscientific research. I begin with a brief overview of the brain-based capacities and functions that are widely viewed as ethically relevant in humans. Similar capacities are found in numerous animal species, including those used in neuroscientific research. I then turn to the overexamined problem of "humanizing" chimeric animals, and why attention to this problem has collapsed into anthropocentric speciesism and human exceptionalism, while overlooking the morally relevant human-like traits and capacities of animals. I then turn to issues of justice in research, and the need for consistency in the scientific and ethical justification for research, one that harmonizes the ethical imperatives of both human subjects research and animal research. Finally, I call on neuroethics to live up to its commitments to rigorously explore, in an empirically informed way, the implications of neuroscientific research, and interrogate research that threatens important and fundamental ethical values.

15.2. Brains, minds, consciousness and moral status

The capacities that are widely thought to make humans the subjects of respect, rights, and moral consideration are functions of the brain and mind (Warren 1997;

DeGrazia 2020). It is a commonly held and rarely questioned premise that consciousness not only matters for moral status, but that it "elevates one's moral status in the world" (Hardcastle 2016), such that humans who are no longer conscious, or creatures thought to not be conscious (or that are conscious in some "lesser" way than paradigmatic humans) have diminished moral status. According to Utilitarian moral theory, for example, what confers moral status on a creature is that it is sentient, or capable of phenomenal experiences (Bentham 1789), although some Utilitarians also argue that the value of an individual life depends on the psychological and cognitive complexity, and the capacity for interests of the individual human or non-human in question (Singer 1979; Levy and Savulescu 2009). Deontological and contractarian theories can consider other cognitive capacities to be relevant to moral status. These can include capacities such as autonomy, moral agency, reason, intelligence, and self-awareness, all of which require consciousness (Warren 1997; Regan 2004; Andrews et al. 2019). Although it is widely assumed, it is not obvious that consciousness is *necessary* for a human or non-human to matter morally (Johnson 2022), but it is surely *sufficient*. Being self-aware, phenomenally aware, or capable of experience makes us a subject, and a subject of moral concern, even if the details of what that means in terms of how we must be treated and regarded remain morally contested. What undergirds consciousness, how brains give rise to consciousness, what consciousness is, and what the significance of consciousness is are all fundamental scientific and philosophical questions that remain unanswered (Dehaene and Naccache 2001; Crick and Koch 2003; Seth et al. 2005; Edelman and Seth 2009; Sinnott-Armstrong 2016; Tononi et al. 2016). Nonetheless, if it is accepted on more than inference and assumption that humans are conscious, then the consciousness of many animals – mammals to be sure, as well as fish, cetaceans and birds, and likely others – cannot be contested without special pleading or ignorance of the evidence accumulated by the sciences. If the consciousness of humans is a sufficient reason for concluding they are subjects of moral concern, then it must be so for non-humans as well. If it is instead other brain-related capacities – emotionality, sociality, agency, autonomy, language use, problem-solving, learning, culture – that matter morally, then they must also matter when we find them in non-humans.

Many ethical concerns related to the development and use of new technologies are within the domain of neuroethics because they relate to brains and affect brain-based capacities. They include concerns related to cognitive and moral capacities and the potential for the neuro-enhancement of both, and questions about consciousness and moral significance regarding cerebral organoids, ex vivo human brain tissue, engineered neural circuitry, AI, and human-animal chimeras (Streiffer 2007; Di Lullo and Kriegstein 2017; Greely et al. 2018). It is noteworthy that while neuroethics has frequently raised ethical concerns about speculative entities that might exist in the near or far future (such as conscious organoids or AIs), it has paid so little attention to the creatures already in our midst, the many conscious, sentient, thinking, intelligent animals used in neuroscientific research. Ironically, some of the speculative entities of possible moral concern, like cerebral organoids and ex vivo brain tissue might, once developed, serve as valuable, humane, ethical, and human-relevant alternatives to the use of living animals in research (Qian et al. 2017). If we are concerned about one, moral consistency demands that we be equally concerned about the other, and should be even more concerned about the entities that already exist, that are already being used. A neuroethics seriously committed to probing and responding to the ethical conduct and implications of brain research must acknowledge the elephant in the room, even if it is as small as a mouse.

15.3. Chimeras and humanization: the overexamined problem

The creation of human/non-human chimeras, particularly when the chimeras are engrafted with human neural cells and tissues, is the rare animal-related issue that has received significant ethical attention within the scientific and neuroethics literature. The expressed concern is the moral humanization of chimeric animals. The National Academies of Sciences, Engineering, and Medicine, in its report on organoids and chimeras, describes some of the ethical issues:

> In some of these conceptions, human-nonhuman animal neural transplants and chimeras involving the brain arouse stronger concerns relative to those involving other organs because many of the capacities associated with higher moral status, such as consciousness, complex problem solving, self-awareness and emotions, are "located" in the brain. From a more introspective view, the brain more than any other organ is believed to define who a person is. As the physical instantiation of characteristics that many people commonly associate with their humanness and individuality, the brain evokes greater concern relative to other organs (National Academies of Sciences Medicine 2021).

In 2005, a report by the United States National Research Council cautioned that the introduction of human tissue into animals warranted attention, especially if the brain is implicated:

> Experiments in which [human embryonic stem] cells, their derivatives or other pluripotent cells are introduced into nonhuman fetuses and allowed to develop into adult chimeras need more careful consideration because the extent of human contribution to the resulting animal may be higher. Consideration of any major functional

contributions to the brain should be a main focus of review (Institutes of Medicine 2005).

The German Ethics Council (Deutscher Ethikrat) notes similar concerns:

> A particular ethical issue is whether the transfer of individual human genes might sometimes alter important characteristics of the receiving species in such a way as possibly to affect the animal's moral status. Drastic modifications of this kind are at least conceivable at [a] biological level.
>
> Particular ethical issues are raised by the possibility that the transplant of human nerve cells or their precursors into the brains of animals – in particular, primates – might give rise to human capabilities in the animal that could in certain circumstances alter its moral status (German Ethics Council 2011).

The worry is that human/non-human chimeras might have enhanced moral status as a result of their "humanized" brains. If that moral status approaches the status of humans, then proceeding with research on those chimeras could be judged as morally impermissible. This poses a dilemma: creating a humanized animal model, in the hopes of optimizing it for human-relevant research that cannot be ethically carried out in humans, might also make using that animal ethically impermissible.

The Danish Council on Ethics, in its report *Man or Mouse?*, notes the potentially wide-ranging problems that human–non-human chimeras might pose:

> The problem would therefore consist of us insisting, in practice, on classifying humans and animals in biological terms and continuing to treat the crossbreed as an animal. If, instead, we introduced a moral classification, in which an individual that has a sufficient number of cognitive characteristics is given status and a claim to protection as a person, there would be no problem for the individual. There again, a number of other problems would arise, and not just for the research projects in hand. Such problems would concern the whole way in which we have organized and marshalled our society (Danish Council of Ethics 2008).

They noted the particular concerns related to the creation of neuro-chimeras.

> Some members consider experiments or therapies capable of altering the brain, i.e. the cognitive characteristics, of animals and humans

primarily capable of altering – or at any rate leading to doubt about – the ethical status ascribable to a particular modified individual, and hence what debt of protection we owe that individual. In the opinion of these members, therefore, experiments potentially capable of assuming such implications should not be performed (Danish Council of Ethics 2008).

They recommended legislative action to grant approval bodies the authority to reject experiments that fall under certain categories, including any experiment that "crucially affects an animal's cognitive functions in a human direction" or that brings "into the world experimental chimeras or hybrids that have been so crucially altered that justified doubt can arise as to whether the crossbreed can still be classified as an animal and can thus be put down if the experiment has an adverse outcome" (Danish Council of Ethics 2008).

The Danish Council thus draws a bright moral line between human and animal, and highlights the moral and social implications of creating humanized chimeras: that their cognitive capacities might become more human, and that such animals might cross over the threshold into the moral domain previously reserved for humans. Should that happen, it would no longer be permissible to treat them "like animals" in research (or elsewhere). This concern also highlights how the worry about humanizing chimeras is specifically brain and cognition related.

The possibility of inducing brain-related traits of moral concern is not farfetched. In 1988, Balaban et al. demonstrated that the creation of quail-chicken chimeras with chimeric brains resulted in the transfer of species-typical vocalizations. The "cross-species behavioral transfer brought about by neuronal transplantation [...] demonstrate[s] that donor cells are functional in the host brain" (Balaban et al. 1988). In embryonic rats, human fetal neural precursor cells grafted into the telencephalic vesicle distributed widely across the rat's brains, generating widespread central nervous system chimerism (Brüstle et al. 1998). When neonatal mice were engrafted with human glial progenitor cells (hGPCs), the

> hGPCs exhibit a competitive dominance when xenografted into the mouse brain that results in the effective, and often complete, replacement of mouse glial progenitors by their human counterparts, with subsequent astrocytic differentiation, thereby yielding murine brains in which human glial cells predominate (Windrem et al. 2014).

Engrafting human glia in neonatal mice has been shown to enhance learning. These "humanized chimeric mice" (as the researchers themselves called them) demonstrated superior discrimination and contextual learning, spatial learning, enhanced memory, and greater synaptic plasticity compared to their non-chimeric peers, with widespread integration of human glial cells in the adult mouse brain.

> By the time these mice reached adulthood, a large proportion of their forebrain glia were replaced by human cells. The chimerization was slowly progressive, so that extensive infiltration of cortex and hippocampus by human cells was evident by 4–12 months (Han et al. 2013).

One researcher described the effects as "whopping". "We can say they were statistically and significantly smarter than control mice" (Coghlan 2014).

Some neuroethicists have dismissed the possibility that commonly used research animals, such as mice and rats, can be humanized to an extent that would warrant moral concern. Greely et al. for example, consider it unlikely given the size and architecture of mouse brains:

> The mouse brain is significantly smaller than the human brain. In volume, it is less than one-thousandth the size of the human brain. Even apart from their smaller size, mouse brains are organized differently from human brains. The proportion of a brain composed of the neocortex, the region most associated in human brains with consciousness, is hugely greater in humans than in mice. The brain is an incredibly complex network of connections. Neuroscientists believe that it is the architecture of the brain that produces consciousness, not the precise nature of the neurons that make it up [...]. A mouse brain made up entirely of human neurons would still be a mouse brain, in size and architecture, and thus could not have human attributes, including consciousness (Greely et al. 2007).

It would be absurd to think that mice, other mammals or other small-brained animals like birds are not conscious. The consciousness of concern, then, is human consciousness, which might be defined either as "consciousness in a human" or as a "human-like" consciousness. The former would categorically exclude all non-human animals, but the unique value of such consciousness could be justified only by anthropocentric speciesism, given the robust and extensive scientific evidence of consciousness in mammals and other classes of animals. A charitable reading of the use of the concept in the literature on chimeras is that what is actually meant is "human-like" consciousness. Human-like, however, is no more well defined than consciousness itself. Greely et al. note, with a reference to Gregor Samsa, "Of course, human consciousness trapped in a mouse's body would truly be cruel treatment, but... this possibility seems extremely unlikely" (Greely et al. 2007).

A human mind in a cockroach body is, of course, the horror of Gregor Samsa's plight. This does not suggest "human consciousness" but rather something akin to awareness of oneself *as human*, but in a non-human body. Greely explicitly says this elsewhere:

> what we should care about, morally, is an organism's mental capabilities. And I would further specify that it is not all mental capabilities but the kind of consciousness, or self-consciousness, that we consider typical of humans – whatever that is and however we try define it (Greely 2014).

The worry, of course, is that "however we try to define it" will be in completely self-serving ways, unsupported by scientific reality. As Capps points out, the *cruelty* for the chimeric human neuron mouse would be not so much that it is embodied in a mouse body (for we are all embodied in our respective bodies), but that

> It *will* suffer as an object of experimentation, of exceptionality, of vulnerability; and it will *know* that we are the cause of its suffering as it comprehends its pitiful existence confined to a laboratory and experimentation (and eventual autopsy) (Capps 2017).

The horror of a human consciousness trapped in a mouse body, then, is not the embodiment itself, but what is done to that body. It would be cruel for the same reasons that it would be cruel to treat humans as experimental mice are currently treated. Greely, for one, thinks this cruelty will not come to pass, again emphasizing the uniqueness of human consciousness:

> Yet I do suspect that we are unlikely, within this century at least, to see human–nonhuman chimeras with a substantially human-like consciousness.

> My skepticism comes partly from our vast ignorance of the daunting neurobiology. We have little, if any understanding, of how the brain generates human consciousness and human mental characteristics. We know even less about what one would have to do to create such consciousness or characteristics in nonhuman animals (Greely 2014).

It is certainly true that consciousness remains a mystery, and we do not know how it is instantiated in the human brain (or any brain, for that matter). And indeed, even detecting consciousness in non-paradigmatically behaving and non-reporting humans remains a considerable challenge (Johnson 2016). What is true of human consciousness is true of animal consciousness as well, or rather, what is true of consciousness *in* humans is also true of consciousness in animals. The idea that the consciousness of humans is unique or special has no scientific basis (Brown 2015; Johnson 2019; Comstock 2020; DeGrazia 2020; Godfrey-Smith 2020; Jones 2020; Sneddon and Brown 2020). It is merely familiar and accessible to conscious humans, just as consciousness in bats is familiar and accessible to bats. Our evidence for animal consciousness is in important ways similar to our evidence of consciousness in non-reporting humans (e.g. infants, young children, persons with advanced dementia). Research to date, however, indicates that the possibility of human-like consciousness (however we define it) or cognitive capacities is not so remote or fantastical that it can be ignored or dismissed as improbable. The acknowledged moral risk is creating a *chimeric human* who can be a subject of morality and justice. Calling such a creature a chimeric mouse or monkey permits merely skeptical replies to the moral concerns, but the skeptical solution to the dilemma posed by human/non-human chimeras dissolves under scrutiny.

15.3.1. *Chimeric non-human primates*

Human/non-human primate (H-NHP) chimeras, because of their genetic and phenotypic proximity to humans, have generated more ethical concern than other chimeric animals, along with numerous proposals to mitigate the likelihood of creating chimeras that would be morally off-limits.

Globally, the use of NHPs in brain research is expanding rapidly. For example, Japan Brain/MINDS project is focused on developing transgenic marmosets for research on dementia and neurodegenerative disorders (Japan Brain/MINDS 2021), and the US Brain Research through Advancing Innovative Neurotechnologies (BRAIN) Initiative funds research on cell and circuit processes, neuromodulation, and functional neuroimaging in NHP brains, and is also expanding its facilities and breeding capacity for marmosets (National Institutes of Health 2019). Before the global Covid-19 pandemic, China was a leading exporter of monkeys for research, and its government has invested significantly in developing capacity for NHP research. Researchers there have created genetically modified and cloned macaques bred for research on brain and eye disorders (Cyranoski 2019).

The BRAIN Initiative of the US National Institutes of Health, in its *Neuroethics Roadmap*, notes in particular the value of NHPs as research animals, with an eye towards the development of more "humanized" models:

> Each model system brings addressable well-developed "species-shared" biologies to experimental paradigms. This is likely true for research with NHPs as experiments move toward understanding traits that are

more human, such as particular aspects of consciousness that inform concepts, such as personhood. As biological aspects and their resulting characteristics are added to non-human species such as NHPs to make them more biologically similar to humans, might they become more morally similar, and in the process, raise unique animal welfare issues. As neuroscience research yields greater scientific insights into the structure and function of animal brains, insights will follow that could inform our understanding of sentience, consciousness, the experience of pain and suffering – and more generally, what we understand about the inner lives of animals. These conceptual aspects of animal experiences are closely connected to the ethical use and treatment of animals (National Institutes of Health 2020).

This statement again recognizes the possibility that humanizing NHPs to optimize their utility in research might generate enhanced ethical concerns. It fails to recognize that the use of NHPs in invasive neuroscientific research already generates enhanced ethical concerns and not merely "welfare issues", and that there is already much that "we understand about the inner lives of animals" because of the valuable contributions made by other sciences that study free-living animals (Boesch 2020).

Moral concerns notwithstanding, research creating chimeras with neural cells has proceeded in NHPs, for example, engrafting human-derived cells in monkey models of Parkinson's disease (Kriks et al. 2011), and injecting fetal human neural stem/precursor cells into experimental autoimmune encephalomyelitis (EAE) marmosets that are used to model multiple sclerosis (Pluchino et al. 2009). Monkey–human chimeric embryos were recently created by researchers who injected human pluripotent stem cells into cynomolgus macaque blastocysts. They found that

> hEPSCs [(human Expanded Potential Stem Cells)] could integrate into the inner cell masses (ICMs) of late monkey blastocysts and contributed to both embryonic and extra-embryonic lineages in peri- and early post-implantation stages during prolonged embryo culture (Tan et al. 2021).

The engrafting of human cells into animal blastocysts, and in particular NHP embryos, is discouraged and prohibited internationally by many funding agencies and guidelines (Institutes of Medicine 2005; National Institutes of Health 2016), while others call for enhanced oversight of such research (International Society for Stem Cell Research 2016). The potential for widespread integration and influence of the human cells in the chimeric embryos (and the resulting animal, if implanted and born), including in the gonads, germline and brain, raises caution flags. Evidence from mouse and rat experiments demonstrate that concern and caution are more than warranted.

The UK Academy of Medical Sciences rightly recognizes that the moral issue is not just *humanizing* NHPs (which they also think unlikely), but rather with augmenting them to resemble Great Apes (including chimpanzees and bonobos), the closest genetic and evolutionary relatives to humans, and animals that are widely considered to merit special status that excludes their permissible use in invasive research. The Academy thinks the possibility is plausible enough to warrant attention:

> One can be confident that the introduction of some human neural stem cells would not endow a monkey with a human-type self-consciousness, since that requires a capacity for higher-order thoughts associated with language, and it is fanciful to suppose that this capacity might be produced in a monkey simply by the introduction of some human neural stem cells into its brain. But once one recognises that the important comparison here is with Great Apes, then the uncertainties that affect our understanding of their cognitive abilities also affect procedures for comparing their abilities to those of enhanced monkeys. Hence if work of this kind with monkeys proceeds it would be important to study some neurally humanised monkeys before potentially damaging medical research on them is undertaken so that an informed assessment of their abilities can be undertaken (Academy of Medical Sciences 2011).

The Academy flags heightened welfare concerns that might result from neural humanization:

> If it turned out that the monkeys were seriously impaired by their neural adaptation, or that the quality of life of breeding colonies of transgenic humanised monkeys were significantly impaired by their humanisation (perhaps by their becoming more aware of their confinement), then these would be powerful reasons for halting the research (Academy of Medical Sciences 2011).

Moreover, they assert that ape-like chimeric monkeys would have to be treated like apes, which "would imply that the reasons we have for not licensing medical research which uses chimpanzees and other Great Apes apply also to research which uses these genetically enhanced monkeys" (Academy of Medical Sciences 2011). In landing on this more subtle distinction, the Academy endorses a kind of ape exceptionalism, while also recognizing that the moral humanization of NHPs is not the only ethical concern raised by such research. We might reasonably wonder, as well, if mice and rats engrafted with human neural cells might also be readily "primatized" rather than specifically "humanized", with implications for how they can permissibly be treated.

15.4. Already human-like: the overlooked problem

On close examination, it is clear that the creation of human/non-human chimeras does not raise *new* ethical concerns. It raises the same concerns that have always existed regarding the use of animals in research.

What is neglected in the discussion of humanizing chimeras is the likelihood that many species already possess the characteristics that effectively "humanize" them – that is, they already share with humans the kinds of capacities and traits that make humans purportedly unique entities of moral concern (Crozier et al. 2020). There are simply no uniquely human capacities or traits, with the exception of having a human genome, and being able to create human offspring, and neither of those traits confer a unique moral status on humans without relying on nonscientific anthropocentric speciesism as a justification. Traits commonly associated with moral status in humans, including psychological, emotional and social complexity, culture, the use of language and tools, consciousness, intelligence, problem-solving, autonomy, moral agency, and even concepts of/rituals associated with death are found in numerous species. Importantly, any morally valued trait found in humans is not found in *all* humans, and is not found *only* in humans. A neuroethics that takes seriously the role of the mind and brain in shaping identity, values, and moral considerability must thus move beyond mere anthropocentric speciesism and human exceptionalism in thinking about the implications of creating human/non-human chimeras.

15.5. Implications: justice in neuroscientific research

In a statement on "unwarranted stem cell exceptionalism", a committee of the International Society for Stem Cell Research (ISSCR) argues for a principle of justice when analyzing the ethics of creating chimeras:

> [O]ne should use existing ethical standards for research, unless something specific to stem cell research drives a need for additional ethical standards. Accepting this basic principle of justice (treating like cases alike) means that one should adhere as much as possible to ethical analytic structures used in relevantly similar contexts (Hyun et al. 2007).

This call to "treat like cases alike" is ironic, for that is precisely what many animal ethicists have long called for – treating all animals, humans included, alike for research purposes (Ferdowsian et al. 2020). That does not mean, as the committee says, merely "building on existing animal welfare structures for animal research" (Hyun et al. 2007) that carve out separate and unequal rules for different species. The kind of exceptionalism that is dominant in the neuroethical discourse on human/non-human chimeras is not stem cell exceptionalism but human exceptionalism, hence the concern with creating an animal that is more human-like than is comfortable in a context where the animal will be experimented on and killed.

Within research ethics, the way the benefits and burdens of research are distributed has long been recognized as a matter of justice. Historically, the exploitation of disadvantaged, captive, and easily coerced populations has prompted the promulgation of research guidelines and codes. The Nuremberg Code responded to research atrocities committed against prisoners and internees by Nazi doctors (Nuremberg Tribunals 1949); internationally, research codes and guidelines have adopted strong protections for those considered vulnerable to exploitation and coercion, including the impoverished, incarcerated, and institutionalized persons who need protection against unreasonable and inequitable exploitation in research (Council for International Organizations of Medical Sciences 2002). As the *Belmont Report* states, justice in the selection of human research subjects requires avoiding the selection of those who are easily exploited and coerced, or those who are unprotected – vulnerable subjects – without strong scientific justification for using them (Belmont Commission 1979). In other words, the vulnerable should not be used in research for reasons of convenience, or because they are easily accessible and unprotected. Much of the burden of protecting *human* research subjects, however, has fallen on animals, and restrictions on human subjects research have increased in number and diversity the animals used in their stead. Non-human animals, compared to humans, are more easily accessible and conveniently used in research – indeed, they are considered commodities that are bred and bought for that purpose. They are also relatively unprotected, their use governed by animal welfare regulations that provide scant protections and do not preclude lifelong captivity, or causing pain, distress, suffering, permanent injury, and death (Pound and Nicol 2018).

Scientists have for millennia used animals for anatomical studies and for research when using humans was considered taboo. Early research codes explicitly endorsed and even required the use of animal experiments as a prerequisite to ethical research with humans. The Nuremberg Code (Nuremberg Military Tribunals 1949), for example, states that human experiments *must* be based on prior research with animals, creating a model for the progression of biomedical research from animal to human that has scarcely been challenged since:

> The experiment should be so designed and based on the results of animal experimentation and a knowledge of the natural history of the disease or other problem under study that the anticipated results justify the performance of the experiment (Nuremberg Tribunals 1949).

The first Declaration of Helsinki in 1964, similarly, linked the ethical permissibility of human subjects research to animal experiments:

> Clinical research must conform to the moral and scientific principles that justify medical research and should be based on laboratory and animal experiments or other scientifically established facts (World Medical Association 1964).

Seven decades later, some regulatory agencies, such as the US Food and Drug Administration, still require animal testing prior to clinical trials in humans (Food and Drug Administration 2021), despite mounting evidence of the failure of animal research to translate to humans (Garner 2014; Johnson 2020b).

In neuroscientific research, NHPs are considered bridges between experiments on animals like rodents and experiments on humans. Marmosets, macaques and other small monkeys are the NHPs most commonly used, in part because they are smaller, easier to breed, easier to house in small cages, and easier to handle than larger NHPs. They are not, however, the NHPs whose brains most resemble human brains – those would be the brains of apes, including chimpanzees and bonobos, whose use is morally frowned upon. Marmosets are commonly used to model human brain disorders like multiple sclerosis, Parkinson's, and autism. Marmosets are tiny primates that can fit in a human hand. They breed twice a year, they grow and mature quickly, and are readily genetically engineered (Servick 2018). Marmosets have smooth, lissencephalic brains quite unlike the furrowed gyrencephalic brains of humans and some other NHPs, but their tiny lissencephalic brains are easy to image. Their frequent use despite significant neuroanatomical differences can be attributed to their easy access, low cost, and convenience rather than to their scientific suitability as substitutes for humans. But the selection of an animal model for the sake of convenience rather than scientific fidelity or discrimination makes their use scientifically unjustified. What is not scientifically justified cannot be morally justified when it results in harm to a sentient creature. If we treat like cases alike, then the selection of NHPs or other animals for non-scientific reasons is unjust, much like the frequent use of vulnerable, incarcerated, and institutionalized persons in mid-20th century research was unjust.

Another way that subject selection can be unjust is when those who are research subjects are not the beneficiaries of research. This includes humans who are economically, socially, and geographically disadvantaged. Within neuroethics, when justice in the context of research is considered it is frequently framed in terms of disparities in access to healthcare and neurotechnologies, where the concern is rightly that those least likely to benefit from the fruits of neuroscientific research bear a disproportionate share of the burdens. The NIH BRAIN Initiative Neuroethics Working Group, considering justice in the development of novel neurotechnologies, states:

> Limited access to safe and effective neural technologies should not exacerbate existing health disparities or inequalities, but neither should the burdens of research fall disproportionately on those who lack access to established interventions (Greely et al. 2018).

Animals *can* benefit incidentally from biomedical research – drugs developed for humans are sometimes adapted for veterinary use, for example. However, they are generally not the beneficiaries of most neuroscientific research. If, like the ISSCR committee (Hyun et al. 2007), we accept as a principle of justice treating like cases alike, then consistently extending that principle to animals that share the capacities and characteristics that make humans subjects of justice can lead to only one conclusion. The animals that are like humans in morally relevant ways must be treated in morally similar ways for purposes of neuroscientific research. This would not exclude outright the use of animals in neuroscientific research (after all, research with humans is permitted), but it would require a harmonization of human and animal research ethics and regulations, which would result in a radical revision of current research practices. It would also require a cultural shift within neuroethics away from anthropocentrism and human exceptionalism when examining the conduct and justification for brain research.

The traditional framing of justice in research as the fair and scientifically justified selection of subjects, and the fair and equitable distribution of the benefits and burdens of research, has profound implications for neuroscientific research with non-human animals that have the capacities and characteristics that make them relevantly similar to humans, and thus subjects of justice.

15.6. Interrogating the anthropocentric culture and human exceptionalism of neuroethics

Non-human entities like robots and AIs have long been recognized to be within the sphere of neuroethical concern, despite having nonbiological "brains" (Johnson 2019). Similarly, human-origin entities and organisms like cerebral organoids, cultured neural tissues and neural cells, and human embryonic stem cells have generated much discussion. Neuroethics has thus long considered entities and organisms other than fully developed humans to be within its purview. The exclusion of non-human animals, with the exception of repeating oft-rehearsed claims about the need to attend to animal welfare, then, is notable and worthy of scrutiny. Recognizing animals as subjects of neuroethical consideration would, to be sure, create discomfort and dissonance, for unlike sentient robots and AIs, these animals already exist, and many are used in neuroscientific research in ways that, were they human, would be prohibited and almost universally acknowledged to be unethical.

In its BRAIN 2025 report, the National Institute of Health (NIH) expressed the desire to change the culture of neuroscience, making it more open to cross-disciplinary collaboration and diverse ideas:

> we hope through the BRAIN Initiative to create a culture of neuroscience research that emphasizes worldwide collaboration, open sharing of results and tools, mutual education across disciplines and added value that comes from having many minds address the same questions from different angles (National Institutes of Health BRAIN Working Group 2014).

Indeed, neuroethics as a discipline has often preached the same message, pushing for greater and earlier integration of neuroethics in neuroscientific work, including the embedding of neuroethicists within the lab for timely, on-the-ground contributions to novel and developing work with the potential to generate unique ethical concerns.

But elsewhere in BRAIN 2025, the authors hype the value of using animals, and NHPs in particular, in neuroscience:

> Although we must be constantly on guard against facile anthropomorphism, the continuity of brain structure and organization across species provides confidence that some cognitive processes analogous to ours are likely to exist in the brains of animals other than humans. Improving the behavioral analysis of these cognitive processes, both in experimental animals and in humans, should be an emphasis of the BRAIN Initiative.

> We expect the BRAIN Initiative to include nonhuman primates such as rhesus macaques, because they are evolutionarily the closest animal model for humans, and this will be reflected in their behavioral and cognitive abilities, genetics, anatomy and physiology (National Institutes of Health BRAIN Working Group 2014).

It is not surprising that an agency dedicated to fostering and funding health-related neuroscientific research would view the evolutionary, behavioral, structural, and cognitive continuities across species as reasons for pursuing research with animals, rather than as reasons against doing so. Indeed, those similarities and continuities provide the scientific justification for using animals, if there is one. But it is not "facile anthropomorphism" that neuroscience needs to guard against, but facile *anthropocentrism* and human exceptionalism, and the view that human/non-human similarities and continuities matter for scientific but not ethical justification simply because it favors human interests to adopt that inconsistency. Neuroethicists have neither sufficiently called out that inconsistency in neuroscience nor encouraged meaningful discussion of the scientific and ethical justifications for experimenting on animals, frequently accepting the proffered scientific rationale: the *possible* benefits for humans. Meanwhile, as a field, neuroethics has largely neglected what is not merely possible, but *actual* in neuroscientific research – significant harms to sentient animals (Pound and Nicol 2018).

An intellectually and ethically rigorous neuroethics must think more critically about the value of using animals as models for humans in brain research (Pound et al. 2004), and think more expansively about the development of human-relevant alternatives to using animals. The discipline of neuroethics has frequently been at the forefront of examining potential but largely speculative ethical concerns about new developments and technologies, scanning the horizon for possibilities that may not yet be in view, as well as possibilities that may never be realized. The moral humanization of human/non-human chimeras is an instructive example of how that analysis is excessively framed and constrained by human exceptionalism. Problematically, there has been scant critical engagement with human exceptionalism and anthropocentrism embedded in the very concept of moral humanization, or with deep questions about what "moral humanization" actually means, both for non-humans and for those humans who have also long been relegated to the margins of moral considerability. There has been undue emphasis on speculative possibilities, such as the development of consciousness in simple entities like organoids, or the retention of consciousness in tiny cubes of ex vivo human brain tissue – with an uncritical acceptance of the assumption that bits of human-origin tissue could plausibly undergird human-like consciousness. Consequently, there has been too little consideration of actual consciousness and its implications in the animals currently used in neuroscientific research, and biomedical research more generally.

Despite the inter- and multidisciplinary nature of neuroethics, it remains for the most part narrowly anthropocentric in its focus. Expanding that focus requires a culture change analogous to the one BRAIN 2025 proposes for neuroscience: a widening of disciplinary boundaries to incorporate more moral perspectives as well as insights from scholarship in other sciences. Maintaining a rigid ethical boundary between humans and non-human animals requires ignoring the implications of the accumulated knowledge and discoveries of the brain sciences, and numerous cognate fields, as well as humanistic disciplines including philosophy and human-animal studies. Scholarship in animal ethics, neuroethics and neuroscience have heretofore advanced on parallel tracks. An empirically informed neuroethics should bend them towards convergence and intersection. A neuroethics that is committed to exploring the implications of neuroscientific research, and interrogating research that threatens important and fundamental ethical values, cannot maintain a rigid doctrine of human exceptionalism that is in significant tension with scientific reality. By adopting a less anthropocentric focus, and a more inclusive ethics, neuroethics can enhance its intellectual rigor, and expand its scope and role in critical and forward-thinking discussions of how developing knowledge of non-human minds challenges human exceptionalism, which currently impedes scientific, social, and moral progress.

15.7. References

Academy of Medical Sciences (2011). Animals containing human material. Report, Academy of Medical Sciences, London.

Andrews, K. and Beck, J. (eds) (2019). *The Routledge Handbook of Philosophy of Animal Minds*. Routledge, New York.

Andrews, K., Comstock, G., Crozier, G., Donaldson, S., Fenton, A., John, T., Johnson, L.S.M., Jones, R., Kymlicka, W., Meynell, L. (2019). *Chimpanzee Rights: The Philosophers' Brief*. Routledge, London.

Balaban, E., Teillet, M.-A., Le Douarin, N. (1988). Application of the quail-chick chimera system to the study of brain development and behavior. *Science*, 241(4871), 1339–1342.

Belmont Commission (1979). The Belmont Report: Ethical principles and guidelines for the protection of human subjects of research. United States Department of Health, Education and Welfare, Washington, DC.

Bentham, J. (1789). An introduction to the principles of morals and legislation [Online]. Available at: https://www.earlymoderntexts.com/assets/pdfs/bentham1780.pdf [Accessed 16 June 2021].

Boesch, C. (2020). The human challenge in understanding animal cognition. In *Neuroethics and Nonhuman Animals*, Johnson, L.S.M., Fenton, A., Shriver, A. (eds). Springer, Cham.

Brown, C. (2015). Fish intelligence, sentience and ethics. *Animal Cognition*, 18(1), 1–17.

Brüstle, O., Choudhary, K., Karram, K., Hüttner, A., Murray, K., Dubois-Dalcq, M., Mckay, R.D.G. (1998). Chimeric brains generated by intraventricular transplantation of fetal human brain cells into embryonic rats. *Nature Biotechnology*, 16(11), 1040–1044.

Capps, B. (2017). Do chimeras have minds? The ethics of clinical research on a human-animal brain model. *Cambridge Q. Healthcare Ethics*, 26(577).

Coghlan, A. (2014). Human cells stage coup d'etat in mouse brains. *New Scientist*, 224(2998), 15.

Comstock, G.L. (2020). Bovine prospection, the Mesocorticolimbic pathways, and Neuroethics: Is a cow's future like ours? In *Neuroethics and Nonhuman Animals*, Johnson, L.S.M., Fenton, A., Shriver, A. (eds). Springer, Cham.

Council for International Organizations of Medical Sciences (2002). International ethical guidelines for biomedical research involving human subjects. Report, Council for International Organizations of Medical Sciences. Geneva, Cham.

Crick, F. and Koch, C. (2003). A framework for consciousness. *Nature Neuroscience*, 6(2), 119.

Crozier, G., Fenton, A., Meynell, L., Peña-Guzmán, D.M. (2020). Nonhuman, all too human: Towards developing policies for ethical chimera research. In *Neuroethics and Nonhuman Animals*, Johnson, L.S.M., Fenton, A., Shriver, A. (eds). Springer, Cham.

Cyranoski, D. (2019). Chinese effort to clone gene-edited monkeys kicks off. *Nature* [Online]. Available at: www.nature.com/articles/d41586-019-00292-w [Accessed 16 June 2021].

Danish Council of Ethics (2008). Man or mouse? Ethical aspects of chimaera research. Report, Danish Council of Ethics, Copenhagen.

Degrazia, D. (2020). Sentience and consciousness as bases for attributing interests and moral status: Considering the evidence – and speculating slightly beyond. In *Neuroethics and Nonhuman Animals*, Johnson, L.S.M., Fenton, A., Shriver, A. (eds). Springer, Cham.

Dehaene, S. and Naccache, L. (2001). Towards a cognitive neuroscience of consciousness: Basic evidence and a workspace framework. *Cognition*, 79(1), 1–37.

Di Lullo, E. and Kriegstein, A.R. (2017). The use of brain organoids to investigate neural development and disease. *Nature Reviews Neuroscience*, 18, 573–584.

Edelman, D.B. and Seth, A.K. (2009). Animal consciousness: A synthetic approach. *Trends in Neurosciences*, 32(9), 476–484.

Ferdowsian, H.R. and Gluck, J.P. (2015). The ethical challenges of animal research: Honoring Henry Beecher's approach to moral problems. *Cambridge Quarterly of Healthcare Ethics*, 24(4), 391–406.

Ferdowsian, H.R., Johnson, L.S.M., Johnson, J., Fenton, A., Shriver, A., Gluck, J.P. (2020). A Belmont Report for animals? *Cambridge Quarterly of Healthcare Ethics*, 29(1), 19–37.

Food and Drug Administration (2021). 21CFR312 investigational new drug application. Report, FDA, United States.

Garner, J.P. (2014). The significance of meaning: Why do over 90% of behavioral neuroscience results fail to translate to humans, and what can we do to fix it? *ILAR Journal*, 55(3), 438–456.

German Ethics Council (2011), Human–animal mixtures in research. Report, Deutscher Ethikrat, Berlin.

Godfrey-Smith, P. (2020). *Metazoa: Animal Life and the Birth of the Mind*. Farrar, Straus and Giroux, New York.

Greely, H.T. (2014). Academic chimeras? *The American Journal of Bioethics*, 14(2), 13–14.

Greely, H.T., Cho, M.K., Hogle, L.F., Satz, D.M. (2007). Thinking about the human neuron mouse. *The American Journal of Bioethics*, 7(5), 27–40.

Greely, H.T., Grady, C., Ramos, K.M., Chiong, W., Eberwine, J., Farahany, N.A., Johnson, L.S.M., Hyman, B.T., Hyman, S.E., Rommelfanger, K.S. (2018). Neuroethics guiding principles for the NIH BRAIN Initiative. *Journal of Neuroscience*, 38(50), 10586–10588.

Gruen, L. (2014). *The Ethics of Captivity*. Oxford University Press, Oxford.

Han, X., Chen, M., Wang, F., Windrem, M., Wang, S., Shanz, S., Xu, Q., Oberheim, N.A., Bekar, L., Betstadt, S. et al. (2013). Forebrain engraftment by human glial progenitor cells enhances synaptic plasticity and learning in adult mice. *Cell Stem Cell*, 12(3), 342–353.

Hardcastle, V.G. (2016). Minimally Conscious States and pain. In *Finding Consciousness: The Neuroscience, Ethics, and Law of Severe Brain Damage*, Sinnott-Armstrong, W. (ed.). Oxford University Press, Oxford.

Hyun, I., Taylor, P., Testa, G., Dickens, B., Jung, K.W., McNab, A., Robertson, J., Skene, L. Zoloth, L. (2007). Ethical standards for human-to-animal chimera experiments in stem cell research. *Cell Stem Cell*, 1(2), 159–163.

Institutes of Medicine and National Research Council (2005). Guidelines for human embryonic stem cell research. Report, The National Academies Press, Washington DC.

Internatonal Society for Stem Cell Research (2016). Guidelines for stem cell research and clinical translation [Online]. Available at: http://www.isscr.org/docs/default-source/guidelines/isscr-guidelines-for-stem-cell-research-and-clinical-translation.pdf?sfvrsn=2 [Accessed 16 June 2021].

Japan Brain/MINDS (2021). Marmoset research [Online]. Available at: https://brainminds.jp/en/research/kind/marmoset-research-en [Accessed 16 June 2021].

Johnson, L.S.M. (2016). Inference and inductive risk in disorders of consciousness. *AJOB Neuroscience*, 7(1), 35–43.

Johnson, L.S.M. (2019). Neuroethics of the nonhuman. *AJOB Neuroscience*, 10(3), 111–113.

Johnson, L.S.M. (2020a). Introduction to animal neuroethics: What and why? In *Neuroethics and Nonhuman Animals*, Johnson, L.S.M., Fenton, A., Shriver, A. (eds). Springer, Cham.

Johnson, L.S.M. (2020b). The trouble with animal models in brain research. In *Neuroethics and Nonhuman Animals*, Johnson, L.S.M., Fenton, A., Shriver, A. (eds). Springer, Cham.

Johnson, L.S.M. (2022). *The Ethics of Uncertainty: Entangled Ethical and Epistemic Risks in Disorders of Consciousness*. Oxford University Press, New York.

Jones, R.C. (2020). Speciesism and human supremacy in animal neuroscience. In *Neuroethics and Nonhuman Animals*, Johnson, L.S.M., Fenton, A., Shriver, A. (eds). Springer, Cham.

Kriks, S., Shim, J.-W., Piao, J., Ganat, Y.M., Wakeman, D.R., Xie, Z., Carrillo-Reid, L., Auyeung, G., Antonacci, C., Buch, A. et al. (2011). Dopamine neurons derived from human ES cells efficiently engraft in animal models of Parkinson's disease. *Nature*, 480(7378), 547–551.

Levy, N. and Savulescu, J. (2009). Moral significance of phenomenal consciousness. *Progress in Brain Research*, 177(361–370).

National Academies of Sciences, Engineering, and Medicine (2021). *The Emerging Field of Human Neural Organoids, Transplants, and Chimeras: Science, Ethics, and Governance*. The National Academies Press, Washington, DC.

National Institutes of Health (2016). Guidelines for human stem cell research [Online]. Available at: https://stemcells.nih.gov/policy/2009-guidelines.htm [Accessed 16 June 2021].

National Institutes of Health (2019). Funding opportunities: Marmoset coordination center (U24 Clinical Trials Not Allowed) [Online]. Available at: https://braininitiative.nih.gov/funding-opportunies/marmoset-coordination-center-u24-clinical-trials-not-allowed [Accessed 16 June 2021].

National Institutes of Health (2020). The BRAIN Initiative and neuroethics: Enabling and enhancing neuroscience advances for society [Online]. Available at: https://braininitiative.nih.gov/strategic-planning/acd-working-groups/brain-initiative®-and-neuroethics-enabling-and-enhancing [Accessed 16 June 2021].

National Institutes of Health BRAIN Working Group (2014). BRAIN 2025 a scientific vision [Online]. Available at: https://braininitiative.nih.gov/pdf/BRAIN2025_508C.pdf [Accessed 16 June 2021].

Nuremberg Military Tribunals (1949). The Nuremberg Code [Online]. Available at: history.nig.gov/display/history/Nuremberg+Code [Accessed 16 June 2021].

Pluchino, S., Gritti, A., Blezer, E., Amadio, S., Brambilla, E., Borsellino, G., Cossetti, C., Del Carro, U., Comi, G., Hart, B. et al. (2009). Human neural stem cells ameliorate autoimmune encephalomyelitis in non-human primates. *Annals of Neurology*, 66(3), 343–354.

Pound, P. and Nicol, C.J. (2018). Retrospective harm benefit analysis of pre-clinical animal research for six treatment interventions. *PloS One*, 13(3).

Pound, P., Ebrahim, S., Sandercock, P., Bracken, M.B., Roberts, I. (2004). Where is the evidence that animal research benefits humans? *BMJ*, 328(7438), 514–517.

Qian, X., Nguyen, H.N., Jacob, F., Song, H., Ming, G.-L. (2017). Using brain organoids to understand Zika virus-induced microcephaly. *Development*, 144(6), 952–957.

Regan, T. (2004). *The Case for Animal Rights*. University of California Press, Berkeley, CA.

Servick, K. (2018). Why are U.S. neuroscientists clamoring for marmosets? *Science* [Online]. Available at: https://www.sciencemag.org/news/2018/10/why-are-us-neuroscientists-are-clamoring-marmosets [Accessed 16 June 2021].

Seth, A.K., Baars, B.J., Edelman, D.B. (2005). Criteria for consciousness in humans and other mammals. *Consciousness and Cognition*, 14(1), 119–139.

Singer, P. (1979). Killing humans and killing animals. *Inquiry*, 22(1–4), 145–156.

Sinnott-Armstrong, W. (2016). *Finding Consciousness: The Neuroscience, Ethics, and Law of Severe Brain Damage*. Oxford University Press, Oxford.

Sneddon, L.U. and Brown, C. (2020). Mental capacities of fishes. In *Neuroethics and Nonhuman Animals*, Johnson, L.S.M., Fenton, A., Shriver, A. (eds). Springer, Cham.

Streiffer, R. (2007). At the edge of humanity: Human stem cells, chimeras, and moral status. *Journal of Philosophical Research*, 32(Supplement), 63–83.

Tan, T., Wu, J., Si, C., Dai, S., Zhang, Y., Sun, N., Zhang, E., Shao, H., Si, W., Yang, P. et al. (2021). Chimeric contribution of human extended pluripotent stem cells to monkey embryos. *Cell*, 184(8), 2020–2032.e2014.

Tononi, G., Boly, M., Massimini, M., Koch, C. (2016). Integrated information theory: From consciousness to its physical substrate. *Nature Reviews Neuroscience*, 17(7), 450–461.

Warren, M.A. (1997). *Moral Status: Obligations to Persons and Other Living Things*. Clarendon Press, Oxford.

Windrem, M.S., Schanz, S.J., Morrow, C., Munir, J., Chandler-Militello, D., Wang, S., Goldman, S.A. (2014). A competitive advantage by neonatally engrafted human glial progenitors yields mice whose brains are chimeric for human glia. *Journal of Neuroscience*, 34(48), 16153–16161.

World Medical Association (1964), Declaration of Helsinki. Ethical Principles for Medical Research Involving Human Subjects. Report, World Medical Association, Helsinki.

16

Cultural Neuroethics in Practice – Human Rights Law and Brain Death

Jennifer A. CHANDLER

Centre for Health Law, Policy and Ethics, University of Ottawa, Canada

16.1. Introduction

Well before there was a recognized neuroethics, science and technology were generating brain-related ethical issues to resolve (Stahnisch 2015; Shen 2016). Twentieth-century advances in resuscitation and mechanical ventilation produced the situation in which a person's cardio-respiratory function could continue despite the irreversible loss of all brain function. These technological changes gave us a metaphysical question with important ethical and legal implications: is a human in this state alive or dead? The concept of brain death has achieved widespread but not universal acceptance in many parts of the world since the first published clinical guideline (Ad Hoc Committee Harvard Medical School 1968). Despite this, the idea that the irreversible loss of all functions of the brain constitutes death continues to be debated, and this debate periodically emerges into the legal domain.

The now 50+-year-old and still simmering problem of brain death offers an excellent case example of the importance of a cultural lens in neuroethics, as well as of how the law in multicultural Western liberal democracies attempts to grapple with cultural diversity in fundamental conceptual and ethical positions. Brain death is ethically and legally significant at two levels. First, it is important precisely because the moral and legal duties owed to living people cease upon death, and so the definition of death has direct consequences for people. Second, social disagreement on a fundamental question like the meaning of death raises the issue of how to balance the benefits and costs of enforcing uniformity, as well as the scope of moral and legal duties to respect and accommodate minority views.

This chapter examines the role of culture in the debate over brain death, noting that the courts are particularly at risk of cultural myopia in the context of the definition of death. This is because legal personhood is extinguished upon death, so claims brought on behalf of brain-dead people are sometimes quickly dismissed on the basis that the claimant is not a legal rights-holder. This means the main issue in question – the rights-based claim of the brain-dead person – is not adjudicated. For example, in the 2018 Ontario case of *McKitty v Hayani*, a claim was brought on behalf of a woman diagnosed brain dead that her religion rejected brain death and the application of that definition to her violated her constitutional right to freedom of religion (Ontario Superior Court of Justice 2018). The court ruled that there was no reason to consider whether her constitutional rights were violated because she was dead and thus had no right to religious freedom. This is clearly an unsatisfying example of circular reasoning, but one that is easier to commit in a context where there is an overwhelming social consensus in favor of brain death as an objective scientific fact, rather than representing a philosophical interpretation of biological facts. The seeming obviousness and indisputability of the position make it easy to dismiss dissenting views in this way. Fortunately, the Ontario Court of Appeal (2019) corrected this error on appeal.

The focus here is on cultural diversity among social groups, including diversity in religious beliefs, but another sometimes closely related dimension is that of ethnic and racial diversity. Reasons for family resistance to brain death are not solely matters of religious or cultural beliefs and practice, but may be due to emotional and psychological factors like grief, anger and mistrust in the health care system. There is evidence of worse health care outcomes and higher levels of mistrust of the medical system amongst racial minorities in some Western countries. Much needs to be done to address this at multiple levels, but it is also relevant to discussions of how to address religious objection to brain death. If a society chooses to accommodate only religious objection, it will induce people to frame their objections in religious terms, whether or not this is the main underlying concern. This is unfortunate as it impedes the recognition of and more direct and helpful response to those underlying concerns.

16.2. The concept of brain death

Brain death has been described as an "artifact of technology" because people in this state could not exist prior to the development of mechanical ventilation to support respiratory and cardiac function in severely brain-injured individuals (Siminoff et al. 2004, p. 2326). The concept was broached several times starting in the 1950s, but the start of its gradual acceptance was the first published clinical definition in the 1968 Report "A Definition of Irreversible Coma" by a committee at the Harvard Medical School (Ad Hoc Committee Harvard Medical School 1968; DeVita et al. 1993; Greer et al. 2020). Various motivations have been suggested for this proposition: to allow for the removal of viable organs for transplantation from deceased ventilated patients, to avoid the futile and wasteful use of medical resources, to address the "burden" to families and society of maintaining people who had "permanent loss of intellect, and perhaps to allow for medical experimentation on brain-dead bodies so as to reduce the use of live human subjects" (Pernick 1999; Truog et al. 2018). Today, brain death is widely accepted around the world, including by medicine, most religious authorities and the legislatures and courts of many countries (Greer et al. 2020; Lewis et al. 2020a). Nevertheless, reasonably sizable minorities of survey respondents in Western countries like Latvia and the US (Siminoff et al. 2004; Neiders and Dranseika 2020) reject brain death and endorse cardio-respiratory definitions of death.

Persistent minority views continue to challenge brain death on conceptual grounds. The challenge with brain death is that the privileging of the brain as *essential* to life reflects a cultural neuroessentialism that is not necessarily universally shared. Some functions performed by the brain may be replaced through technological means (e.g. ventilation to replace lost respiratory drive), but there remains the question of whether other brain functions that cannot be replaced in this way are in fact essential to life. Veatch has identified three stable classes of views on this point: whole-brain definitions, higher-brain definitions and cardiocirculatory or somatic definitions (Veatch 2019, p. 387).

The dominant whole-brain definition holds that death means the irreversible loss of all functions of the whole brain, including the brainstem. This view therefore regards the continuation of any brain function as sufficient for life. The ambiguity of the term *function* has led to problems for this definition. The awkwardness of the American Academy of Neurology's recent declaration (Russell et al. 2019) that continued neuroendocrine *function* is consistent with brain death has led to calls to revise the definition (Lewis et al. 2020b).

The higher-brain definition holds that certain functions such as capacity for consciousness are essential to life, and so the presence of some other brain functions will not be enough to qualify as living in the absence of the essential brain functions (Veatch 2019, p. 392). In this view, a person in a permanent vegetative state (PVS) would be dead.

The cardiocirculatory definition holds that an individual is alive until the heart and circulation stop irreversibly (Veatch 2019, p. 390). Current practice where there is no intention to attempt resuscitation is to wait five minutes to determine whether to declare cardiocirculatory death (Dhanani et al. 2021). The somatic definitions regard the somatic functioning of the organism as life and note that various complex somatic functions continue in medically supported brain-dead bodies (Veatch 2019, p. 390). On this point, Shewmon has noted a whole range of functions can continue in brain-dead bodies including homeostasis of multiple physiological and chemical parameters; assimilation of nutrients; elimination, detoxification and recycling of cellular waste; energy balance; maintenance of body temperature (albeit subnormal); wound healing; fighting of infections; development of febrile response to infection (albeit rarely); cardiovascular and hormonal stress responses to incision; successful gestation of a fetus; sexual maturation and proportional growth (Shewmon 1998).

Many have noted that the choice amongst these views is a philosophical or moral question, not a biomedical question – all three can accept the same underlying biological facts and still hold their view of what functions are essential for the status of life (Shewmon 1998; Siminoff et al. 2004, p. 2326; Ross 2018). The Ontario Court of Appeal recently stated that,

> [t]he determination of legal death is not simply, or even primarily, a medical or biological question. The question of who the law recognizes as a human being – entitled to all of the benefits and protections of the law – cannot be answered by medical knowledge alone. Facts about the physiology of the brain-dead patient are needed to determine what obligations are owed to the brain-dead patient, but the enquiry is not ultimately technical or scientific: it is evaluative. Who the common law ought to regard as a human being – a bearer of legal rights – is inescapably a question of justice, informed but not ultimately determined by current medical practice, bioethics, moral philosophy, and other disciplines (Ontario Court of Appeal 2019, p. 29).

There may be very good policy reasons to select a particular legal definition of death, and societies can stipulate legal definitions for policy purposes, whether or not they have a solid philosophical or biological basis. The legal approach to the beginning of life furnishes an example. In Canada, legal personhood begins when a fetus has been born alive. A similar approach to stipulating a definition for death is likely needed, given the arbitrariness of selecting which bodily functions signify life within a context of increasing technological capacity to support disaggregated bodily functions. The essential problem is thus revealed – the legal definition of death is a social choice, and all of the problems of fairness in adjudicating amongst differences of opinion within multicultural societies come to the fore.

16.3. Objections to brain death: culture, religion and demographic minorities

Given the metaphysical nature of the question of defining death and the social consequences of its deployment, it is not surprising that there is cultural variation in acceptance of the concept of brain death. Most major religions accept brain death as death, although several religions (or subgroups within those religions) do not, including Shintoism and some Orthodox Jews, Buddhists, Muslims and Indigenous people in North America (Chamsi-Pasha and Albar 2017; Pope 2017b, p. 291; Veatch 2019, p. 392).

Yang and Miller note the speed with which brain death was accepted and legally endorsed in Western societies as compared to the slower and more conservative approach in Eastern countries (Yang and Miller 2015, p. 214). Empirical studies of public opinion show lower levels of public acceptance of the idea that brain death is death in Japan and China than in Western countries like the United States, the United Kingdom, Germany and France (Yang and Miller 2015, p. 216). Yang and Miller suggest that Western openness to brain death was underpinned by philosophical ideas of the dichotomy between body and soul, the soul or conscious mind as the essence of personhood and the relationship of soul and body as one of master and tool. Conversely, Eastern philosophies like Shintoism and Taoism do not separate mind from body, and they tend to view the boundary between life and death as blurred and gradual (Yang and Miller 2015, p. 219). The Japanese see the mind and body as inseparable, and so brain death or cessation of brain function independent of other body functions is inconsistent with traditional Japanese values (Terunuma and Mathis 2021). Buddhism regards consciousness, representing personal and collective identity, as distributed throughout the body and not exclusively located in the brain, such that consciousness may subsist elsewhere in the body despite a lack of brain activity (Yang and Miller 2015, p. 220). In summary, "philosophical and religious backgrounds appear to reflect the ease or difficulty with which societies have adapted to brain death" (Yang and Miller 2015, p. 220).

However, a key point is that objections to brain death are not solely based in religious or cultural ideas, but may also reflect psychological factors like grief, anger and mistrust. Racial minorities may be less likely to trust medical prognoses and diagnoses for catastrophically brain-injured family members because of historical medical abuses, current systematic disparities in health care outcomes, together with personal experience of social discrimination.

In their research on attitudes towards organ donation, Siminoff et al. (2006) found that when compared to White Americans, "African-Americans express greater concerns about the trustworthiness of the health care system, both in general and in terms of the donation system specifically". For example, they were less likely to believe doctors can be trusted to pronounce death correctly for patients eligible to donate organs.

There may be racial disparities in health care outcomes for the very class of patients at issue in brain death – those with catastrophic brain injury. Bowman et al. investigated the outcomes by race or ethnicity for people hospitalized for moderate to severe traumatic brain injury between 2000 and 2003. They found that compared to White patients, Black and Asian patients had higher in-hospital mortality and Black and Hispanic patients were less likely to be discharged to rehabilitation centers. They could not conclusively attribute this to biased medical decision-making as opposed to other potential contributing factors, and recommended further research on this point. There is also evidence that racial minorities in the United States receive more life-prolonging care (e.g. resuscitation, intensive care unit hospitalization, feeding tube use). It is not clear why this is so, but one suggestion is a lack of trust in health care providers. Other suggested explanations are social, religious and cultural values related to care, less information about advance directives, or discrepancies in patient–physician communication (Mack et al. 2010).

16.4. What is at stake with the definition of death?

Death is clearly of profound social, emotional and psychological significance for humans, and it is usually associated with important religious and cultural practices (Ohnuki-Tierney et al. 1994). However, it is also of critical legal importance since the moment of death signifies the extinction of legal personhood, and with it a whole range of legal rights and duties. Thus, the definition of death has major consequences for a person. We may object by saying that this person is dead, and so how can there be any consequences to suffer? But this prejudges the issue in question, of course. The point is that on this metaphysical question, the line can be drawn in various places and groups of people will lose or acquire legal personhood on the basis of where it is drawn.

The definition of death is not a purely private matter, as important social and legal consequences flow from a person's status as alive or dead (Lewis et al. 2019, p. 12; Ross 2018).

– Criminal law: Is a person guilty of assault or homicide?

– Family law: When does a spouse become a widow or widower?

– Insurance: When are life insurance policies payable? When do health or disability insurance benefits end?

– Wills and estates: At what point should an estate be distributed?

– Medical care: Is there any obligation to provide continuing medical care, nutrition and hydration?

– Organ donation: When may vital organs be removed for transplant?

Criminal law is understood as public in nature, and so the public accordingly has an interest in the proper application of the criminal law. A clear and uniform definition of death is therefore needed for this purpose. On occasion, those objecting to brain death may be in conflict of interest. In the 2014 case of Issac Lopez, the two-month-old baby's father admitted injuring him and was arrested for child abuse. The mother launched a challenge to the diagnosis of brain death on her own, given the father's possible motivation to avoid homicide charges (Pope 2014).

Disputes over brain death frequently revolve around the most immediate consequence of a brain death diagnosis, namely the termination of medical support, nutrition and hydration. The rules on withdrawing life support vary among jurisdictions, but many either explicitly insist that a physician obtain surrogate consent to the withdrawal of life-sustaining therapies or leave it unclear, in which case risk-averse physicians often seek consent (Pope 2013). However, if a person is dead, no surrogate consent is required to withdraw life support (with certain jurisdictions like New Jersey that allow exemptions from brain death). Thus, the definition of death determines which end-of-life decision-making regime is applicable and so has important consequences for the patient and family.

The definition is also very important to health care providers because the obligation to provide ongoing care that they regard as a futile can bring about considerable moral distress for them. What is in fact futile or undignified is a matter of disagreement, but many health care providers experience moral distress in the end-of-life context. Much of the research on moral distress over futile treatment concerns dying patients, as opposed to those who are brain dead (Dzeng et al. 2016; St Ledger et al. 2021). It is possible that distress about exacerbating and prolonging suffering is greater in the case of these still-living patients, rather than brain-dead patients who are beyond suffering. However, there is evidence that even health care providers struggle to regard a warm body with a beating heart as dead (White 2003), and moral distress associated with imposing suffering and indignity on a brain-dead patient may thus still arise. In any event, we can also imagine moral distress associated with the diversion of resources from others in need, the demoralization associated with the perceived waste of time and expertise, and a feeling that continued interventions do not express proper respect for a person's remains (Russell et al. 2019, p. 231).

The definition chosen also matters to society since many would consider health care resources to be wasted in the case of brain-dead bodies. The resource implications would depend upon how often brain death is rejected. The temporary accommodation of objecting families (i.e. a delay in the removal of ventilation) is very common even when not legally required, because this delay allows families to come to terms with their loss and resolves at least some objections. Indeed, the American Academy of Neurology appears to endorse limited accommodation, stating that "[t]he AAN is respectful of and sympathetic toward requests for limited accommodation based on reasonable and sincere social, moral, cultural, and religious considerations [… but] such requests must be based on the values of the patient and not those of loved ones or other surrogate decision-makers" (Russell et al. 2019, p. 231). New Jersey allows legal exemption from brain death, and there is some evidence that extended medical maintenance of dead bodies is not widespread there (Son and Setta 2018).

Organ donation and transplantation is also critically affected by the definition of death because the dead donor rule precludes causing the death of the donor by removal of vital organs. Prior to the adoption of brain death, organ donation was only possible following cardio-respiratory failure or from living donors, and a significant limit on the success of posthumous donation was damage to donated organs from lack of oxygen (DeVita et al. 1993). The acceptance of brain death allowed organs to be removed from dead donors, while their hearts were still beating, protecting organs from anoxic damage. Indeed, one of the reasons for recognizing brain death cited in the 1968 Harvard Medical School report was to allow this. For many years, brain-dead donors were the main source of organs for transplants as the use of organs after cardio-respiratory failure led to poorer transplant outcomes (DeVita et al. 1993, pp. 123–124). More recently, there has been a return to DCD (donation after cardio-respiratory death) due to the scarcity of brain-dead and living donors, and the unmet need for transplantation. Nonetheless, brain-dead donors remain an important source of organs for transplantation.

16.5. Legal responses to brain death objection

All societies face the question of how to address divergence between the individual and the society or among cultural groups on important philosophical, religious and moral questions. Where the views of different groups largely affect members of those groups, without substantial impact on others or on important public policies, mutual accommodation is easiest.

In the case of brain death, the stakes are high for the individuals and families concerned, but there are also consequences for health care providers and society. Whether the consequences for the latter groups are sufficient to justify the imposition of the majority view of brain death upon dissenting minorities is a critical question. One of the legal structures available to balance the interests at stake in cases of conflict between fundamental values and interests are constitutional human rights guarantees that are codified in many liberal democracies. These documents declare a set of fundamental human rights and freedoms, and they constrain legislative and government action that contravenes those rights and freedoms. Rights are not absolute, and mechanisms to try to fix the appropriate limits and make trade-offs are needed for these constitutional structures. In the case of the Canadian Charter of Rights and Freedoms, certain rights and freedoms are declared, but they are subject to an explicit "reasonable limits" clause (Canadian Charter of Rights and Freedoms 1982, sec. 1). This clause allows the government to argue that the infringement of a right is necessary to achieve a pressing governmental objective, that there is no less infringing way to achieve that objective, and that the harm to those whose rights are infringed is proportionate to the objective the government is seeking to achieve.

A recent Ontario case illustrates the operation of this mechanism in Canada, even though the court did not reach a final decision on the essential underlying question of whether the constitutional right to religious freedom was violated by the application of the brain death diagnosis.

In 2017, 27-year-old Taquisha McKitty was declared brain dead by two critical care physicians in a hospital in Brampton, Ontario (Ontario Superior Court of Justice 2018). Three additional physicians, including a neurologist, repeated the testing and confirmed the diagnosis. The results of a nuclear brain blood flow study and a somatosensory evoked potential test were also consistent with brain death. The McKitty family objected that her Christian beliefs held that death occurs only when cardiorespiratory function ceases, and so the application of the brain death diagnosis violated her right to freedom of religion under the Canadian Charter of Rights and Freedoms. The family also argued that she was not dead, but was instead a vulnerable disabled person, and that the definition of brain death violated her constitutional rights to life and the equal protection and benefit of the law without discrimination on the basis of disability.

Several difficulties confronted the McKitty claim, one of which was that the Charter applies to government action, and in the absence of a legislated definition of brain death in Ontario, it was necessary to find a sufficient connection to governmental actions or purposes to support the claim. Another key challenge facing the claim was whether, as a dead person, she had Charter rights at all. The trial judge wrote:

> I find that brain death extinguishes personhood, and with it, the right to assert Charter protection. [The question of whether] personhood extends beyond brain death and is to include physiological or biological function as life is engaging in a philosophical and theological discussion – a discussion that is beyond the role of this court. Just as the courts have not engaged in a theological debate on when life begins, so too should the court not become involved in a debate about when life ends. Decisions that involve social, political, moral, bioethical, and philosophical considerations are appropriately left to the legislature (Ontario Superior Court of Justice 2018, para. 205).

This is profoundly unsatisfying reasoning, given that the very issue in dispute is side-stepped. We can imagine a thought experiment to make the problem stand out. If a law were to declare patients in a permanent vegetative state to be dead (as some advocates of higher brain definitions of death support, but no jurisdiction accepts), it would be impossible to argue that this violates the right to life of people with these disabilities because they would not have Charter rights. Similar problems of losing the ability to dispute brain death upon being declared dead have arisen in other legal proceedings. In Ontario, disputes over capacity and consent to medical treatment are resolved initially by a tribunal called the Consent and Capacity Board. In several cases involving family objection to the removal of ventilation from brain-dead individuals, that Board has held that it does not have the ability to resolve the matter because its jurisdiction is limited to matters involving the "treatment" of "persons". Since the patient is dead, there is no "person", and mechanical ventilation of a dead body is not "treatment" as defined under the legislation applied by the Board (Pope 2017b, p. 275; Ontario Superior Court of Justice 2018, paras. 277–287).

Fortunately, the appellate court addressed the trial-level problem of circular reasoning in *McKitty v Hayani*. It held that where the rule that limits access to the protections of the Charter is the very rule whose constitutionality is questioned, it is necessary to presume the application of the Charter for that specific purpose. To do otherwise, "begs the question that is in dispute" (Ontario Court of Appeal 2019, para. 47).

The Ontario Court of Appeal did not, however, decide on the main issue of whether the application of a brain death diagnosis violated Taquisha McKitty's religious rights. This was because there had been inadequate evidence available on McKitty's religious beliefs and on the likely impacts of recognizing a religious accommodation. The Court stated that these questions must be left to another case and so, in Ontario, we await another case on this question.

It is noteworthy that the McKitty family pointed to some of the accommodations provided in other jurisdictions, such as in the state of New Jersey, which allows a religious exemption from the application of brain death. Should a future Canadian claimant succeed in persuading a court that brain death infringes their Charter rights, then the government would likely be confronted with the question of whether similar accommodations are constitutionally required in Canada despite its different health care funding model.

16.6. Accommodation of dissenting views

There is a fairly broad spectrum of potential accommodations available for minorities who object to the idea of brain death. The most limited accommodation is to delay the removal of ventilation and other medical support. This is legally obligatory in the US states of New York and California. Even where it is not legally required, it appears common for health care facilities to provide delays as a compassionate measure for a day or two, in order to let families come to terms with the situation (Pope 2017a).

A more extensive accommodation is offered in the US state of New Jersey, which recognizes brain death as death, but allows an indefinite exemption for those who object on religious grounds (Pope 2017a). Israel's *Brain-Respiratory Death Law* of 2008 also accepts brain death as death, but states that life support may not be withdrawn if the patient's family indicates that the patient opposed brain death on religious or other grounds (a translation is offered by the Halachic Organ Donor Society: www.hods.org/pdf/law51%20Braindead.pdf).

A narrow exemption that fixes the time of death on the basis of neurological criteria, but precludes the removal of medical support would allow for a uniform approach to other important legal questions such as criminal law, family status and succession. However, this would leave the major underlying question of how to provide continuing care and support for brain-dead patients. This is an issue for both privately and publicly funded systems. If the numbers of such accommodations are small, the problem remains manageable. Veatch suggests that if the costs expand and "strain the interests or rights of fellow members of an insurance pool, they can explore the possibility of excluding coverage of life support for people who choose to be considered living when they have dead brains" (Veatch 2019, p. 398). While this may be possible for private insurance systems, it is less obvious how to constrain the costs if brain death exemptions are allowed in a publicly funded system.

16.7. Conclusion

The problem of defining death arose because of advances in resuscitation and life-sustaining technology. Cultures and religions vary on how to interpret the meaning and moral significance of a state in which brain functions have been lost, but other functions dependent upon the brain like respiration are supported artificially. Societies with liberal democratic governments continue to grapple with how to balance individual freedoms against collective interests. The question of whether and how to accommodate diverging views on questions as fundamental as the definitions of life and death have continued to arise in the 50+ years since the first clinical definition of brain death was proposed.

It is likely that the future will present new iterations of this problem. Resuscitation technologies are continuously advancing. Recently, cellular function was restored in some neurons four hours after circulatory arrest in pigs (Vrselja et al. 2019). While this study achieved limited restoration of cellular function, and did not restore brain function, it suggests that "in the future it may be possible to reperfuse the brain and restore brain functions" (Gardiner et al. 2020). In addition, new technologies to support or replace lost functions have arisen, such as extra-corporeal membrane oxygenation (ECMO), which replaces cardio-respiratory function and allows a person to be sustained without a heartbeat, or even a heart (Gardiner et al. 2020).

Cultural neuroethics will be important in analyzing the conceptual and moral questions about the meaning and significance of new biological states created through these evolving technologies, and in directing attention to the varying views associated with cultural differences. Legal structures such as the constitutional human rights available in liberal democracies can help to resolve the potential conflicts that flow in practice from these varying views. However, there are limitations to the use of these structures, which apply to government action only. Furthermore, to the extent the focus is on freedom of religion alone, this obscures the fact that there are other philosophical, emotional and social reasons for rejecting a brain death diagnosis, particularly for demographic groups with greater levels of mistrust of medical authorities. How to respond to these other reasons is another important concern for cultural neuroethics and the law.

16.8. Acknowledgments

The author gratefully acknowledges financial support for this research from the Canadian New Frontiers in Research Fund (grant number: NFRFE-2019-00759).

16.9. References

Ad Hoc Committee Harvard Medical School (1968). A definition of irreversible coma: Report of the ad hoc committee of the harvard medical school to examine the definition of brain death. *JAMA*, 205(6), 337. doi: 10.1001/jama.1968.03140320031009.

Bowman, S.M., Martin, D.P., Sharar, S.R., Zimmerman, F.J. (2007). Racial disparities in outcomes of persons with moderate to severe traumatic brain injury. *Medical Care*, 45(7), 686–690.

Canadian Charter of Rights and Freedoms (1982). Part 1 of the Constitution Act. Being Schedule B to the Canada Act 1982 (UK), 11.

Chamsi-Pasha, H. and Albar, M.A. (2017). Do not resuscitate, brain death, and organ transplantation: Islamic perspective. *Avicenna Journal of Medicine*, 7(2), 35–45.

DeVita, M.A., Snyder, J.V., Grenvik, A. (1993). History of organ donation by patients with cardiac death. *Kennedy Institute of Ethics Journal*, 3(2), 113–129. doi: 10.1353/ken.0.0147.

Dhanani, S., Hornby, L., van Beinum, A., Scales, N.B., Hogue, M., Baker, A., Beed, S., Boyd, J.G., Chandler, J.A., Chassé, M. et al. (2021). Resumption of cardiac activity after withdrawal of life-sustaining measures. *New England Journal of Medicine*, 384(4), 345–352. doi: 10.1056/NEJMoa2022713.

Dzeng, E., Colaianni, A., Roland, M., Levine, D., Kelly, M.P., Barclay, S., Smith, T.J. (2016). Moral distress amongst American physician trainees regarding futile treatments at the end of life: A qualitative study. *Journal of General Internal Medicine*, 31(1), 93–99. doi: 10.1007/s11606-015-3505-1.

Gardiner, D., McGee, A., Bernat, J.L. (2020). Permanent brain arrest as the sole criterion of death in systemic circulatory arrest. *Anaesthesia*, 75(9), 1223–1228. doi: 10.1111/anae.15050.

Greer, D.M., Shemie, S.D., Lewis, A., Torrance, S., Varelas, P., Goldenberg, F.D., Bernat, J.L., Souter, M., Topcuoglu, M.A., Alexandrov, A.W. et al. (2020). Determination of brain death/death by neurologic criteria: The world brain death project. *JAMA*, 324(11), 1078. doi: 10.1001/jama.2020.11586.

Lewis, A., Bonnie, R.J., Pope, T., Epstein, L.G., Greer, D.M., Kirschen, M.P., Rubin, M., Russell, J.A. (2019). Determination of death by neurologic criteria in the United States: The case for revising the uniform determination of death act. *Journal of Law, Medicine & Ethics*, 47(S4), 9–24. doi: 10.1177/1073110519898039.

Lewis, A., Bakkar, A., Kreiger-Benson, E., Kumpfbeck, A., Liebman, J., Shemie, S.D., Sung, G., Torrance, S., Greer, D. (2020a). Determination of death by neurologic criteria around the world. *Neurology*, 95(3), e299–e309. doi: 10.1212/WNL.0000000000009888.

Lewis, A., Bonnie, R.J., Pope, T. (2020b). It's time to revise the uniform determination of death act. *Annals of Internal Medicine*, 172(2), 143. doi: 10.7326/M19-2731.

Mack, J.W., Paulk, M.E., Viswanath, K., Prigerson, H.G. (2010). Racial disparities in the outcomes of communication on medical care received near death. *Archives of Internal Medicine*, 170(17). doi: 10.1001/archinternmed.2010.322.

Neiders, I. and Dranseika, V. (2020). Minds, brains, and hearts: An empirical study on pluralism concerning death determination. *Monash Bioethics Review*, 38(1), 35–48. doi: 10.1007/s40592-020-00114-0.

Ontario Court of Appeal (2019). McKitty v. Hayani, ONCA 805.

Ontario Superior Court of Justice (2018). McKitty v. Hayani, ONSC 4015.

Ohnuki-Tierney, E., Angrosino, M.V., Becker, C., Daar, A.S., Funabiki, T., Lorber, M.I. (1994). Brain death and organ transplantation: Cultural bases of medical technology [and comments and reply]. *Current Anthropology*, 35(3), 233–254. doi: 10.1086/204269.

Pernick, M.S. (1999). Brain death in a cultural context: The reconstruction of death, 1967–1981. In *The Definition of Death: Contemporary Controversies*, Youngner, S.J., Arnold, R.M., Schapiro, R. (eds). Johns Hopkins University Press, Baltimore, MD.

Pope, TM. (2013). Dispute resolution mechanisms for intractable medical futility disputes. *New York Law School Law Review*, 58(2), 347–369.

Pope, T.M. (2014) Legal briefing: Brain death and total brain failure. *Journal of Clinical Ethics*, 25(3), 245–57.

Pope, T.M. (2017a). Brain death forsaken: Growing conflict and new legal challenges. *Journal of Legal Medicine*, 37(3–4), 265–324. doi: 10.1080/01947648.2017.1385041.

Pope, T.M. (2017b). Brain death forsaken: Growing conflict and new legal challenges. *Journal of Legal Medicine*, 37(3–4), 265–324. doi: 10.1080/01947648.2017.1385041.

Ross, L.F. (2018). Respecting choice in definitions of death. *Hastings Center Report*, 48, S53–S55. doi: 10.1002/hast.956.

Russell, J.A., Epstein, L.G., Greer, D.M., Kirschen, M., Rubin, M.A., Lewis, A. (2019). Brain death, the determination of brain death, and member guidance for brain death accommodation requests: AAN position statement. *Neurology*, 92(5), 228–232. doi: 10.1212/WNL.0000000000006750.

Shen, F.X. (2016). The overlooked history of neurolaw. *Fordham Law Review*, 85, 667–695.

Shewmon, D.A. (1998). "Brainstem death," "brain death" and death: A critical re-evaluation of the purported equivalence. *Issues in Law & Medicine*, 14(2), 125–145.

Siminoff, L.A., Burant, C., Youngner, S.J. (2004). Death and organ procurement: Public beliefs and attitudes. *Social Science & Medicine*, 59, 2325–2334.

Siminoff, L.A., Burant, C.J., Ibrahim, S.A. (2006). Racial disparities in preferences and perceptions regarding organ donation. *Journal of General Internal Medicine*, 21(9), 995–1000.

Son, R.G. and Setta, S.M. (2018). Frequency of use of the religious exemption in New Jersey cases of determination of brain death. *BMC Medical Ethics*, 19(1), 76. doi: 10.1186/s12910-018-0315-0.

St Ledger, U., Reid, J., Begley, A., Dodek, P., McAuley, D.F., Prior, L., Blackwood, B. (2021). Moral distress in end-of-life decisions: A qualitative study of intensive care physicians. *Journal of Critical Care*, 62, 185–189. doi: 10.1016/j.jcrc.2020.12.019.

Stahnisch, F.W. (2015). History of neuroscience and neuroethics: Introduction. In *Handbook of Neuroethics*, Clausen, J. and Levy, N. (eds). Springer, Dordrecht. doi: 10.1007/978-94-007-4707-4_22.

Terunuma, Y. and Mathis, B.J. (2021). Cultural sensitivity in brain death determination: A necessity in end-of-life decisions in Japan. *BMC Medical Ethics*, 22(1), 58. doi: 10.1186/s12910-021-00626-2.

Truog, R.D., Pope, T.M., Jones, D.S. (2018). The 50-year legacy of the Harvard report on brain death. *JAMA*, 320(4), 335. doi: 10.1001/jama.2018.6990.

Veatch, R.M. (2019). Controversies in defining death: A case for choice. *Theoretical Medicine and Bioethics*, 40, 381–401. doi: 10.1007/s11017-019-09505-9.

Vrselja, Z., Daniele, S.G., Silbereis, J., Talpo, F., Morozov, Y.M., Sousa, A.M.M., Tanaka, B.S., Skarica, M., Pletikos, M., Kaur, N. et al. (2019). Restoration of brain circulation and cellular functions hours post-mortem. *Nature*, 568(7752), 336–343. doi: 10.1038/s41586-019-1099-1.

White, G. (2003). Intensive care nurses' perceptions of brain death. *Australian Critical Care*, 16(1), 7–14. doi: 10.1016/S1036-7314(03)80023-1.

Yang, Q. and Miller, G. (2015). East–west differences in perception of brain death: Review of history, current understandings, and directions for future research. *Journal of Bioethical Inquiry*, 12(2), 211–225. doi: 10.1007/s11673-014-9564-x.

17

Neuroscientific Research, Neurotechnologies and Minors: Ethical Aspects

Laura PALAZZANI
LUMSA Univesity, Rome, Italy

17.1. The importance of neuroscientific and neurotechnological research on minors

The participation of minors in scientific research always raises specific ethical issues in general, due to their age (which covers a long time span, from 0 to 18 years) and different degrees of maturity (variable according to the different existential, social and cultural circumstances), their specific vulnerability related to conditions of dependence when they are very small and not fully aware of the possible risks of certain choices, with a progressive acquisition of the ability to participate in choices. Furthermore, children are often treated as "little adults", whereas it is always necessary to consider that it does not suffice to make the application of technologies or drugs proportional on a quantitative level in order to reduce the impact and potential damage/risks. It should always be taken into account that the body's response mechanisms in pediatric age are also qualitatively different from those of adults, given that they are in the phase of growth and development.

In this sense, research on children always requires specific scientific and ethical precautions. This prudence is already outlined on the level of pharmacological research. When the risks are high, evidence that it is not possible to carry out the research on adults (who are able to take risks consciously) is required and when the presumptive risks are limited, it always prioritizes research with direct benefits for those who participate and carefully evaluates when the benefits are indirect, making it an essential condition that the benefits must at least be aimed at children in the same condition in the face of minimal risks and inconveniences[1].

On the basis of the principle of equality and justice, children, like any other humans, have a right to have research carried out on them in order to achieve possible conditions of health in the same way as adults: it would be unethical to exclude children from scientific research because it would mean to discriminate against them with respect to other subjects in the protection of fundamental interests and rights, such as life and health. But scientific research on minors, although necessary, raises some ethical issues.

In the field of neuroscience and neurotechnology, the discussion on research on minors must generally be included within the ethics of pharmacological research (with particular reference to risks/benefits, informed consent/assent, i.e. participation in relation to the age and maturity of the minor) and research in the field of genetics (with particular reference to the right to know/not to know and privacy), but it must also be articulated in a specific way, since it presents some peculiar ethical issues. Some national and international bioethics committees have taken and are taking stand on some aspects and the regulations also highlight the need for specific implementations.

Today, there are many diverse neurotechnologies available for research in the field of neuroscience[2]. It is important to preliminarily distinguish the different levels; in fact, sometimes they are also intertwined and superimposed: the level of research into therapies on sick subjects both in terms of prevention, diagnosis, treatment and rehabilitation, the level of advancement of knowledge (even non-therapeutic) in both sick and healthy subjects, the level of cognitive enhancement on healthy subjects in order to monitor abilities or alter/increase cognitive abilities in view of intellectual-emotional improvements.

In this context, specific attention should be paid to children and adolescents due to the uncertainty of even long-term or irreversible risks, also given the plasticity of

---

1. Regulation (EU) No 536/2014 of the European Parliament and the Council of 16 April 2014 on clinical trials on medicinal products for human use, and repealing Directive 2001/20/EC.
2. For example: neuroimaging or brain, electroencephalography, computed tomography, magnetic resonance imaging, magnetoencephalography, positron emission tomography, cranial ultrasound, functional magnetic resonance imaging, neurodevices implanted, deep brain stimulation, brain computer interfaces.

brain development. It is important to distinguish the ethics of research in relation to the modality of research with invasive and non-invasive technologies, recording, measurement, monitoring and stimulation, superficial or deep, for the therapeutic, non-therapeutic or potential purpose of monitoring or alteration and relating to the "scope of application in the medical and non-medical sector" (Chneiweiss 2006).

A specific field of research is developmental cultural neuroscience (Chiao and Ambady 2007; Kim and Sasaki 2014), which focuses on the role played by culture in neurobiological processes related to infancy, childhood and adolescence. Theories in developmental psychology outline the role of social inputs in shaping children's cognition, emotion and behavior (Collins and Steinberg 2006). Along this line, neuroscience is researching how diverse cultural environments influence the development of children's brains, cognition and behaviors in the intersection of cross-cultural psychology and neuroscience. Developmental cultural neuroscience investigates similarities and differences in the brain, and behavioral development across the entire life span, starting from childhood, using a neuroimaging technology along with observation, and experimental approaches. Developmental cultural neuroscience can provide new insights into cultural socialization, the interaction between neurobiological systems, psychological development and culture, as social environment. This line of research will provide a comprehensive understanding of the brain's plasticity in the dynamic process of cultural transmission. Developmental cultural neuroscience provides researchers with a unique approach to investigating when, how and why children in different cultures show divergent neural, psychological and behavioral trajectories over the course of development. Advances in this promising field may provide valuable insights into cultural transmission and neuroplasticity, with implications for promoting children's learning and mental health in diverse cultures.

In this chapter, we will deal with the ethical issues arising in the field of neuroscientific research and the use of neurotechnologies on minors, in general, and also all of the general/customary ethical requirements being applied to developmental cultural neuroscientific research. We will also identify some/several specific problems in this specific context.

17.2. Ethical criteria of neuroscientific research on minors in the medical field

The ethical criteria for research are similar to those of pharmacological research and genetic research on minors in some respects: these criteria require specific articulation in the field of neuroscience and neurotechnology. Some ethical criteria, on the other hand, are specific to the field of neuroscience and neurotechnology (Illes 2010).

17.2.1. Justification of scientific relevance of the research

A first ethical requirement, as with any research, is the justification of its scientific relevance, that is, to demonstrate that research can achieve an advancement of scientific knowledge (an intrinsic objective of research itself) with a beneficial impact in general for individuals and society. The scientific relevance must also be calibrated based on the expertise of the researchers and qualifications of the facility. These requirements must always be met, above all in the specific context of research involving minors, given their particular vulnerability.

In the case of minors, it is essential that the research team also has specific expertise in studies on children, that there are adequate guarantees regarding the appropriateness of the study design and study methodology with respect to the specific needs of children (e.g. research that minimizes physical and mental invasiveness and minimize discomfort and risks) and the competency of the researcher. It should also be verified that the study is necessary for the minor and is not feasible for adults, who are able to have a greater awareness of the potential risks.

It is essential for the research to be scientifically evaluated by an ethics committee, which has specialist expertise in both pediatrics and neuroscience, as well as in ethics, in order to define the relevance of the research and priority with respect to the needs of those participating. In the specific area of cultural neuroscience, consideration should be given to the importance of having an expert in this field with professional competence involving children.

17.2.2. The benefit/risk balance and the protection of physical and mental integrity

The ethical evaluation of research concerns the balancing of risks and benefits, in relation to the invasiveness of the interventions and the purposes, in relation to the protection of the physical and mental integrity of the minor, which is the priority ethical value (Oviedo Convention 1997).

The increased invasiveness is justified only for therapeutic purposes. To the extent that neurotechnologies are invasive, and therefore expose children to particular risks both physically and psychologically, as well as discomfort (which can also cause pain in the physical, psychological and social sense), the research is justifiable only if there is a proportion between the expected benefits for the treatment of the disease and the foreseeable risks. Such research, if invasive, is only justified in cases of severe disease and in the absence of less invasive non-experimental alternative therapies, applying the principle of the best interest of the child and the minimization of risks and inconveniences. Preclinical animal studies and the results of adult studies should be carefully evaluated for a more accurate scientific basis, and the participation of the smallest number of children should always be encouraged to the extent possible (National French Consultative Ethics Committee for Health and Life Sciences 2012, 2013).

More ethically delicate is the problem of involving children, be they sick or healthy, as control groups and subjects in non-therapeutic research, with only an indirect benefit for children suffering from the same pathology. This type of research is not excluded but must enable a significant advancement of knowledge with minimum risk and discomfort (Italian Bioethics Committee 2010).

17.2.3. Personal identity

Research with neurotechnologies may be invasive both on a physical and psychological level. Some neurotechnologies may alter personality and personal identity that are the core of the character, traits and preferences present throughout the person's life, related to the notion of authenticity. Authenticity means that a person is fully themselves when they act according to their desires and preferences. The application of some neurotechnologies may be a challenge to personal identity and the alteration of personality, and this is specifically problematic with minors.

Changes in mood and anxiety are reported after deep brain stimulation treatment. Memory modification techniques may use pharmacological means and turn to implanting chips in the brain. It can completely erase a memory or reduce the emotional impact of a painful memory. The perception of past experiences may be distorted and affect personal identity and authenticity. Deep brain stimulation can pose a threat to the individual's integrity: the control exerted by the technology on certain parts of the body is experienced as a form of subjugation of the person to the technical device. There is the possibility that the device can be controlled remotely by a clinician, and this, perhaps, without the patient's knowledge. Depending on the type of pathology, the perception of the impact of neurotechnologies on personal identity varies (changes in behavior, total change), and also in regards to duration. All the emerging issues related to personal identity and neurotechnologies raise the ethical need for precaution in front of the risks and uncertainties, above all with regard to children and adolescents having a developing brain.

The development of the brain is strictly related to other biological systems in the prenatal, early childhood and adolescent periods, in the complex interaction between genetics (nature) and environment (nurture, culture) shaping the growing child, physically and emotionally, socially and behaviorally. The complex interaction between genetics and life experience (including culture) that shapes the development of neurobiological systems, particularly in the early childhood and adolescent period, may impact the structure and function of the developing brain, and consequently health and well-being. The brain develops rapidly during this period, a time during which experiences shape the developing brain, with implications for health, learning and behavior throughout the lifespan.

Neurotechnology has the potential to transform children and adolescents' plastic, still developing brain in a myriad of ways that will shape their future identity with long-lasting, if not permanent, effects. Neurodevices and brain–computer interfaces embedded while an individual is still undergoing significant neurodevelopment (thought to extend until at least 25 years of age) make it difficult to distinguish traits and behavior, which can be attributed to the neurodevice versus the "normal" maturation of the brain.

In this context, neuroscientific and neurotechnological research should apply the principle of precaution, in the meaning of abstension when the potential risks are high, uncertain and unavoidable. Research on minors is justified only if the technologies used are less invasive.

17.2.4. *Autonomy: informed consent of minors and informed consent of parents*

One of the most critical elements of research in children is consent.

The need to obtain the consent of both parents is a consolidated bioethical line of thought in particular cases of social hardship (e.g. neediness, poor education, immigration). It is furthermore important that, with regard to the general information prior to consent, the researcher evaluates the real motivations leading the parents to accept the recruitment of their child in the research project, so as to exclude the existence of ethically unacceptable reasons for this: for example, to benefit from medical treatment otherwise not guaranteed or obtain greater attention from the doctors in the treatment of the children.

The consent of the parents should be accompanied by the assent of the child, which is proof of their actual involvement in the medical decisions, together with their parents (Wendler 2006). Such assent should be obtained through appropriate information and communication with the child, suitable for their age and their intellectual and emotional capacity to understand. It is impossible to establish time limits in the formulation of assent. There should be an approximate distinction between pre-school age (with communication through pictures) and early school age (with pictures accompanied by cartoons with short simple explanations), to progressively develop a more complex elaboration up to adolescence or the so-called "mature minors".

The appropriateness of the information will be evaluated case by case according to the cultural and social context, as well as in relation to the existential context, since each child has a different evolution and maturity and can react differently to illness or pain. Minors should receive information, from expert personnel, that is proportional to their capacity to understand the risks and benefits, and furthermore the investigator is called upon to take into consideration the desire expressed by the minor to take part in the research or to withdraw from it at any moment. The child should be told that their desires will be considered important in the decision, although making it clear that they alone cannot be decisive. Specific attention should be paid so that the involvement of the child is not an indirect insistence on participation, which should always be free and unconditioned by external factors (verifying the absence of so-called "undue inducement"). The conditioning of minors is particularly problematic in a pediatric phase, given the child's vulnerability from the external influences of adults, members of the family and doctors.

In the context of assent, the children should be helped by the researcher to understand the aim of the research, the procedures foreseen and the experiences that they will have. The researcher should always attempt to perceive how much the children have actually understood and what their often unexpressed concerns are, in order to help participants to overcome them. It is important that the researchers, together with the parents, always act in the best interests of the children, helping them to develop their awareness and choice, whenever possible. In this sense, the informed consent/assent cannot be reduced to a mere procedure, but must be carried out in an interaction between researcher, parents and children, to be realized over time so that there is room for clarifications and the reaching of possible shared decisions. Both consent and assent should be in written form. It should be made clear that these records can be withdrawn at any time, without having to give any justification. In the case of conflicts and disputes, suitable psychological assistance and ethical consultation could be guaranteed.

In neurosciences, very specific ethical issues emerge.

Fast advances in neuroscience and the possibility of identifying the neural correlates of decision-making open up the possibility of acquiring accurate information about people's competence to consent to medical clinical procedures and medical research. The possibility of creating a reliable neural test of competence and decision-making competence to consent opens up many questions: who will choose and set the threshold between what may be a competent or incompetent patient? This is still an open question that has a specific relevance for minors, because of their development, and requires further bioethical and biolegal analysis.

There is the dilemma of valid informed consent for deep brain stimulation treatment in depressed patients: this specific consent/assent calls for additional steps to be taken to ensure that subjects have understood the information necessary for informed consent because of their psychological status. The problem of surrogate consent for patients in a minimally conscious state or those who are incompetent is particularly problematic: research can only be done with the consent of the representatives (parents or tutor) and the so-called "deferred" consent/assent, expressed as soon as consciousness and awareness are recovered. The problem of communication with patients in a minimally conscious state and for end-of-life decisions can be solved by neurotechnologies, through brain–computer interface communication

Cultural belonging may be a challenge in the informed consent process, because of possible miscomprehension and cultural barriers. An intercultural approach to communication and a participatory approach to the informed consent process (e.g. taking into account the perspectives of different cultural groups in the development of information materials, etc.) can empower culturally diverse subjects to make autonomous decisions with regard to their participation/non-participation in neuroscientific research. During the consent process, a "cultural insider" or "cultural mediator" should also be involved, that is, a person who has knowledge of the language and familiarity with the culture of a particular group through their membership in that group.

The "cultural insider" should facilitate researchers in interactions with members of the community and the participants in the research. Involving cultural insiders helps to gain a deeper understanding of the socio-cultural contexts of the research setting and enables researchers to conduct research using culturally sensitive methods, as local people will be more inclined to share their feelings and perceptions when they trust people recognized as insiders. In this way, research can be culturally informed at each step of the research process, through cultural community engagement. This is particularly important with parents, as well as with children, due to their fragility. An adequate training of researchers regarding a knowledge of the main cultural patterns involved in interaction with potential participants is another key aspect for a culturally sensitive informed consent/assent process. In a context of cultural diversity, it may be significantly challenging to ensure effective benefit–risk communication between researchers and participants, in order to prevent misconceptions arising from overestimation of the envisaged benefits deriving from inclusion in research.

17.2.5. *Incidental findings*

In research with neurotechnologies, the problem of incidental findings or unexpected brain anomalies in functional magnetic resonance imaging (fMRI) emerges. The frequent discovery of unexpected anomalies (or the difficulties in interpreting them) may pose ethical challenges such as determining appropriate strategies for communication (before and after) and how to handle clinically relevant incidental findings, above all concerning minors.

The principle of respect for the autonomy of the parents may be in conflict with the principle of beneficence of the children, and their best interest, when people subjected to brain imaging research renounce their right to know and they claim the right not to know. In seeking to do good (beneficence), a researcher may deem it a responsibility both at the ethical and deontological levels to disclose the potential consequences of incidental findings discovered during brain imaging research because they may be beneficial to health (at the level of prevention/treatment/care), but doing so may be a breach of respect for a person's right to autonomy.

Research teams in the field of neurosciences should always include clinical medical personnel to interpret incidental findings and give research participants the information. Extra caution should therefore be taken in the formulation of the informed consent process, which needs to explicitly include the reference to incidental findings. A specific circumstance regards children: if the incidental findings reveal or may reveal information of clinical relevance for them, parents have a duty to know, in their best interest. This problem is analogous to the discussion surrounding genetic testing. The handling of incidental findings is especially problematic in healthy research subjects.

Communication strategies are particularly complex and delicate, and require specific education and training, above all in the case of the fMRI scans, which are difficult to interpret, even by specialists. In all research, there should be a discussion on how to handle clinically relevant incidental findings and the communication of results (communication strategies). There is a possible impact of false positive and false negative. Few empirical data and normative analyses exist for research with minors: the importance of involving minors in disclosure emerges (Di Pietro and Illes 2013).

Furthermore, in this area, specific attention needs to be given to cultural differences.

17.2.6. *Predictive value of brain images*

The question of the predictive value of brain images and the possibility of diagnosing certain dispositions in the brain (e.g. the likelihood of getting a certain disease) is also of ethical concern in research in neurosciences. During research, the subject should be aware of the possibility of ascertaining the probability or certainty of such a disposition, since the possibilities of false positives (diagnosing a pathology that is not there) or the possibilities of false negatives (the failing to identify or communicate a possibly life-threatening condition) may have major consequences on patients. The negative impact of false positives weighed against the potential consequences associated with failing to identify or communicate a possibly life-threatening condition needs to be considered in neuroresearch.

This aspect is extremely delicate with minors, in a similar way to predictive genetic tests, because of the difficulty to manage an existentially problematic outcome, which may cause anxiety in a scenario of predictive incurable illness, considering the uncertainty.

Respect for the child's autonomy and future decision-making (the so-called right to an open future) sometimes call for a special management of the information that has no immediate relevance in the child's health or health management, above all when no treatment or preventive interactions are available. More robust protective measures may be required in order to preserve basic human rights, including the autonomy of vulnerable humans, as children are.

This predictive scenario may have various implications in different cultural contexts: some cultures may better accept illness or the risk of illness than others. This should be taken into consideration during the communication process, paying specific attention to children.

17.2.7. Single-patient reports

In the novel field of neurotechnology, there is a clear risk of an excessive reliance on "single-patient" case reports. There is a tendency of "selective reporting", which may be highly problematic. This implies a possible over-reporting of positive results, but it can also be the basis for the duplication of efforts. Research groups may therefore reproduce studies, not knowing that similar studies have already been done and failed, which is highly problematic in the field of deep brain stimulation due to the risks associated with invasive technologies. Consequently, it is critical that all such studies are registered in a public database (e.g. the WHO International Clinical Trials Registry Platform).

The problem of selective publication, which is a general scientific integrity problem, becomes very urgent in neuroscience. It is ethically necessary to establish a comprehensive case registry and quality outcome reporting, above all for minors, in order to any avoid duplication, which translates into useless research exposed to futile risks.

17.2.8. Privacy: the use of neural data

The new possibilities of monitoring and manipulating the human mind through neuroimaging open up new challenges to the right to privacy of individuals, accessing not only their behaviors, but also their thoughts (the so-called "mind reading/reading the mind"). Neurotechnology may map, scan and record brain activity and transmit the digital data of the users in implanted neurodevices, as in deep brain stimulation. Information collected from neurodevices can be used to identify someone or reveal their brain activity. These results may be used in the stigmatization of neurological or mental health conditions. Research on "mind reading", especially in relation to unexpectedly found data or detection, based on the neuroimaging of psychological states, may expose the subject of research to risks of detection and may have personal implications. Some of those consequences have already been raised by genetic research, mainly those regarding the access to such information by third parties (employers, insurance companies).

Also, the use of neuronal activity monitoring or testing in order to check the neural states of patients for the purposes of determining competence to consent may be used to violate "cognitive privacy", "neural privacy" or "informational privacy". The right to privacy also includes the protection of intimate thoughts, in such terms that some claim "cognitive freedom" related with exercising autonomy over brain conditions.

Concerning the consent to share data, alternative systems, such as the opting out system, have also been suggested, with the consideration to treat neural data in the same way that organs and tissues are treated in some jurisdictions, for example, for transplant purposes, where individuals would need to explicitly opt in to share neural data from any device. This would involve a safe and secure process, including a consent procedure that clearly specifies who will use the data, for what purposes and for how long. This is particularly problematic for minors, because of the absence or limited capacity to be aware of the possible risks.

Research outcomes such as the utility of "brain data" should be subjected to ethical considerations such as non-discrimination and privacy protection (e.g. to prevent re-identification and unauthorized re-use), and ensuring benefits that also accrue to those who participated in the research. This is especially the case for children, whose brain data could be defining not only to shape them (their perception as a human), but also to have an impact on their future opportunities (employment, insurance, etc.).

The evaluation of privacy may differ from culture to culture. Nevertheless, the main aim of researchers should be to correctly inform the participants about the possible consequences of this kind of research, above all when parents or children do not consider privacy relevant.

17.2.9. *Justice*

The exclusion and inclusion criteria of research participation should be justified on an objective and scientific basis. The distribution of benefit sharing to research participants should also be considered, paying specific attention to the most vulnerable, such as minors. The access to novel neurotechnology treatment will not always be possible. Disproportional distribution of benefit/risks may affect vulnerable groups due to age, socio-economic status, geographical location or cultural belonging, in less developed regions.

17.3. Research for neuro-enhancement purposes

Research may also develop towards neuro-cognitive enhancement that refers to interventions designed to improve mental and emotional performance, considered "normal", with psychotropic drugs affecting mental processes, neuro-imaging technologies to assess or alter brain function, such as deep transcranial magnetic stimulation over the cortex, using brain implants and employing a brain–computer interface. Pediatric neuro-enhancement is an upcoming topic, evaluating tools to shape children and ways to improve children's capacities. These issues appear in a different light given new paths in technological development, and changing demands and expectations regarding parenting.

Research on neurotechnologies for enhancement purposes is very problematic in general and especially in minors, and it is the subject of discussion, because of the non-proportionality between the risks (which may be long term and irreversible), compared to the benefits, which are non-recovery but improvement of "normal" functions, considering that – in the case of minors – the brain is still in development.

In the case of small/young children and neuroenhancement, the choice would lie with the parents, therefore, without the children having a clear awareness of the possible risks; in the possible participation of older children who consent to research, the consent may be induced and not autonomous, being a consequence of the hidden coercion of their parents. There is a debate on explicit and implicit coercion in regard to neuroenhancement. In case a minor is being forced by their parents to take pills to perform better in school, this would be a case of explicit coercion, whereas it is implicit coercion to perform better. Both scenarios pose a serious risk to the autonomy of the individual. There is a possible infringement of autonomy through coercive pressure to improve, either by educators or parents.

What is also very controversial is the use of helmet electrode recording to monitor children's brainwaves, in order to survey awareness and attention, by teachers and parents, in order to improve concentration. This opens up challenges to children, who can be easily manipulated because of a specific weakness or dependence. There is a profile of responsibility of the medical specialist who must ensure the appropriacy of the prescription and therefore prevent an "improper" non-therapeutic use of these drugs or technologies, which are not proven for safety and efficacy (Health Council of the Netherlands 2003; The Danish Council of Bioethics 2011).

Uses of non-invasive neurostimulation or brain–computer interfaces, either for "enhancement" purposes or for gaming or teaching/learning, because of the lack of any clear associated health benefits and because of the uncertainties regarding risks, mean that it is important to reflect on several ethical concerns with prudence and caution. In particular, to minimize the pursuit of unnecessary brain interventions, there is a need to ensure the originality and rigor of research investigating non-therapeutic uses in humans and also to disseminate existing evidence. There is a particular concern in children, in whom the effects of neurostimulation or brain–computer interface on the developing brain are not well known. There is a need for observational research with children who are already using neurotechnologies to address this concern, and also for advice to be issued to teachers and parents about the current evidence of the efficacy of neurofeedback as an educational enhancement tool.

It should also be outlined that the cognitive improvements (e.g. in memory) supposedly achieved through neurostimulation are more effectively realized through conventional educational means. While education aims to develop capacities for global improvements in learning, these kinds of studies may reach a transient improvement in relation to specific tasks. It should also be considered that cognitive function can be improved in a more lasting manner through instruction and continuous training, a rich social life and relationships; from study, learning, continuous stimulation of interests and healthy lifestyles (nutrition, physical activity, regular sleep). It is a path that clearly requires a lot of time, but (perhaps) it is more respectful of the opportunities for growth and development of personal and relational identity (Italian Bioethics Committee 2014).

17.4. Use of neurotechnologies in non-medical field without any research

It should also be said that the research and results are applied in a medical context, as well as in a non-medical one (e.g. in the context of neuromarketing, teaching, gaming and entertainment/recreational sector). Often, neurotechnologies are also used in non-medical sectors without adequate research according to the ethical and legal rules agreed by the international community for experimentation, exposing subjects to considerable risks, without having adequate information and awareness.

An example are the games already popular on the market based on non-invasive electroencephalography and brain–computer interface technology for recreational purposes. An adequate debate is essential for the bioethical and bio-juridical profiles in this sector, with particular regard to minors and the use of technologies for ludic and recreational purposes. A large number of people use these applications, despite the lack of any clear evidence on benefits/risks, and the fact that they are not necessary and used without medical monitoring (in private settings[3]). A delicate ethical problem emerges with neuromarketing companies, which may run studies involving humans without clear informed consent and approval from an ethics committee, outside the medical context. The collection of consumer data in the non-medical field (as neuromarketing) without consent may lead to manipulation to sell unhealthy products to a certain vulnerable population, such as minors. There

---

3. See: https://neuroscape.ucsf.edu.

should be an acknowledgment of the limits of the scientists' understanding of the brain and the limited capacity to cure or remove all suffering. The virtue of humility goes beyond the principle of caution and suggests a need for permanent deliberation about the right action in a given situation. Responsibility is the virtue that strives for a balance between the principles of beneficence and caution. On the one hand, it calls for ethical formation of researchers; on the other hand, it also connects with the social responsibility of researchers regarding the translation of the researchers' work into the public sphere[4] (Ienca and Andorno 2017; OECD 2019).

17.5. Some conclusive reflections

It is particularly important to emphasize the relevance of ethics in the field of neuroscience and recent developments in neurotechnology. The International Bioethics Committee of UNESCO is about to publish a document, the first such document of global bioethics, which addresses the issue emphasizing, in a special section dedicated to research, the sensitivity of ethical issues, with specific reference to minors. There is no specific literature on the topic, in particular on developmental cultural neuroscience with reference to minors.

This contribution has identified the main ethical requirements of research in the field of neuroscience and the use of new neurotechnologies with specific reference to minors, adapting these where necessary, with regard to cultural specificity. The ethical elements of particular importance are the scientific justification of the research, the balance of benefits/risks burden, the protection of personal identity, autonomy, specifically through informed consent/assent to participate in the research, the protection of privacy, and justice. Specific issues arise from incidental findings, predictions (from genetic tests) and the use of single case reports.

---

4. International frameworks and protocols that apply exclusively to neurosciences are rare. However, several regulatory frameworks are applicable to neuroscience. These include the latest version of the World Medical Association's Declaration of Helsinki (2013), the Universal Declaration on Bioethics and Human Rights of UNESCO (2005), the Convention on Human Rights and Biomedicine of the Council of Europe (1997), and several Directives of the European Union (such as Clinical Trials Regulation 2014 and Medical Device 2017), which are relevant for clinical trials in the field of neuroscience and for medical devices. Regarding medical/pharmaceutical enhancement, there are no instruments which directly address neuroenhancement. Only rules of professional standards are applicable to medical/pharmaceutical neuroenhancement, because the problem lies in the "off label" use of medication prescribed in another medical context. It is therefore very difficult to regulate the phenomenon as such. Particular issues in the licensing procedure of non-invasive novel neurotechnologies for gaming purposes are at stake.

17.6. References

Chiao, J.Y. and Ambady, N. (2007). Cultural neuroscience: Parsing universality and diversity across levels of analysis. In *Handbook of Cultural Psychology*, Kitayama, S. and Cohen, D. (eds). Guilford Press, New York.

Chneiweiss, H. (2006). *Neurosciences et neuroéthique. Des cerveaux libres et heureux*. Alvik éditions, Paris.

Collins, W.A. and Steinberg, L. (2006). Adolescent development in interpersonal context. In *Handbook of Child Psychology: Social, Emotional, and Personality Development*, 6th edition, Eisenberg, N. (ed.). Wiley, Hoboken, NJ.

Deutscher, E. (2013). Informationen und Nachrichten aus dem Deutschen Ethikrat. *Infobrief*, 14, December.

Di Pietro, N.C. and Illes, J. (2013). Disclosing incidental findings in brain research: The rights of minors in decision making. *Journal of Magnetic Resonance Imaging*, 38, 1009–1013.

Dibljevic, V. and Racine, E. (2019). Pediatric neuro-enhancement, best interest and autonomy: A case of normative reversal. In *Shaping Children, Advances in Neuroethics*, Nagel, S.K. (ed.). Springer.

Dubljević, V., Jox, R.J., Racine, E. (2017). Neuroethics: Neuroscience's contributions to bioethics. *Bioethics*, 31(5), 326–327.

Farah, M.J. (2002). Emergent ethical issues in neuroscience. *Nature Neuroscience*, 5, 1123–1129.

Farah, M.J. and Heberlein, A.S. (2007). Personhood and neuroscience: Naturalizing or nihilating. *The American Journal of Bioethics*, 71, 37–48.

Farisco, M. and Evers, K. (eds) (2017). *Neurotechnology and Direct Brain Communication. New Insights and Responsabilities Concerning Speechless But Communicative Subjects*. Routledge, New York.

Figueroa, G. (2016). Neuroethics: The pursuit of transforming medical ethics in scientific ethics. *Biological Research*, 49, 11.

Garnett, A., Whiteley, L., Piwowar, H., Rasmussen, E., Illes, J. (2011). Neuroethics and fMRI: Mapping a fledgling relationship. *PloS One*, 6(4), e18537.

Goering, S. and Yuste, R. (2016). On the necessity of ethical guidelines for novel neurotechnogy. *Cell*, 167, November 3.

Health Council of the Netherlands (2003). Human enhancement. Report, Health Council of the Netherlands.

Ienca, M. and Andorno, R. (2017). Towards new human rights in the age of neuroscience and neurotechnology. *Life Sciences, Society and Policy*, 13, 5.

Illes, J. (ed.) (2006). *Neuroethics: Defining the Issues in Theory, Practice, and Policy*. Oxford University Press, New York.

Illes, J. (2010). Empowering brain science with neuroethics. *Lancet*, 376, 1294–1295.

Italian National Bioethics Committee (2010). Neuroscience and human experimentation: Bioethical problems. Report, Italian National Bioethics Committee.

Italian National Bioethics Committee (2014). Neuroscience and pharmacological cognitive enhancement: Bioethical aspects. Report, Italian National Bioethics Committee.

Jarchum, I. (2019). The ethics of neurotechnology. *Nature Biotechnology*, 37, 993–996.

Kim, H.S. and Sasaki, J.Y. (2014). Cultural neuroscience: Biology of the mind in cultural contexts. *Annual Review of Psychology*, 65, 487–514. doi: 10.1146/annurev-psych-010213-115040.

National French Consultative Ethics Committee for Health and Life Sciences (2012). Ethical Issues Arising out of Functional Neuroimaging. *Opinion*, 116.

National French Consultative Ethics Committee for Health and Life Sciences (2013). Recours aux techniques biomédicales en vue de "neuro-amélioration" chez la personne non malade : enjeux éthiques. *Opinion*, 122.

Nuffield Council on Bioethics (2013). Novel neurotechnogy: Intervening in the brain. Report, Nuffield Council on Bioethics, UK.

OECD (2019). Recommendation of the Council on Responsible Innovation in Neurotechnology. Report, OECD.

Palazzani, L. (2019). *Innovation in Scientific Research and Emerging Technologies: A Challenge to Ethics and Governance*. Springer Nature Switzerland, Cham.

Palmerini, E. (2015). A legal perspective on body implants for therapy and enhancement. *International Review of Law, Computers & Technology*, 29(2–3), 226–244.

Sandel, M.J. (2011). The case against perfection: What's wrong with designer children. In *Atlantic Monthly*, Jecker, N. (ed.). Jones & Bartlett Learning, Mississauga, ON.

The Danish Council of Ethics (2011). Medical enhancement. Report, The Danish Council of Ethics.

U.S. President's Council on Bioethics (2003). Beyond therapy: Biotechnology and the pursuit of human improvement. Report, U.S. President's Council on Bioethics.

Conclusion

Michele FARISCO[1,2]

[1] *Centre for Research Ethics and Bioethics, Uppsala University, Sweden*
[2] *Biogem, Biology and Molecular Genetics Research Institute, Ariano Irpino, Italy*

Even if it is still young, neuroethics is an established field of research with recognizable contents and methods. However, this does not mean that neuroethics has a rigid and definite identity: different approaches have been introduced and refined, and they can arguably be considered as expressions of different "schools of thought" and traditions, including cultural traditions.

In this sense, neuroethics is not an exception within the ethical and more general philosophical debates, which are composite, multi-voiced and dialectic (i.e. informed by lively debates). Empirical neuroethics, neurobioethics, fundamental neuroethics and political neuroethics are illustrative examples of the contemporary understanding of the discipline, which also includes other expressions. These different approaches share a common reference point (i.e. the brain and related research) and a shared general goal (i.e. to clarify the connection between brain research/technology and morality/ethics) to then analyze the emerging issues through different methods pursuing different specific aims (e.g. to investigate the practical implications of brain research rather than identifying the impact of brain research on traditional concepts, including morally salient notions such as responsibility and free will).

The debate about the epistemological statute of neuroethics is not new, as confirmed by the chapters of the present volume. What is new is the urgency to include cultural diversity in this debate. The topic of culture is complex and multi-faceted, and certainly deserves more in-depth and circumscribed analyses. The contributions in this volume provide a first systematic reflection on the connection between culture and neuroethics, which is extremely needed, especially if neuroethics (whatever it is its understanding) aims to maximize its impact on society. In order to do so and to avoid being conceived only as an academic discipline or a discussion among experts in the field, neuroethics needs to create a link (or to improve it if already in place) with a number of stakeholders (including decision-makers, politicians, citizens' organizations, patients' associations, industries, etc.). For this link to be effective, a shared conceptual and linguistic platform is crucial, otherwise no communication and no mutual understanding are possible. Furthermore, in order to build such a "common ground" it is necessary to take into account cultural variation as a key factor. Importantly, cultural variation, like culture itself, is a multi-faceted concept: it relates to disciplinary as well as anthropological, sociological and political differences. All these variations significantly impact the way people perceive, value, accept and ultimately have access to brain research findings and related technological products. In fact, culture is a kind of filter that biases the perception of science and technology, affects how many people know about it and can access it, and determines whether and how much society can take advantage of their potential benefits.

Here lies a *fundamental risk of creating new inequalities*, not only in the economic ability to have access to the most updated scientific and technological achievements, but also in the preliminary possibility *to know, appreciate, and freely decide* about them. In fact, as illustrated by the contributions to this volume, cultural identity (including all its explicit and more subtle variations) results from different factors and has a number of ethically relevant implications. *Neuroethics has to face this challenge, which implies different specific tasks*. First, *a clarification of the best strategy to embrace* appears necessary. Global, cross-cultural, multicultural neuroethics are some of the proposed approaches to assessing the issue, all with their own justifications, shortcomings and promises. Second, *an agreement about the connection between the different understandings of neuroethics is necessary*. Notwithstanding the different specific forms, including those represented in this volume, neuroethics is a consistent discipline that stands as a general framework. Therefore, it is crucial to clarify how different understandings of the discipline interact with each other, in order to avoid conflicts and improve reciprocal collaboration. Third, *clear recommendations for including cultural variation within the neuroethics scope are necessary*.

On the basis of the analyses included in this volume, as a provisional conclusion, I would like to formulate the following recommendations addressed to three main stakeholders: researchers; developers and providers of technological services; neuroethicists.

– To the people involved in scientific and technological research:

To provide the necessary conceptual clarity, particularly about those terms that play a key role in their explanations/theories and those expressions that are prone to misleading interpretation by people from other disciplines or that tend to raise hyped expectations from the public. For example, when the issue at stake is how to improve the health of loved ones or how to avoid new forms of mental privacy infringement, the risk of misinterpretation due to excessive emotional involvement is very high.

To pay attention to the language used when communicating findings so that it is adapted to different publics. For example, sharing the results of cutting-edge research on the neuronal underpinnings of conscious perception with a public of neuroscientists is not the same as it is with the members of patients' associations: the respective biases and needs are different and should be reflected in the language used.

To create communication strategies that are sensitive to the needs and possibilities, including epistemic limitations, of different publics, which entails that they must be multi-faceted and multi-channel. To illustrate this, while a scientific paper might be sufficient to communicate with peers, when a researcher wants to reach out to public society at large (e.g. patients' families/caregivers), other strategies are necessary, such as direct communication or other forms of public engagement.

To raise awareness of the *multidimensional* ways in which *cultural differences*, including disciplinary differences, express themselves. Ideally, as part of their curricula, researchers should be informed about how culture creates diversity at multiple levels, within the broad spectrum of what constitutes the personal *Weltanschauung*.

To be encouraged to develop measures *to minimize culturally based biases* in research and relative communication. From the awareness of the risk of being biased by their own culture, researchers should take steps to minimize it.

To recognize and pay attention to the mutual interaction between science (e.g. neuroscience) and society, considering the potential social impact of research not as an additional point that should possibly be included in scientific reflection, but as an essential component of it. It is time to recognize that scientific research is an integral part of society, reflecting its values and contributing to sharing them. Sociological and philosophical analyses of science revealed this point a while ago, but its translation in actual scientific practice is still lacking.

To recognize the potential impact of extra-scientific factors on how scientific research is defined and conducted. Priorities and goals of scientific and technological research are not defined autonomously, but rather result from a complex interplay between science, technology and external factors, including politics, economy, ideologies, religion, etc. In a word: cultural diversity.

To recognize the social tendency to attribute epistemic authority to science and to use it to maximize public benefit rather than to manipulate society. Notwithstanding some anti-scientific and technophobic trends (as recently shown by the Covid-19 pandemic), the main public tendency is still to consider science and technology as sources of reliable knowledge. Researchers should take this as a social responsibility for improving social benefits and avoiding any kind of negative exploitation.

To promote engagement activities with the public, in order to raise awareness of the role played by culture and to identify priorities for research and ethically sustainable methods to achieve relevant goals. To include public society not only as a passive repository of information but as an active participant in the process of scientific and technological research (from topic selection to goal setting) should be part of the social responsibility of scientists and technology developers.

– To the people involved in providing technological services:

To facilitate the identification, understanding and assessment of societal issues emerging from the technology offered and its cultural situatedness, for example by including neuroethics and RRI services. Importantly, these services should be conceived not only as regulatory and compliance tools, but also as instruments to inform the services ethically, including considerations on cultural diversity.

To include a training program intended to raise awareness of the cultural diversity and how it impacts the social exploitation of technological services. This kind of awareness can play an important role in increasing the number of people who eventually have access to the services.

To create a virtual space, for example in the form of discussion forums or capacity development activities, *for mutual understanding between developers and users* on the impact of cultural diversity on research and service offerings. Implementing a dialogue with different stakeholders from different backgrounds is crucial to align the services with actual social needs.

To promote awareness of the gap between research findings and public benefits by identifying different stakeholders, understanding relevant needs and benefits, and engaging with them. The social benefit derived from research findings is not straightforward, and actual public gaps need to be identified and overcome.

– To the people involved in neuroethics reflection:

To promote an intradisciplinary clarification about the mutual connection between the different forms of neuroethics reflection. Identifying points of possible collaboration would be more constructive than emphasizing the differences.

To engage in multi- and interdisciplinary collaboration, particularly with researchers in neuroscience and developers of neurotechnologies, in order to identify emerging ethical issues, set priorities among them and collaborate in defining an effective strategy for assessing them. How this can be achieved concretely is an open and big challenge. It is likely that different strategies are needed depending on the different disciplines involved (e.g. neuroscience rather than AI).

To think about and promote concrete engagement activities with the publics, in order to make ethical reflection as concrete as possible and to maximize its positive impact on society. Ethical reflection should be done starting from and with the publics to make it more effective.

To include cultural differences as an important factor in shaping the ethical analysis, both in the identification of issues and in their assessment. Cultural diversity should be kept as a point of reference, and attempts to find shared ethical values and agreed-upon evaluations should be inspired by it.

To pay special attention to underrepresented or minority cultural groups. New kinds of inequalities are possible, especially new forms of "technological illiteracy" that preclude any possibility of benefiting from new technologies. This happens both within the same country (depending on different cultural backgrounds including different levels of instruction) and in some countries more than in others (depending on financial, political, social and cultural differences).

List of Authors

Jennifer A. Chandler
Centre for Health Law, Policy and Ethics
University of Ottawa
Canada

Veljko Dubljević
Department of Philosophy and
Religious Studies
North Carolina State University
Raleigh
USA

Kathinka Evers
Centre for Research Ethics and Bioethics
Uppsala University
Sweden

Michele Farisco
Centre for Research Ethics and Bioethics
Uppsala University
Sweden
and
Biogem, Biology and Molecular Genetics
Research Institute
Ariano Irpino
Italy

Cynthia Forlini
School of Medicine
Faculty of Health
Deakin University
Geelong
Australia

Karin Grasenick
Convelop Cooperative Knowledge
Design GmbH
Graz
Austria

Manuel Guerrero
Center for Research Ethics and Bioethics
Uppsala University
Sweden
and
Department of Bioethics and
Medical Humanities
Faculty of Medicine
University of Chile
Santiago
Chile

Karen Herrera-Ferrá
Asociación Mexicana de Neuroética
Mexico City
Mexico

Mai IBRAHIM
Department of Philosophy and
Religious Studies
North Carolina State University
Raleigh
USA

L. Syd M JOHNSON
Center for Bioethics and Humanities
SUNY Upstate Medical University
New York
USA

Fabrice JOTTERAND
Medical College of Wisconsin
Milwaukee
USA
and
University of Basel
Switzerland

Denis LARRIVEE
Mind and Brain Institute
University of Navarra Medical School
Pamplona
Spain
and
Arts and Sciences Department
Loyola University Chicago
USA

Amal MATAR
Center for Research Ethics and Bioethics
Uppsala University
Sweden

Georg NORTHOFF
Faculty of Medicine
Centre for Neural Dynamics
The Royal's Institute of Mental Health
Research
Brain and Mind Research Institute
University of Ottawa
Ontario
Canada
and
Mental Health Centre
Zhejiang University School of Medicine
Hangzhou
and
Centre for Cognition and Brain Disorders
Hangzhou Normal University
China

Laura PALAZZANI
LUMSA Univesity
Rome
Italy

Eric RACINE
Pragmatic Health Ethics Research Unit
Institut de recherches cliniques
de Montréal
and
Université de Montréal
and
McGill University
Montreal
Quebec
Canada

Karen S. ROMMELFANGER
Neuroethics and Neurotech Innovation
Collaboratory
Center for Ethics Neuroethics Program
Emory University
and
Institute of Neuroethics Think and Do Tank
Atlanta
USA

Arleen SALLES
Center for Research Ethics and Bioethics
Uppsala University
Sweden
and
Neuroetica Buenos Aires (NEBA) BsAs
Argentina

Abdou Simon SENGHOR
Pragmatic Health Ethics Research Unit
Institut de recherches cliniques
de Montréal
and
McGill University
Montreal
Quebec
Canada

Laura SPECKER SULLIVAN
Philosophy Department
Fordham University
Bronx
USA

Jie YIN
School of Philosophy
Fudan University
Shanghai
China

Index

A, B

agency, 5, 22, 27, 29, 31, 32
animal
  brains, 250, 258
  research, 249, 250, 260, 262, 263
anthropology, 95, 96, 99–101, 104
applied
  ethics, 190
  neuroethics, 40, 43
artificial intelligence, 43
autonomy, 164, 165, 167–169, 171, 172, 174, 292, 295–297, 299, 301
biases, 217, 223
bioethics, 1, 4–13
bipolar disorder (BD), 26, 30, 31
body, 159–162, 164, 165, 168–170, 172, 173, 175
brain (*see also* animal *and* deep brain stimulation *and* Human Brain Project), 22–30, 32, 33, 95–104, 110–120, 127, 128, 130–137, 139, 249–260, 262–265
  -computer interfaces (BCI), 77, 79, 80, 86
  -to-brain interfaces (BBI), 77, 79–81, 85–91, 93
  activity, 24
  death, 271–273, 275–283
  enhancement, 237, 239
  research, 96, 97, 99, 100
  science, 3
BRAIN Initiative, 257, 263, 264

C, D

Chile, 235, 236, 238–244
chimeras, 251–258, 260, 261, 265
Chinese philosophy, 184–186, 189, 190
Christian religion, 162, 175
cognitive enhancement, 181, 182, 191
colonialism, 59, 60, 68
community-based participatory research (CBPR), 150
conceptual neuroethics, 109, 118–120
Confucius, 188
consciousness, 24–26, 39, 41–43, 46, 250–252, 255–260, 265
contingency, 160–167, 172–174
cooperation, 148
cross-cultural neuroethics, 119, 144–146, 148, 149, 151, 153, 155
cultural diversity, 110–114
culture, 109–111, 113–120, 143–148, 150–153, 195–198, 200, 202, 204–206, 209–211, 218, 220, 225, 228, 272, 275, 282
deep brain stimulation, 22, 30
demographic minorities, 275
dimorphism, 221, 231
diversity, 55–60, 62–69
  anthropological, 56–58, 62, 64, 68
  biological, 58
  cultural diversity, 56–60, 62–66, 68

E, F

electrocardiogram (ECG), 199
electroencephalogram (EEG), 199–202, 208–211
engagement, 10, 12, 14–16
enhancement, 21, 22, 30, 288, 298, 299, 301
ethics (*see also* neuroscience), 56–59, 61–67, 69, 70, 180, 188, 189, 191, 195–198, 202, 206–214, 288–291, 300–303
  of neuroscience, 4, 6–9, 12, 14, 16, 21, 22
free will, 48–50
freedom, 165, 166, 174
fundamental neuroethics, 40–44, 46–48, 51

G, H

gender (*see also* sex/gender), 78, 82–85
  medicine (GM), 218, 222–224, 226
global
  market, 125, 127, 128, 131, 132
  neuroethics, 113
globalization, 125–127, 129, 131, 137
gongfu, 179, 184, 187, 190
guideline, 227
healthcare, 4, 5, 7–9, 14
human
  exceptionalism, 249, 250, 260, 261, 263, 265, 266
  flourishing, 57, 61, 65–67
  identity, 42, 48
  nature, 180, 183, 184, 186–188
  rights, 235–239, 241–245, 279, 282
Human Brain Project (HBP), 41, 42, 51
Hume's law, 159, 161, 162, 175

I, J

identity, 59, 96, 98–103
incidental findings, 295, 296
informed consent, 288, 292–295, 300, 301
innovations, 4, 7, 8, 13, 15, 16, 41, 42, 47, 48, 51
integrity, 290, 291, 297
intercultural understanding, 146–149, 151
interdisciplinary, 38, 39, 41, 46, 48, 51, 109
international neuroethics, 111, 113
intersectionality, 224
intracultural creativity, 146, 148
Islamic religion, 198, 199, 202–204, 206
justice, 250, 257, 260–263, 288, 298, 301

L, M

language, 196–198, 200–202, 204–207
Latin America, 128, 133, 135–137
law, 271, 274, 277, 280–283
metaphysics, 164, 165
Mexico, 126, 128, 130, 132–137
mind, 24, 95, 96, 99–104, 127, 129–132, 134, 136, 137, 249, 250, 264, 266
minors, 287–301
moral
  neuroenhancement, 179–181, 184, 189, 190
  status, 250–254, 259, 260
multidisciplinary, 41, 51

N, O

nature, 160–162, 165, 166, 169, 172, 174, 175
neo-Confucianism, 190
neuro-ecological model, 22, 26, 29, 32, 33
neuroessentialism, 102
neuroethical
  legal social and cultural issues (NELSCI), 131, 132, 135
  reflection, 217, 218, 228, 229

neuroethics, 21, 32, 35, 37, 39–44, 46–48, 51, 55–57, 59, 62–69, 71, 75, 77–79, 81, 89–92, 95–102, 104, 109–114, 118–120, 127–137, 143–151, 153–156, 159, 160, 162, 173, 175, 189, 195–197, 202–207, 249–252, 257, 260, 262–266, 271, 282
neuroexceptionalism, 95, 102, 104
neurofeminism, 77, 78, 89
neurogendering, 222
neurology, 4, 5, 8, 97, 98, 101
neuroplasticity, 78, 80, 84, 85, 91
neuroprotection, 235, 236, 238–241, 243, 244
neurorights, 236, 243–246
neuroscience (*see also* ethics), 3–16, 22–24, 27, 31, 32, 96, 97, 99–101, 127–137, 143, 145, 146, 148, 149, 154, 160–162, 164, 170, 171
of ethics, xv, xix, 6, 8, 9, 14, 21, 22
neuroscientific research, 250, 252, 258, 260, 262–266, 289, 294
neurosexism, 77–79, 81–85, 89, 91, 222
neurotechnology, 7, 8, 10, 56, 62, 63, 65, 66, 68, 69, 127–137, 235–240, 243, 288–290, 292, 297, 298, 301
non-reductionism, 22, 26

Nuremberg Code, 238
organ donation, 276–278
organism, 167, 168, 170, 173, 174

P, R

person, 162, 165
personal identity, 291, 301
policy, 3–5, 14
political neuroethics, 79, 89–91
privacy, 288, 297, 298, 301
psychiatric disorders, 26, 29, 33
religion, 159, 160, 162–164, 166, 169, 175, 205, 272, 275, 280, 282, 283

S, T, U

sex, 78, 82–84
sex/gender, 217, 218, 220, 222, 225–229
simulation, 41, 42, 46
single photon emission computed tomography (SPECT), 199, 200
time perception, 24, 32
transcranial magnetic stimulation (TMS), 81, 87, 88
Universal Declaration of Human Rights, 236, 237, 241, 243–245